Lecture Notes in Control and Information Sciences

Edited by A. V. Balakrishnan and M. Thoma

For information about Vols. 1–21 please contact your bookseller or Springer-Verlag.

Lecture Notes in Control and Information Sciences

Edited by M. Thoma and A. Wyner

89

G. K. H. Pang
A. G. J. MacFarlane

An Expert Systems Approach to Computer-Aided Design of Multivariable Systems

Springer-Verlag Berlin Heidelberg GmbH

Authors
Dr. Grantham K. H. Pang
Department of Electrical Engineering
University of Waterloo
Waterloo
Ontario N2L 3G1
Canada

Professor Alistair G. J. MacFarlane
Engineering Department
University of Cambridge
Trumpington Street
Cambrigde CB2 1PZ
England

ISBN 978-3-540-17356-4   ISBN 978-3-540-47438-8 (eBook)
DOI 10.1007/978-3-540-47438-8

2161/3020-543210

ACKNOWLEDGEMENTS

G.K.H. Pang would like to thank his wife Fanny for helping with the preparation of this manuscript, and the Croucher Foundation for financial support. We are grateful to J-M. Boyle for many helpful discussions and detailed comments, and to Cambridge University Engineering Department for the use of its computing facilities.

CONTENTS

Tables and Figures are placed at the end of each chapter except Fig. 6.12.

NOTATION

Unless otherwise stated, the following notation will be adopted:

$a \simeq b$ means a is approximately equal to b

$a := b$ means a is defined to be b or a denotes b

$\mathbb{R}, \mathbb{C}$:= field of real and complex numbers, respectively

$\mathbb{C}_+$:= $\{ z \in \mathbb{C} \mid \mathrm{Re}\, z \geq 0 \}$, the closed right-half complex plane

For $z \in \mathbb{C}$

$| z |$:= modulus (or magnitude) of z

$\angle z$, $\arg z$:= argument of z

$\mathrm{Re}\,z, \mathrm{Im}\,z$:= real, imaginary part of z, respectively

For $k \in \mathbb{R}$

$\sqrt{k}$:= square root of k ; also written as $k^{1/2}$; unless otherwise stated, the value is taken to be positive

E^k := k dimensional Euclidean space

$\max_{k}$:= the maxmium with respect to k

$\mathbb{R}(s), \mathbb{C}(s)$:= field of rational functions in s with coefficients in $\mathbb{R}, \mathbb{C}$

$O(s^i)$:= a quantity of order s^i (or less)

Let F be any one of $\mathbb{R}, \mathbb{C}, \mathbb{R}(s)$ or $\mathbb{C}(s)$, then :

$F^{m \times \ell}$:= set of $m \times \ell$ matrices with elements in F

$F^{m \times \ell}(s)$:= set of $m \times \ell$ matrices with elements in F(s)

F^n := vector space of $n \times 1$ column vectors with elements in F, over an appropriate field

Let $M \in F^{m \times \ell}$ where F is either $\mathbb{R}$ or $\mathbb{C}$, then :

m_{ij} := (i,j)th entry of M ; we also write $M = (m_{ij})$

$\{ g_i \}$:= set of eigenvalues (spectrum) of M ; also known as characteristic values or gains ; generally, g_i are arranged in descending order of their magnitude

$\{ \sigma_i \}$:= set of singular values of M ; also known as principal gains ; generally, σ_i are arranged in descending order of their magnitude

$\bar{\sigma}(M)$:= maximum singular value of M

$\underline{\sigma}(M)$:= minimum singular value of M

M^t := transpose of M

M^{-1} := inverse of M

M^* := conjugate transpose of M

$|M|$:= (x_{ij}) where $x_{ij} = |m_{ij}|$

$argM$:= (x_{ij}) where $x_{ij} = arg\ m_{ij}$

$\| M \|_F$:= $(\sum_{j=1}^{\ell} \sum_{i=1}^{m} |m_{ij}|^2)^{1/2}$, the Frobenius norm of M

$\| M \|_2$:= $\bar{\sigma}(M)$, spectral norm or maximum singular value of M

I_m := mxm unit matrix

Let $u \in F^{\ell}$ where F is either $\mathbb{R}$ or $\mathbb{C}$, then

$\| u \|_2$:= $(u^* \cdot u)^{1/2} = (\sum_{i=1}^{\ell} |u_{ij}|^2)^{1/2}$, the Euclidean vector norm of u

u^t := transpose of the vector u

$diag\{d_i\}_{i=1}^{n}$:= nxn diagonal matrix with $d_1,\ldots,d_n$ along the diagonal ; also written as $diag\{d_1,\ldots,d_n\}$ or $diag\{d_i\}$

Let $A \in \mathbb{R}^{n \times n}$, $B \in \mathbb{R}^{n \times \ell}$, $C \in \mathbb{R}^{m \times n}$, $D \in \mathbb{R}^{m \times l}$ and s be the frequency variable ($s \in \mathbb{C}$), then :

$G(s)$:= $C(sI_n - A)^{-1}B + D$, the plant open–loop gain (transfer function) matrix

Also, let g be the gain variable ($g \in \mathbb{C}$) and $\ell = m$, then:

$S(g)$:= $B(gI_m - D)^{-1}C + A$, the closed–loop frequency matrix

Let $\Omega \subset \mathbb{C}$ and $G(s) \in \mathbb{R}(s)^{m \times \ell}$, then:

$\text{\#SMP}[G(s),\Omega]$:= number of Smith–McMillan poles of G(s) in Ω

Let ς be a (finite number of) closed curve(s) in $\mathbb{C}$, then:

$\text{\#E}(\varsigma,a)$:= number of encirclements of ς around the point a ; anti–clockwise encirclements are taken as positive

List of Symbols:

0	zero; zero vector; zero matrix
i	integer
j	$\sqrt{-1}$; integer
ω	angular frequency
D_{NYQ}	Nyquist D-contour ; Section 3.2
MS(G)	measure of skewness, a normality indicator ($G \in C^{m \times m}$); Section 3.4.4
MS(k)	MS(G(jk)) where $k \in R$, measure of skewness of G(jk); Section 6.6
$\kappa(G)$	spectral condition number ($G \in C^{m \times m}$); Section 3.4.5
cond(g_i)	condition number for an eigenvalue g_i ; Section 3.4.5
ρ_i	gain ratio; Section 3.4.8
==>,<==	implies, is implied by
$\square$	marks the end of a proof

List of Abbreviations:

AI	Artificial Intelligence; Chapter 2
AIRC	Aircraft Dynamics Model; Section 7.4.3
AUTO	Automobile Gas Turbine Model; Section 6.9.1
CACSD	Computer-Aided Control System Design; Section 2.2
CS, CSi	Misalignment Angle, i^{th} branch of; Section 3.4.7
CVD	Characteristic Value Decomposition; Section 3.2
E, Ei	Eigenloci, Characteristic Gain Loci, i^{th} branch of; Fig. 5.1.1
FLOW	Flow-box Model; Section 5.5.3
FOO	Full-Order Observer; Fig. 8.1
GHEL	Helicopter Model; Section 6.9.2
GROC	Rocket Engine Model; Section 5.5.4

HF   High Frequency; Section 8.1

HFS   High Frequency Sub-controller; Section 5.2.1

IF   Intermediate Frequency; Section 8.1

KBF   Kalman-Bucy Filter; Chapter 2

KEE   Knowledge Engineering Environment; Section 8.8

LF   Low Frequency; Section 8.1

LFS   Low Frequency Sub-controller; Section 5.2.2

LHP   Left-Half Plane; Section 7.2

LQR   Linear Quadratic Regulator; Chapter 2

LTR   Loop Transfer Recovery; Chapter 2

MAID   Multivariable Analytical & Interactive Design; Section 8.4.1

MIMO   Multi-Input, Multi-Output; Section 4.2

NSRE   Non-Square Chemical Reactor Model; Section 6.10.2.1

OBC   Observer-Based Controller; Section 7.2

P, Pi   Principal Gain Loci, i^{th} branch of; Fig. 5.1.1

P+I   Proportional plus Integral; Section 5.4

REAC   Chemical Reactor Model; Section 5.5.5

RFA   Reverse Frame Approximation; Section 6.1

RFAT   Reverse Frame Alignment Technique; Section 8.1

ROO   Reduced-Order Observer; Fig. 8.1

SDT   Simple Design Technique; Section 5.2

SISO   Single-Input, Single-Output; Chapter 2

SVD   Singular Value Decomposition; Section 3.3

STD   Schur Triangular Decomposition; Section 3.4

TGEN   Turbo-Generator Model; Section 6.9.3

w.r.t. with respect to

iff   if and only if

Units:

m	meter
N	newton
rad	radian
s	second
rev	revolution ($2 \cdot \pi$ radians)
min	minute (60 seconds)
kN	kilonewton (1000 N)

CHAPTER ONE

INTRODUCTION

The theory and codified practice of automatic control is an organised body of shareable knowledge, and the importance of developing appropriate interactive computing environments lies largely in making such a specialised body of knowledge easily usable and easily accessible, and therefore easily shareable. Expert and knowledge-based systems have a key role to play in the creation of such environments. The work presented here is concerned with the investigation of expert system techniques for the design of linear multivariable feedback control systems. It is important that the procedures used by such an expert system to manipulate models and their attributes are formulated in terms of a set of individual functions which a designer can cause to be executed on the computer. Only in such circumstances will one be able to formulate any high-level machine-based procedure which would "explain" its actions. For the same reason the procedures used by the machine must be coherent with the principles in terms of which the man thinks about the tasks which are being carried out. In order to achieve these key attributes of referential transparency and coherence, the analysis and design techniques presented here are based on an appropriate generalisation of classical feedback methods. This enables a comprehensive and accurate representation of the behaviour of a multivariable feedback system to be given in terms of a basic set of graphical indicators.

1.1 The Interactive Design Process

The relationship between man and machine in the interactive design process is summarised in Fig. 1.1. We consider data passed from machine to man in terms of <u>indicators</u> and data passed from man to machine in terms of <u>drivers</u>. The man works in terms of a high level conceptual framework and accesses in the machine a powerful manipulative framework. The basic task in creating a satisfactory interactive computing system is to get these two frameworks to mesh together satisfactorily via an appropriate set of indicators and drivers.

It is important to realise, as illustrated by Fig. 1.2, that design is a feedback process, and that in general both the object being created and the specification against which it is being manipulated are being iteratively adjusted in a feedback cycle of dependence as the design proceeds. Design can also be described as a process of instantiation: the progessive generation of a specific fully defined object from an initial incomplete general description. In creating a specific instance of the general class of object desired, the designer is grappling with both uncertainty and complexity, and it is for coping with these twin sources of difficulty that the interactive man/machine combination is well suited: the man to handle uncertainty and the machine to handle complexity. In seeking to define the relative roles of man and machine one must start from a consideration of their strengths and weaknesses in respect of the tasks involved. The man has as strengths:

- ° the ability to abstract, simplify and conceptualise
- ° the ability to handle incomplete and ill-defined descriptions
- ° experience and common sense
- ° adaptability and flexibility
- ° skills in pattern-recognition and association.

He has as weaknesses:

- ° short-term memory limitations
- ° slowness in executing complex procedures
- ° tendency to fatigue and distraction

- ° varying responses to similar stimuli
- ° inability to handle many disparate activities at the same time
- ° difficulties in long-term memory retrieval.

The machine has as strengths:

- ° speed and reliability
- ° extensive and accurate short-term and long-term memory
- ° indifference to fatigue
- ° predicability of response
- ° ability to handle large amounts of data and to perform a number of unrelated tasks simultaneously
- ° ability to accurately execute extremely complex formally-specified procedures.

It has as weaknesses:

- ° inability to generalise
- ° no conceptual level and no common sense
- ° inability to disambiguate and to handle uncertainty
- ° lack of flexibility and adaptability.

As has already been emphasised, design is a feedback process and this feedback is critically important in progressively stripping away the uncertainty in the original design specification. In the following sub-section design methods are considered in three categories associated with different amounts of initial uncertainty about the behaviour of the objects being handled: analytical, procedural and experimental. Although the requirements are markedly different in the three cases, the same general principles apply. The man sets and refines goals, argues from general principles in terms of abstract concepts, and handles ambiguity, conflict of objectives and uncertainties in description and performance. The machine evaluates functions, executes complex procedures, searches through complex data sets, generates and manipulates indicators, and accepts and acts on drivers.

When developing an interactive computing environment, we have to take proper account of the man as well as the machine. In discussing this it is useful to talk in terms of principles and procedures. Principles are the organisers of high-level declarative knowledge, and procedures are the implementors of low-level imperative knowledge; a man thinks in terms of general principles, and a machine functions in terms of formally specified procedures. In the interactive computing context we have to handle both formal and informal knowledge, and also declarative and imperative knowledge, and somehow we have to make them all fit together in an effective and efficient way, as illustrated by Fig. 1.3.

The basic problem of an automatic control system designer is to create or modify a given dynamical system so that it has a specified behaviour or set of attributes. In doing so he wants to use the interactive computing environment to:

- handle formal declarative knowledge by evaluating for him the behaviour and attributes of any given dynamical model

- handle formal imperative knowledge by executing appropriate sequences of procedures in order to attain specified objectives

- handle informal declarative knowledge in the form of textual descriptions of background theory, codes of practice, design data bases etc.

- handle informal imperative knowledge in the form of design guidelines, rules of practice, mandatory design requirements, etc.

To do all these satisfactorily will require a wide range of software, display and interface technology and the use of appropriate declarative and imperative languages. There is tremendous potential for improving the interactive computing environment by making large amounts of informal declarative and imperative knowledge available in carefully designed and integrated "soft" interaction modes, where by soft interaction we mean the sort of interaction involved in using a friendly "help" system, or in using a browser.

1.2 Analytical, Procedural and Experimental Modes of Design

In the design process we are dealing with three things:

° objects

° attributes of objects

° operations on objects.

From a very general point of view there are three main possibilities when considering design methodologies:

° a given attribute determines an appropriate object

° the set of objects can be searched through to find one with the desired attribute

° a given object can be manipulated using the available operations on it to turn it into one having the desired attribute.

Hence, in terms of the amount of analytical knowledge the designer has at his disposal, we can split his possible methods of working into those which are:

° attribute-centred, or analytical, when an exactly computable answer is available

° operation-centred, or procedural, when he is working from an extensive knowledge base but does not have an exact prescription for a solution

° object-centred, or experimental, when he is working from a restricted
   knowledge base and is systematically searching for a solution.

These are not hard and fast distinctions; they are however helpful when
seeking to relate the designer's conceptual framework to the machine's
manipulative framework. These relationships are summarised in Fig. 1.4.

1.3 Outline of Monograph Contents

An outline of the remainder of this monograph is as follows. Chapter
2 discusses the use of expert systems for designing control systems, and gives
a brief summary of some previous work in this field. The indicators required
to assess the three key aspects of the feedback design problem - stability,
performance and robustness - are presented in Chapter 3. In Chapter 4, we
introduce a complete set of indicators (primary indicators) which can be
used for interactive design. A systematic approach to the design problem is
then given in Chapters 5, 6 and 7. The spirit of this approach, which is
conceived in terms of a systematic manipulation of a set of gains and phases
associated with an open-loop transfer function matrix is that of the classical
approach to feedback system design. Chapter 8 discusses the development of an
expert system using the systematic design approach which has been developed.
A frame-based approach is proposed for representing the design knowledge.
Finally, Chapter 9 draws some conclusions about the work done and suggests
some areas for future work.

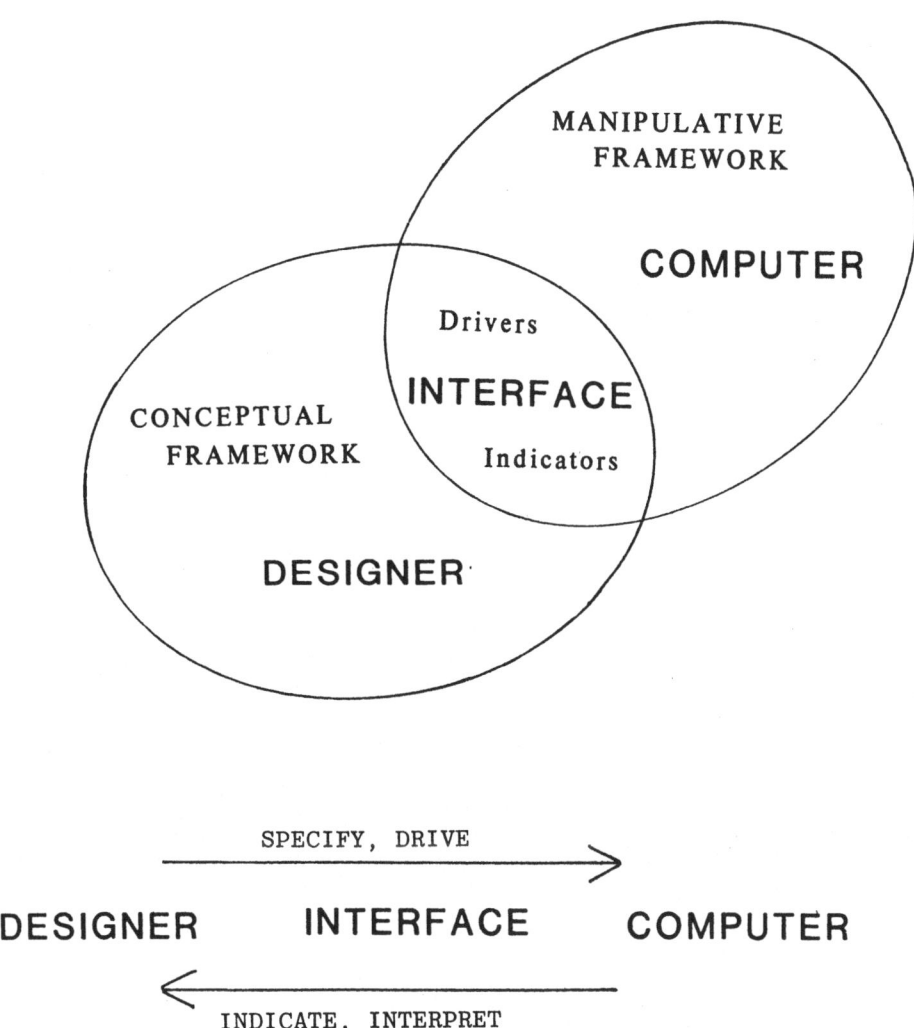

Fig. 1.1 Interactive design process

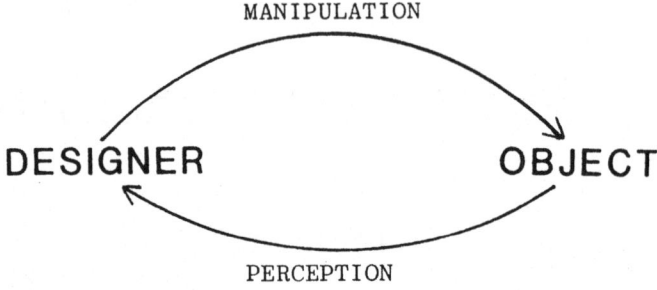

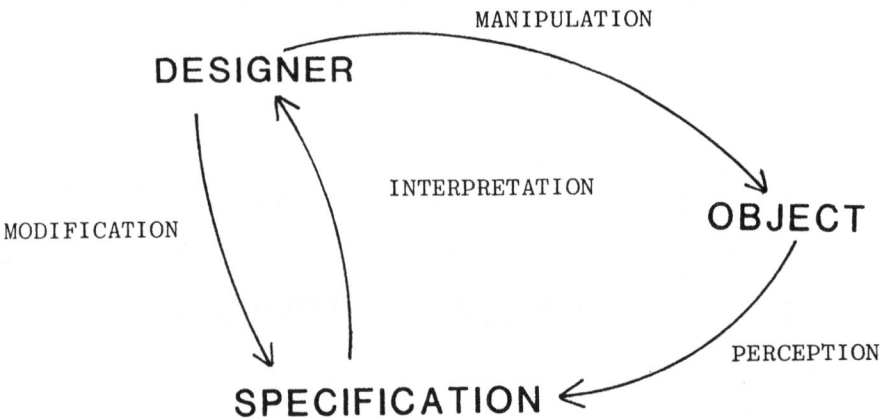

Fig. 1.2 Design is a feedback process

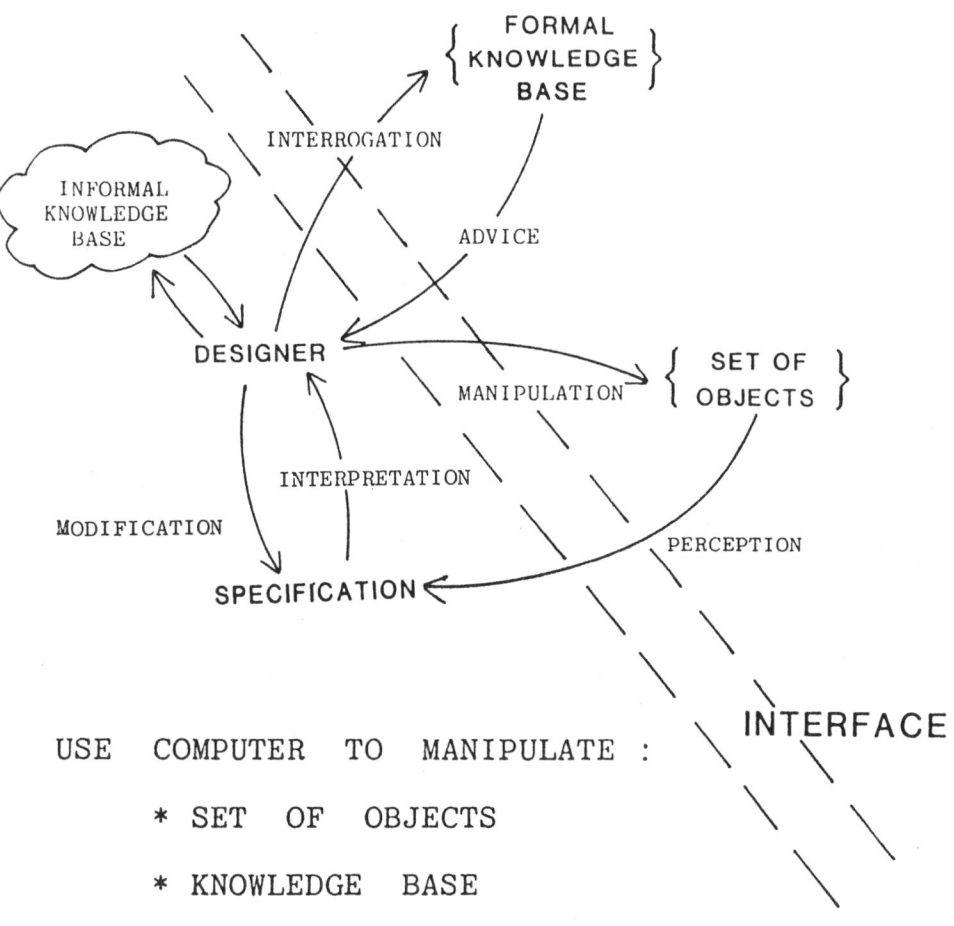

USE COMPUTER TO MANIPULATE :

    * SET OF OBJECTS

    * KNOWLEDGE BASE

INTERACTIVE COMPUTING HELPS TO HANDLE :

    * COMPLEX OBJECTS

    * COMPLEX KNOWLEDGE

Fig. 1.3 Interaction with objects and with
knowledge base

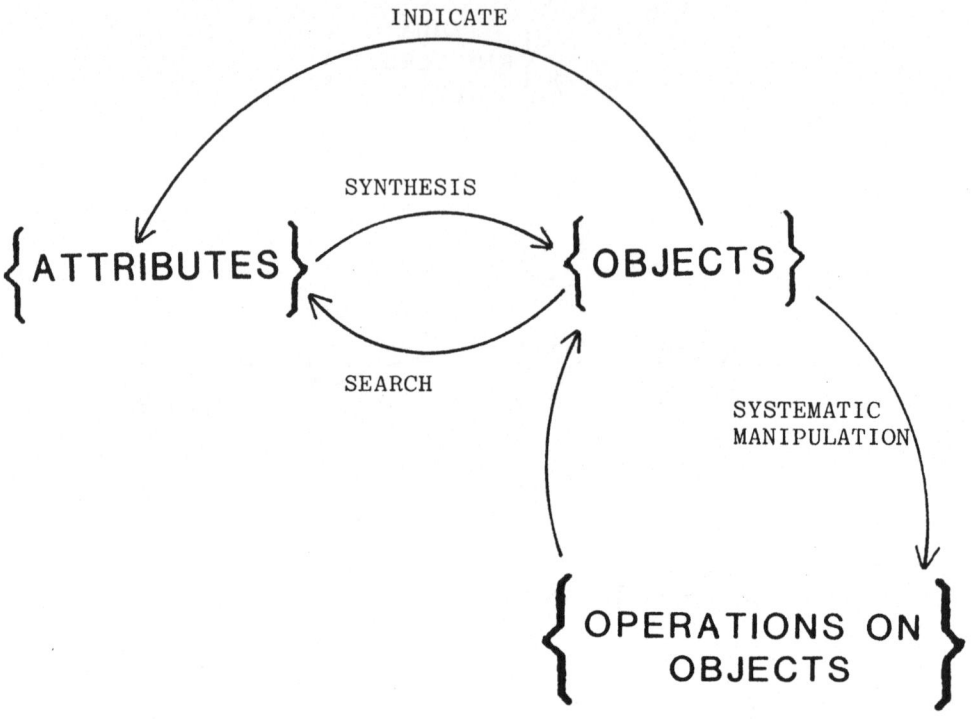

Fig. 1.4   Objects, attributes and operations

CHAPTER TWO

USE OF EXPERT SYSTEMS FOR CONTROL SYSTEM DESIGN

There is rapidly growing interest in the use of artificial intelligence (AI) techniques in control systems work. A good discussion is given by Taylor and Frederick (1984) of the use of expert systems for control system design. They have explored their potential for control design by concentrating on the use of classical techniques for single-input single-output systems. Åström, Anton and Årzén (1986) have discussed the use of expert systems in control system implementation, in particular the replacement of the heuristic logic in PID controller implementations by a rule-based expert system.

Sage (1981) gives a careful and systematic treatment of analysis and design techniques for single-input single-output (SISO) systems, making considerable use of design-decision charts. Although he does not specifically deal with expert systems, his treatment would be a good point of departure for anyone wishing to construct a rule-based description of SISO design techniques. James et al. (1985) and Nolan (1986) have described expert systems for SISO controller design, and have approached the problem in different ways. The expert system designed by James et al. automatically designs a compensator for a given plant, and provides a designer with little scope for manipulating the design. Nolan, in contrast to this, has aimed at building an expert system which functions as a designer's assistant. He considers that the expert system should be able to carry out block diagram analysis of a given system, and provide the designer with assistance in selecting the type of compensation required. His type of expert system would link with a conventional control system CAD package which would provide the appropriate analysis and synthesis software required.

Birdwell and his colleagues (Birdwell et al., 1985) have also described an expert system for the analysis and design of control systems. Their prototype system, CASCADE, deals with the design of controllers for linear multivariable systems using linear quadratic regulator (LQR) and Kalman Bucy filter (KBF) design techniques to satisfy frequency-domain measures of performance. They use the loop transfer recovery (LTR) technique, described by Doyle (1981), to maintain the overall performance characteristics in the final design.

Trankle and Markosian (1985) have investigated the use of real-time expert systems for the adaptive control of multivariable systems. These authors have developed a control system using these techniques which can adapt to significant changes in an original plant model. An expert system performs the task of identifying the current model of the plant and then designing an appropriate new control law. They have approached the re-design problem by developing an expert planner which has the design rules as operators. The design process consists of a sequence of operations obtained by heuristic planning.

Trankle and his co-workers (1986) have also described an expert system for control system design in a further paper. They have again used the ideas of planning systems and organized the execution process on two levels. The upper level is concerned with developing an overall strategy and so setting up a list of goals to be accomplished by the lower level, which then develops specific computer-aided control system design package commands for achieving these goals. Heuristic rules are used to help with the planning at each level. Hence, the overall problem is decomposed into a series of smaller problems and their expert system is constructed in a hierarchical form.

2.1 Expert Systems and Interactive Computing

The emergence and rapid development of interactive computing has important implications for work on all aspects of automatic control. Its importance transcends the mere utility of fast, flexible and responsive computing; a successful interactive computing environment makes a specialised body of knowledge accessible and therefore shareable. If it has an appropriate interface it also allows an investigator easily to create and modify systems and hence easily to experiment with them. A graphics interface is ideal for many engineering applications since research has shown that a wide variety of information is best digested and interpreted in this form (Haber and Wilkinson, 1982).

A striking feature of all the applications of expert systems which are described in the literature is their highly domain-specific nature; powerful applications can only be expected in areas where a highly organized body of knowledge exists. Currently many such areas, and this is certainly the case with control engineering, have seen the development of a variety of interactive design packages which require very considerable skill for their proper use. There is thus a real need for the deployment of expert systems techniques in making such packages accessible and more flexibly useful.

2.2 Need for Expert System Design Environment

The need is particularly acute for interactive design. Design involves a variety of complex trade-offs as well as exploratory and experimental processes. It is always difficult to reconcile the conflicting aspects of initial specifications, and a designer has to decide on the cost he is prepared to pay for any particular feature or constraint. Expert systems have been found useful in solving such problems of judgement based on experience.

The use of expert systems can provide a high-level design environment which is powerful, supportive, flexible, broad in scope, and readily accessible to non-expert users (Taylor and Frederick, 1984). They can guide a user through a complex design process, help him to use the facilities available and provide appropriate explanations at each stage of the design process. A powerful package for computer-aided control system design (CACSD) must have not only a user-friendly interface, reliable numerical software, good interactive graphics capabilities, and good database management, but also an extensive control engineering knowledge base. Bünz and Gütschow (1985) have emphasised this point during their description of their interactive CACSD package. Hence the users of future packages will essentially be supported by a comprehensive knowledge base, accessed through an expert system.

CHAPTER THREE

INDICATORS OF STABILITY, PERFORMANCE AND ROBUSTNESS

3.1 Introduction

In the analysis and design of multivariable feedback systems, the designer has to handle a range of aspects of closed-loop system behaviour. In the context of control system design, any graph generated by the computer and presented to the designer, indicating some aspects of the system behaviour, is termed an indicator. Indicators for stability, performance, robustness, sensitivity and normality are proposed by MacFarlane and Hung (1985). In this section, we discuss the three key aspects of the design problem: stability, performance and robustness. The discussion of stability and performance indicators is not new, but is included for completeness. The main result is the use of gain divergences to monitor the robustness properties of the feedback loop and its relationships to a normality indicator (Hung and MacFarlane, 1982; MacFarlane and Hung, 1985).

3.2 Indicators of Stability

The generalised Nyquist diagram is a good indicator of the closed-loop stability of a system (MacFarlane and Hung, 1985). It is obtained by computing the characteristic value decomposition (CVD) of G(s) at each s = jω within the frequency region of interest. It is well-established that the generalised Nyquist diagram may be used to give a natural generalisation of Nyquist's fundamentally important Stability Criterion (MacFarlane and Postlethewaite, 1977; Desoer and Wang, 1980; Smith, 1981).

Theorem 3.2.1 (Generalised Nyquist Stability Criterion)

Let $G(s) \in \mathbb{R}_p(s)^{m \times m}$ be a proper rational function of order $m \times m$. The closed-loop system $[I_m + G(s)]^{-1} G(s)$ is stable if and only if

$$-1 \notin g \circ D_{NYQ} \text{ and } \$E(g \circ D_{NYQ}, -1) = \$SMP[G(s), \mathbb{C}_+]$$

where

$g \circ D_{NYQ}$ denotes the set of eigenloci,

$\$E(g \circ D_{NYQ}, -1)$ denotes the number of encirclements around the point $(-1+j0)$,

and $\$SMP[G(s), \mathbb{C}_+]$ denotes the number of Smith-McMillan poles of $G(s)$ in the closed right-half frequency plane.

It should be noted that anti-clockwise encirclements are taken as positive.

Proof: See MacFarlane and Postlethwaite (1977).

A corresponding generalisation of the classical Evan's Root Locus diagram by Postlethwaite and MacFarlane (1979) has led to the generalised root locus diagram (multivariable root locus). Presented in this diagram are the characteristic frequency loci which are traced out by the eigenvalues of the closed-loop frequency matrix $S(g)$ as the gain variable g traverses the negative real axis in the gain plane. It has been shown that these loci are the 180° phase contours of the characteristic gain function $g(s)$ in the frequency plane (Postlethwaite and MacFarlane, 1979).

The generalised Nyquist diagram is plotted in the s-plane. A generalised Bode diagram is obtained by plotting the characteristic gains and phases along a logarithmic frequency scale. It has been shown that the well-known Bode relationships between gain and phase behaviour can be extended to the multivariable case (Smith, 1982) with some additional restrictions. For practical consideration of closed-loop stability, the generalised Bode diagram is very useful because it helps to determine the gain-phase trade-offs during design.

3.3 Indicators of Performance

The principal gains can be used as indicators of closed-loop performance of a system (MacFarlane and Hung, 1984). These are obtained by computing the singular value decomposition (SVD) of G(s) at each s = $j\omega$ within the frequency region of interest. They generalise the scalar concept of gain to the multivariable case (MacFarlane and Scott-Jones, 1979). When they are plotted in generalised Bode diagram form, the maximum and minimum singular values give an upper and lower bound on the characteristic gain moduli.

When considering the performance of a closed-loop system, one is concerned with the ability of the system to track commands, and to reject noise and disturbances. For the standard multivariable feedback arrangement of Fig. 3.1, we have that

$$y(s) = L^{-1}(s) \ r(s) + F^{-1}(s) \ d(s) - L^{-1}(s) \ n(s)$$

$$:= y_r(s) + y_d(s) + y_n(s)$$

where

y(s) is the output transform vector, and

r(s), d(s) and n(s) are the reference input, disturbance and output sensor noise transform vectors respectively.

F(s) = I + G(s)K(s) = I + Q(s) is the return-difference operator, and

L(s) = I + Q^{-1}(s) is the inverse return-difference operator,

where Q(s) is the open-loop gain matrix.

Hence, the closed-loop system performance in terms of reference input tracking, disturbance rejection and noise rejection depends on the singular values of F^{-1} and L^{-1}, and these in turn can be related to the singular values of Q. Bounds using singular values which relate F^{-1} and L^{-1} with Q have been given by MacFarlane and Hung (1984). This enables the maximum and minimum principal gains of the open-loop gain matrix Q to be used as indicators of closed-loop performance. Doyle and Stein (1981) have discussed them as σ-plots. In general, we would like to obtain large $\underline{\sigma}(Q)$ over a large frequency range (from the low to the intermediate frequency region) to make the error due to input tracking and disturbances small. At high frequencies, we want small $\bar{\sigma}(Q)$ to make the error due to sensor noise small and the system robust. ($\bar{\sigma}(Q)$, $\underline{\sigma}(Q)$ denote the maximum and minimum singular values of Q respectively.)

3.4 Indicators of Robustness

Another very important aspect of feedback system behaviour is robustness. By robustness, we mean the ability of a system to remain stable despite model inaccuracies and parameter variations. To obtain a robust system is very important because the prime reason for using feedback control is to combat uncertainties. From a performance point of view we are concerned with the effect of perturbations on singular values, and from a stability point of view their effect on characteristic values.

3.4.1 Normality of a matrix

It is well-known that the eigenvalues of a normal matrix are relatively insensitive to perturbations (Wilkinson, 1965). Therefore, we aim to obtain a compensated system which is as normal as possible, especially near the critical crossover frequency.

Let $G \in \mathbb{C}^{m \times m}$ be an m x m complex matrix. In a characteristic value decomposition (CVD)

$$G = W \Lambda W^{-1}$$

where W is the eigenvector frame matrix, and

Λ is a diagonal matrix with eigenvalues on its diagonal.

Theorem 3.4.1

G is normal iff

G has a complete orthonormal system of eigenvectors.

Proof: See Hung and MacFarlane (1982)

In a singular value decomposition (SVD)

$$G = Y \Sigma U^*$$

where Y and U are the output and input gain frame matrices respectively, and

Σ is a diagonal matrix with singular values on its diagonal.

A matrix G is said to be gain balanced if all the singular values are equal. Also, it is said to be aligned if $U^* \cdot Y$ is a diagonal matrix.

Theorem 3.4.2

Let $G \in \mathbb{C}^{m \times m}$

G is normal iff

G is partially gain balanced and partially aligned.

The proof is given in Appendix E.

Let $G \in \mathbb{C}^{m \times m}$ have a Schur Triangular Decomposition (STD)

$$G = S \, T_u \, S^*$$
$$= S \, (D + T) \, S^*$$

where S is unitary,

T_u is upper triangular,

D is diagonal containing the eigenvalues of G, and

T is the strictly upper triangular part of T_u.

Theorem 3.4.3

Let $G = S (D + T) S^*$ be an STD of G,

$$G \text{ is normal iff}$$

$$T \text{ is } 0 \quad .$$

Proof: (==>)

G is normal

==> G has a CVD $G = W \wedge W^*$ where W is the unitary eigenvector frame

==> T = 0

(<==)

T = 0

==> $G = S D S^*$

==> G has a complete orthonormal system of eigenvectors

==> G is normal (by Theorem 3.4.1)   □

Theorem 3.4.4

Let $G \in \mathbb{C}^{mxm}$

$$G \text{ is normal iff}$$

$$g_i = \sigma_i e^{j\theta_i}$$

$$i = 1, \ldots, m$$

where $\{g_i\}$ is the set of characteristic values (eigenvalues),

$\{\sigma_i\}$ is the set of principal gains (singular values), and.

$\{\theta_i\}$ is the set of principal phases.

Proof: (<==)

Let $G \in \mathbb{C}^{mxm}$ have a Schur Triangular Decomposition as

$$G = S (D + T) S^* \qquad \text{where S is unitary}$$

$$= S (\operatorname{diag}\{g_i\} + T) S^*$$

$$= S \ (\ \text{diag}\{\sigma_i e^{j\theta_i}\} + T \ ) \ S^* \qquad\qquad [3.4.1.1]$$

Hence, taking the Frobenius norm and squaring both sides, we have

$$\| G \|_F^2 = \| \text{diag}\{\sigma_i e^{j\theta_i}\} + T \|_F^2 \qquad\qquad [3.4.1.2]$$

since the Frobenius norm of a matrix is invariant under a unitary

transformation. The equation becomes

$$(\ \Sigma \ \sigma_i^2 \ ) = (\ \Sigma \ \sigma_i^2 \ ) + \| \ T \ \|_F^2 \qquad\qquad [3.4.1.3]$$

Hence, T = 0 which implies that G is normal by Theorem 3.4.3.

(==>) (This part of the proof was first given by Dr. Y.S. Hung)

$$G = Y \cdot \text{diag}\{g_i\} \cdot Y^* \qquad \text{where Y is unitary}$$

$$= Y \cdot \text{diag}\{|g_i|\} \cdot \text{diag}\{e^{j\angle g_i}\} \cdot Y^*$$

$$= Y \cdot \text{diag}\{|g_i|\} \cdot U^*$$

where

$$U^* = \text{diag}\{e^{j\angle g_i}\} \cdot Y^* \ .$$

This is in fact an SVD of G with

$$\Sigma = \text{diag}\{|g_i|\} \qquad\qquad [3.4.1.4]$$

and

$$Y \cdot U^* = Y \cdot \text{diag}\{e^{j\angle g_i}\} \cdot Y^*. \qquad\qquad [3.4.1.5]$$

From [3.4.1.4], we have

$$| \ g_i \ | = \sigma_i \qquad\qquad [3.4.1.6]$$

From [3.4.1.5], we have

$$\angle g_i = \theta_i \qquad\qquad [3.4.1.7]$$

Equations [3.4.1.6] and [3.4.1.7] together imply that

$$g_i = \sigma_i e^{j\theta_i} \ . \qquad\qquad \square$$

Theorem 3.4.5

Let G $\in$ $\mathbb{C}^{m \times m}$

G is normal iff

$$|g_i| = \sigma_i \qquad\qquad i = 1, \ldots, m$$

Proof: This follows from Theorem 3.4.4.

3.4.2   Interpretation of normality

a. From Theorem 3.4.1, we observe that a matrix is normal when its eigenvector

   frame is orthogonal i.e. all the eigenvectors are normal to each other.

   Therefore, from the characteristic value decomposition (CVD) point of view,

   a matrix not being normal is associated with the non-orthogonality of its

   eigenvector frame.

b. From Theorem 3.4.2, we observe that a matrix is normal when one of the

   following occurs:

(i)   its input and output gain frames are aligned, or

(ii)   all the singular values are equal i.e. balanced, or

(iii) there is a partial alignment and a partial gain balancing.

   Therefore, from the singular value decomposition (SVD) point of view, a

   matrix not being normal is associated with the degree of misalignment of

   the singular vector frames as well as the amount of imbalance in the

   singular values.

c. From Theorem 3.4.4 & 3.4.5, we observe that a matrix is normal when its

   singular values equal the magnitude of its characteristic values.

   Therefore, the non-normality of a matrix is associated with the divergence

   between the singular values and the magnitude of the characteristic values.

   This emcompasses aspects of both the CVD and SVD (or a Schur Triangular

   Decomposition point of view).

3.4.3   Perturbation bounds for polar factors

          Bounds have been given by Higham (1984) for the changes induced in the polar factors of a matrix by perturbations of the matrix in the following theorem.

Let $G \in \mathbb{C}^{m\times m}$ have polar decomposition

$$G = \phi M_r$$

where ϕ is the phase matrix, and

M_r is the right modulus matrix.

Theorem 3.4.6

If $\epsilon = \dfrac{\| \Delta G \|_F}{\| G \|_F}$ satisfies $K_F(G) \cdot \epsilon < 1$,

then $G + \Delta G$ has the polar decomposition

$$G + \Delta G = (\phi + \Delta\phi)(M_r + \Delta M_r)$$

where $\dfrac{\| \Delta M_r \|_F}{\| M_r \|_F} \leq \sqrt{2} \cdot \epsilon + O(\epsilon^2)$ [3.4.3.1]

and $\dfrac{\| \Delta\phi \|_F}{\| \phi \|_F} \leq (1 + \sqrt{2})K_F(G) \cdot \epsilon + O(\epsilon^2)$ [3.4.3.2]

with $K_F(G) = \| G \|_F \cdot \| G^{-1} \|_F$

Proof: See Higham (1984)

This shows that the modulus part of a polar factorisation is relatively insensitive to the effects of perturbation, but that the sensitivity of the phase part depends on $K_F(G)$. From the feedback point of view, it means that the principal gains, and hence the indicators of performance, are relatively insensitive to the effects of perturbations. Therefore, it is the effect of perturbations on the characteristic values of G with which we must be principally concerned when characterising robustness. Putting it another way, performance indicators are robust but stability indicators are not.

3.4.4 Departure from normality

Although normal matrices have nice spectral properties, they constitute only a relatively small set among general matrices. Also, it would be difficult to obtain a compensated system which is normal over the entire frequency range of interest. Therefore, we have to introduce a measure of departure from normality for a general matrix.

Definition 3.4.7: MS(G) — A Normality Indicator

Let $G \in \mathbb{C}^{m \times m}$ have an STD

$$G = S (D + T) S^*$$

We define

$$MS(G) = \frac{\| T \|_F}{\| G \|_F}$$

where $\| \cdot \|_F$ denotes the Frobenius norm of a matrix (Wilkinson, 1965; Hung and MacFarlane, 1982). Note that

$$\| G \|_F = \| S (D + T) S^* \|_F$$

$$= \| D + T \|_F$$

since the Frobenius norm is invariant under a unitary transformation.
Hence,

$$\| G \|_F^2 = \| D \|_F^2 + \| T \|_F^2$$

and

$$\| T \|_F^2 = \| G \|_F^2 - \sum_{i=1}^{m} | g_i |^2$$

where g_i are the eigenvalues of G (i = 1,...,m). Therefore, although an STD is not unique, $\| T \|_F$ is unique because it is independent of the particular

STD taken.

Definition 3.4.8: Skewness of a matrix (Hung and MacFarlane, 1982)

A matrix is skew iff it is not normal.

MS(G) has been defined as a quantitative measure of departure from normality. In other words, it is a measure of the skewness of a matrix. When the matrix is very skew, $\| T \|_F$ tends to $\| G \|_F$ and MS(G) tends to 1. In the limiting case, MS(G) = 1. On the other hand, MS(G) = 0 when G is normal. Therefore, the normality indicator MS(G) has a range from 0 to 1.

3.4.5 Measures of conditioning of a matrix with respect to its eigenvalue sensitivity

Here, we present some measures of departure from normality which are based on the characteristic value decomposition of a matrix.

Definition 3.4.9: $\kappa(W)$ - Spectral condition number of a matrix with respect to its eigenvalue problem (Wilkinson, 1965)

Let $G \in \mathbb{C}^{m \times m}$ have a CVD

$$G = W \Lambda W^{-1}$$

We define the spectral condition number

$$\kappa(W) = \| W \|_2 \cdot \| W^{-1} \|_2 \qquad [3.4.5.1]$$
$$= \bar{\sigma}(W) / \underline{\sigma}(W)$$

where $\bar{\sigma}$ & $\underline{\sigma}$ denote the maximum & minimum singular value of a matrix respectively.

Definition 3.4.10: $\text{cond}(g_i)$ - Condition number for an individual eigenvalue (Wilkinson, 1965; Stewart, 1973; Kautsky et al., 1985)

Let $G \in \mathbb{C}^{m \times m}$ have a CVD

$$G = W \Lambda V$$

where W = eigenvector frame with all the right eigenvectors,

Λ = diag$\{g_i\}$ with g_i distinct, i = 1,...,m, and

$V = W^{-1}$ with all the left eigenvectors.

Suppose w_i have been scaled so that

$$\| w_i \|_2 = 1$$

and v_i have been scaled so that

$$v_i^* \cdot w_i = 1$$

We define

$$\text{cond}(g_i) = \| v_i \|_2 \qquad \qquad [3.4.5.2]$$

Bounds on the spectral condition number of a matrix are given as follows: (Wilkinson, 1965; Kautsky et al., 1985)

$$\max_i \text{cond}(g_i) \le \kappa(G) \le \sum_{i=1}^{m} \text{cond}(g_i) \qquad [3.4.5.3]$$

Thus, $\kappa(W)$ and $\text{cond}(g_i)$ are both measures of sensitivity of eigenvalues of G which can be used as measures of robustness.

3.4.6 Robustness characterisation of individual characteristic value

Theorem 3.4.11

Let g_i be a simple eigenvalue (characteristic value) of $G \in \mathbb{C}^{m \times m}$ with right eigenvector w_i and left eigenvector v_i. Suppose w_i has been scaled so that

$$\| w_i \|_2 = 1$$

and v_i has been scaled so that

$$v_i^* w_i = 1.$$

Also, let g_i' be an eigenvalue of (G + E) with E being the perturbation to the matrix. Then

$$| g_i - g_i' | \le \| E \|_2 \| v_i \|_2 + O(\| E \|_2^2)$$

$$\leq \epsilon \parallel v_i \parallel_2 + O(\epsilon^2)$$

where $\epsilon = \parallel E \parallel_2$

Proof: See Stewart (1973), pp.295-296, Thm 4.2.

This shows that the perturbed eigenvalue is bounded by the size of the left eigenvector norm for a fixed size of perturbation. Thus if $\parallel v_i \parallel_2$ is large, g_i will be ill-conditioned. Therefore, $\parallel v_i \parallel_2$ can be used as a condition number for an individual eigenvalue of a matrix, which justifies Definition 3.4.10 above. These condition numbers may be useful when the perturbation of one eigenlocus is more important than others.

3.4.7 MS(G) in relation to gain balancing and gain direction alignment

A measure of non-normality based on the singular value decomposition of a matrix is given below.

Definition 3.4.12: CS_i - Misalignment angle (MacFarlane and Hung, 1984)

Let $G \in \mathbb{C}^{m \times m}$ have an SVD

$$G = Y \Sigma U^*$$

Also let $\Psi = U^* \cdot Y$

The quantities $\mid \Psi_{ij} \mid$ are given by $\mid u_i^* \cdot y_j \mid$ where u_i^* is the ith row of U^* and y_j is the jth column of Y. We define

$$CS_i = \cos^{-1} \mid \Psi_{ii} \mid$$
$$= \mid u_i^* \cdot y_i \mid \qquad [3.4.7.1]$$

CS_i represents the angle between the ith input and output principal gain directions. When $CS_i = 0$, the ith principal gain direction is said to be aligned. Also, we then have $cond(g_i) = 1$ and $\sigma_i = \mid g_i \mid$.

If $G \in \mathbb{R}^{2 \times 2}$ have an SVD

$$G = Y \Sigma U^t$$

$$= \begin{bmatrix} \cos\theta_1 & -\sin\theta_1 \\ \sin\theta_1 & \cos\theta_1 \end{bmatrix} \begin{bmatrix} \sigma_1 & \\ & \sigma_2 \end{bmatrix} \begin{bmatrix} \cos\theta_2 & -\sin\theta_2 \\ \sin\theta_2 & \cos\theta_2 \end{bmatrix}^t .$$

we can do a scaling and multiply $1/\sigma_2$ to G without affecting the phases or misalignment angles of G.

Hence,

$$G = Y \text{ diag}\{ \sigma , 1 \} U^t$$

where $\sigma = \sigma_1 / \sigma_2$.

Next, a unitary transformation is performed on G and we let

$$\tilde{G} = Y^t G Y$$

$$= \begin{bmatrix} \sigma & 0 \\ 0 & 1 \end{bmatrix} \begin{bmatrix} \cos\theta & -\sin\theta \\ \sin\theta & \cos\theta \end{bmatrix} \qquad\qquad [3.4.7.2]$$

where $\theta = (\theta_1 - \theta_2)$.

Note that G and $\tilde{G}$ are equivalent matrices and the misalignment angles CS_1 and CS_2 are equal to θ. σ represents the amount of gain imbalance. Fig. 3.2 shows a plot of MS(G) against θ for different values of σ .

3.4.8 MS(G) in relation to divergences between characteristic values and singular values

Definition 3.4.12: Gain ratio

Let $G \in \mathbb{C}^{m \times m}$ have characteristic values g_i and singular values $\sigma_i (i = 1,\ldots,m)$ and they are arranged in descending order of their magnitude.

We define the gain ratios

$$\rho_i = \frac{| g_i |}{\sigma_i} \qquad\qquad i = 1,\ldots,m$$

Note that $\rho_i = 1$ when G is normal ($i = 1,\ldots,m$).

Theorem 3.4.13

The normality indicator MS(G) for a matrix $G \in \mathbb{C}^{m \times m}$ can be expressed explicitly in terms of its gain ratios ρ_i as

$$MS(G) = \left[\frac{\displaystyle\sum_{i=1}^{m} \sigma_i^2 (1 - \rho_i^2)}{\displaystyle\sum_{i=1}^{m} \sigma_i^2} \right]^{1/2} \qquad [3.4.8.1]$$

$$i = 1, \ldots, m$$

Proof:

Let $G \in \mathbb{C}^{m \times m}$ have an SVD

$$G = Y \cdot \Sigma \cdot U^* \qquad [3.4.8.2]$$

where Y and U are unitary, and

$\Sigma = \text{diag}\{\sigma_i\}$ (i=1,...,m)

Also, let G have a Schur Triangular Decomposition

$$G = S \cdot T_u \cdot S^*$$
$$= S \cdot (D + T) \cdot S^* \qquad [3.4.8.3]$$

where S is unitary, and

$D = \text{diag}\{g_i\}$ with i=1,...,m, and

T is the strictly upper triangular part of T_u.

Taking the Frobenius norm of G in [3.4.8.2] and noting that the Frobenius norm is invariant under a unitary transformation gives

$$\| G \|_F = \| Y \cdot \Sigma \cdot U^* \|_F$$
$$= \| \Sigma \|_F$$
$$= (\Sigma \sigma_i^2)^{1/2} \qquad [3.4.8.4]$$

Taking the Frobenius norm of G in [3.4.8.3] gives

$$\| G \|_F = \| S \cdot (D + T) \cdot S^* \|_F$$

$$= \| \ D + T \ \|_F$$

$$= (\ \Sigma \ |g_i|^2 + \| \ T \ \|_F^2 \ )^{1/2} \qquad [3.4.8.5]$$

Equating [3.4.8.4], [3.4.8.5] we then obtain

$$\| \ T \ \|_F^2 = \Sigma \ \sigma_i^2 - \Sigma \ |g_i|^2$$

By definition,

$$MS(G) = \frac{\| \ T \ \|_F}{\| \ G \ \|_F}$$

and so

$$MS(G)^2 = \frac{\| \ T \ \|_F^2}{\| \ G \ \|_F^2}$$

$$= \frac{\Sigma \ \sigma_i^2 - \Sigma \ |g_i|^2}{\Sigma \ \sigma_i^2}$$

$$= \frac{\displaystyle\sum_{i=1}^{m} \ \sigma_i^2 \ (1 - \frac{|g_i|^2}{\sigma_i^2})}{\displaystyle\sum_{i=1}^{m} \ \sigma_i^2}$$

$$= \frac{\displaystyle\sum_{i=1}^{m} \ \sigma_i^2 \ (1 - \rho_i^2)}{\displaystyle\sum_{i=1}^{m} \ \sigma_i^2}$$

Hence,

$$MS(G) = \left[\frac{\displaystyle\sum_{i=1}^{m} \ \sigma_i^2 \ (1 - \rho_i^2)}{\displaystyle\sum_{i=1}^{m} \ \sigma_i^2} \right]^{1/2}$$

□

The characteristic values g_i and singular values σ_i are assumed to have been arranged in descending order of their magnitude so far. However, if the gain ratios ρ_i are obtained for an arbitary ordering of g_i and σ_i,

[3.4.8.1] would still be valid.

From the above analysis, it can be seen that the amount of divergence between the magnitude of the corresponding pairs of characteristic and singular values is neatly related to the normality indicator MS(G). That is, the skewness of a matrix may be expressed as a summation of divergences of magnitude of the corresponding pairs of characteristic and singular values.

If G is written as in [3.4.7.2], we can plot MS(G) against the ratio between the maximum singular value (σ) and the characteristic value of maximum magnitude (see Fig. 3.3). This is done for different gain ratios σ. Fig. 3.3 shows that MS(G) is a monotonic increasing function of the gain ratio. Therefore, it is reasonable to use gain divergences as an indication of robustness. This result leads to the concept of a set of primary indicators which will be defined later.

Theorem 3.4.14

Let $G \in \mathbb{C}^{m \times m}$ have characteristic values g_i, singular values σ_i and singular phases θ_i (i=1,...,m).

$$\sum_{i=1}^{m} \angle g_i = \sum_{i=1}^{m} \theta_i \qquad [3.4.8.6]$$

$$\prod_{i=1}^{m} |g_i| = \prod_{i=1}^{m} \sigma_i \qquad [3.4.8.7]$$

That is, the sum of characteristic phases equals the sum of principal phases and the product of the singular values equals the product of the magnitude of the characteristic values.

Proof:

Let $G \in \mathbb{C}^{m \times m}$ have a CVD

$$G = W \cdot \Lambda \cdot W^{-1}$$

$$= W \cdot diag\{g_i\} \cdot W^{-1}$$

$$= W \cdot \text{diag}\{|g_i|\} \cdot \text{diag}\{e^{j\angle g_i}\} \cdot W^{-1} \qquad [3.4.8.8]$$

Also let G have an SVD

$$G = Y \cdot \Sigma \cdot U^*$$

$$= (Y \cdot \Sigma \cdot Y^*)(Y \cdot U^*)$$

$$= (Y \cdot \text{diag}\{\sigma_i\} \cdot Y^*)(P \cdot \text{diag}\{e^{j\theta_i}\} \cdot P^*) \qquad [3.4.8.9]$$

where $Y \cdot U^*$ is called the phase matrix. The CVD of which gives the singular phases θ_i of G and P is another unitary matrix.

Taking the determinant of G in [3.4.8.8.]

$$\det(G) = \prod_{i=1}^{m} |g_i| \cdot \prod_{i=1}^{m} e^{j\angle g_i} \qquad [3.4.8.10]$$

Taking the determinant of G in [3.4.8.9]

$$\det(G) = \prod_{i=1}^{m} \sigma_i \cdot \prod_{i=1}^{m} e^{j\theta_i} \qquad [3.4.8.11]$$

From [3.4.8.10] & [3.4.8.11]

$$\prod_{i=1}^{m} |g_i| = \prod_{i=1}^{m} \sigma_i$$

and

$$\sum_{i=1}^{m} \angle g_i = \sum_{i=1}^{m} \theta_i$$

$\square$

3.4.9 MS(G) in relation to the variation of the spectrum under perturbations

The variation of the spectrum of a skew matrix G of order m x m when perturbed to $G(I + \Delta)$ can be bounded in terms of MS(G). By a result of Henrici (1962) and Wilkinson (1965), for any eigenvalue g of $G(I + \Delta)$, there exists an eigenvalue g_i of G such that

$$\frac{|\,g - g_i\,|}{1 + \dfrac{MS(G)}{\alpha} + \,\ldots\ldots\, + \dfrac{MS(G)^{m-1}}{\alpha^{m-1}}} \leq \|\,G\,\|_2 \cdot \|\,\Delta\,\|_2 \qquad [3.4.9.1]$$

where

$$\alpha = \frac{|\,g - g_i\,|}{\|\,G\,\|_F} \,. \qquad [3.4.9.2]$$

From Theorem 3.4.3, MS(G) = 0 when the matrix is normal. Therefore, the above equation reduces to

$$|\,g - g_i\,| \leq \|\,G\,\|_2 \cdot \|\,\Delta\,\|_2 \,. \qquad [3.4.9.3]$$

If G is close to normal with a small skewness measure MS(G), the spectrum would be expected to remain reasonably insensitive to perturbations.

From [3.4.9.1],

$$\frac{|\,g - g_i\,|}{1 + \dfrac{MS(G)}{\alpha} + \,\ldots\ldots\, + \dfrac{MS(G)^{m-1}}{\alpha^{m-1}}} \leq \|\,G\,\|_2 \cdot \|\,\Delta\,\|_2 \leq \|\,G\,\|_F \cdot \|\,\Delta\,\|_2$$

$$[3.4.9.4]$$

Substituting [3.4.9.2], we have

$$\frac{\alpha}{1 + \dfrac{MS(G)}{\alpha} + \,\ldots\ldots\, + \dfrac{MS(G)^{m-1}}{\alpha^{m-1}}} \leq \|\,\Delta\,\|_2 = \delta$$

$$[3.4.9.5]$$

Putting $x = \dfrac{MS(G)}{\alpha}$ $\qquad\qquad\qquad\qquad$ [3.4.9.6]

Equation [3.4.9.5] becomes

$$\frac{\alpha}{1 + x + x^2 + \,\ldots\ldots\, + x^{m-1}} \leq \delta$$

$$[3.4.9.7]$$

That is,

$$\frac{MS(G)}{\delta} \le x + x^2 + \dots + x^m \qquad\qquad [3.4.9.8]$$

Let $f(x) = x + x^2 + \dots + x^m \qquad\qquad [3.4.9.9]$

Note that $f(x)$ is a monotonically increasing function for $x > 0$ and f is continuous. Therefore, function f has an inverse and $x = f^{-1}(\frac{MS(G)}{\delta})$ is a unique non-negative solution of [3.4.9.8]. That is,

$$\frac{MS(G)}{\alpha} \ge f^{-1}(\frac{MS(G)}{\delta})$$

and

$$\alpha \le \frac{MS(G)}{f^{-1}(\frac{MS(G)}{\delta})} \qquad\qquad [3.4.9.10]$$

For $m = 2$,

$$\alpha \le \frac{2 \cdot MS(G)}{-1 + \sqrt{1+4 \cdot MS(G)/\delta}} \qquad\qquad [3.4.9.11]$$

and

$$|g - g_i| \le \frac{2 \cdot MS(G) \cdot \|G\|_F}{-1 + \sqrt{1 + 4 \cdot MS(G)/\delta}} \qquad\qquad [3.4.9.12]$$

Hence, we can obtain a graph with α against $MS(G)$ for different amounts of perturbation δ (see Fig.3.4). We observe that α is a concave, montonic increasing function of $MS(G)$. The proof of this for the general case is given in Appendix A. Thus, $MS(G)$ has been related to the susceptibility of the characteristic gains to perturbations. This confirms that $MS(G)$ is an indicator of normality and hence robustness. Therefore, given the amount of perturbation over a certain frequency region, the upper bound of the perturbed eigenloci is known. This is useful for the assessment of robust stability of a system.

3.4.10 MS(G) in relation to the spectral radius expansion factor

From the previous section, we have seen how the perturbed eigenvalue bound (α) is related to the skewness measure (MS(G)) for different amounts of perturbations (δ). Here, we express the perturbed eigenvalue bound in terms of a radius expansion factor and relate it to MS(G).

Definition 3.4.16: Spectral radius expansion factor

Using the same notation as in Section 3.4.9., the spectral radius expansion factor is defined as:

$$\beta = \frac{\alpha}{\| \Delta \|_2} \ . \qquad\qquad [3.4.10.1]$$

From [3.4.10.1] and [3.4.9.2], we obtain

$$| g - g_i | = \beta \| \Delta \|_2 \| G \|_F \ . \qquad\qquad [3.4.10.2]$$

When MS(G) = 0, from [3.4.9.3], we obtain

$$| g - g_i | \leq \| \Delta \|_2 \| G \|_F \ . \qquad\qquad [3.4.10.3]$$

In the limiting case,

$$| g - g_i | = \| \Delta \|_2 \| G \|_F \ . \qquad\qquad [3.4.10.4]$$

Therefore, we let $\beta = 1$ when MS(G) = 0 .

Substituting [3.4.10.2] into [3.4.9.4],

$$\frac{\beta \cdot \| \Delta \|_2 \cdot \| G \|_F}{1 + (\frac{MS(G)}{\alpha}) + \ \ldots\ldots\ + (\frac{MS(G)}{\alpha})^{m-1}} \leq \| \Delta \|_2 \cdot \| G \|_F \qquad\qquad [3.4.10.5]$$

and hence

$$\beta \leq 1 + \frac{MS(G)}{\alpha} + (\frac{MS(G)}{\alpha})^2 + \ \ldots\ + (\frac{MS(G)}{\alpha})^{m-1} \qquad\qquad [3.4.10.6]$$

Substituting [3.4.10.1] into [3.4.10.6] and letting $\delta = \| \Delta \|_2$

$$\beta^m \le \beta^{m-1} + \beta^{m-2}(\frac{MS(G)}{\delta}) + \beta^{m-3}(\frac{MS(G)}{\delta})^2 + \ldots + (\frac{MS(G)}{\delta})^{m-1} \quad [3.4.10.7].$$

For m = 2,

$$\beta \le (1 + \sqrt{1+ 4\cdot MS(G)/\delta})/2 \qquad [3.4.10.8]$$

Figure 3.5 plots β against MS(G) for different amounts of perturbation δ in the limiting case. The radius expansion factor β may be used as a measure of spectral sensitivity.

3.5 Summary

It is well-known that characteristic values (eigenvalues) are of fundamental importance for assessing the closed-loop stability of a system. For the assessment of performance, however, principal gains (singular values) are used for they give a more accurate measure of gain. In this chapter, it has been shown that the divergences between the magnitude of characteristic values and principal gains give an indication of the normality of a system. A normality indicator MS is defined and its relationship with gain divergences is also demonstrated.

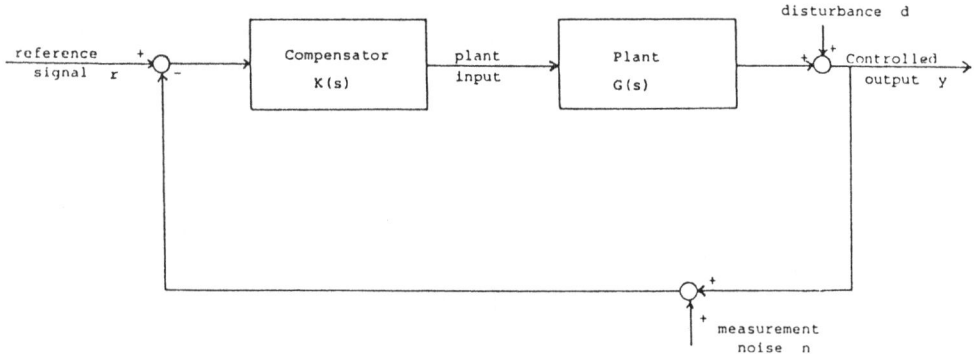

Fig.3.1 Multivariable linear feedback control system

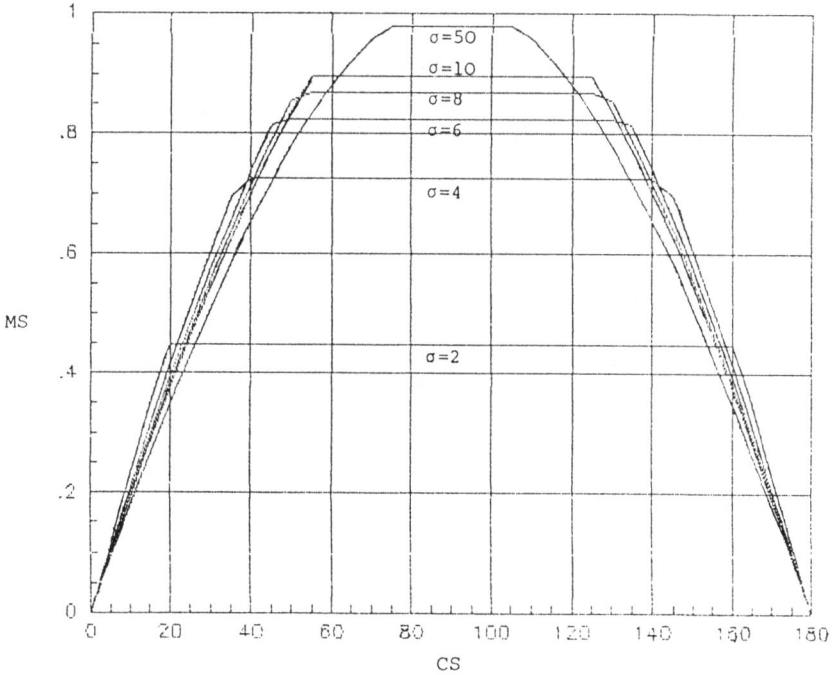

Fig. 3.2 MS(G) against the misalignment angle(CS)
for different gain ratios (σ).

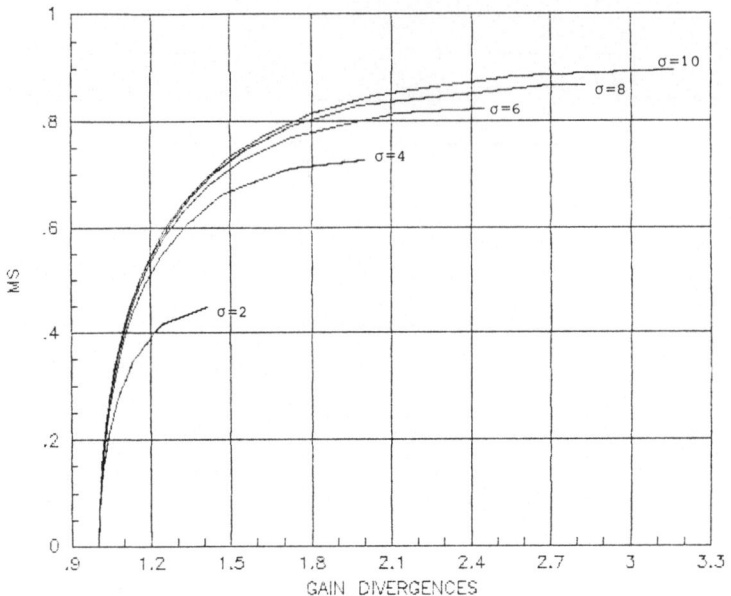

Fig. 3.3 MS(G) against gain divergences for different gain ratios (σ).

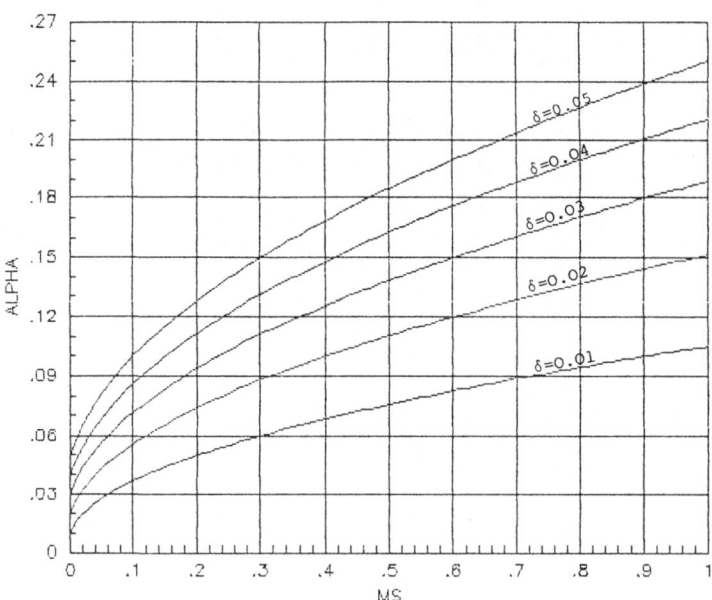

Fig. 3.4 α against MS(G) for different amount of perturbation (δ).

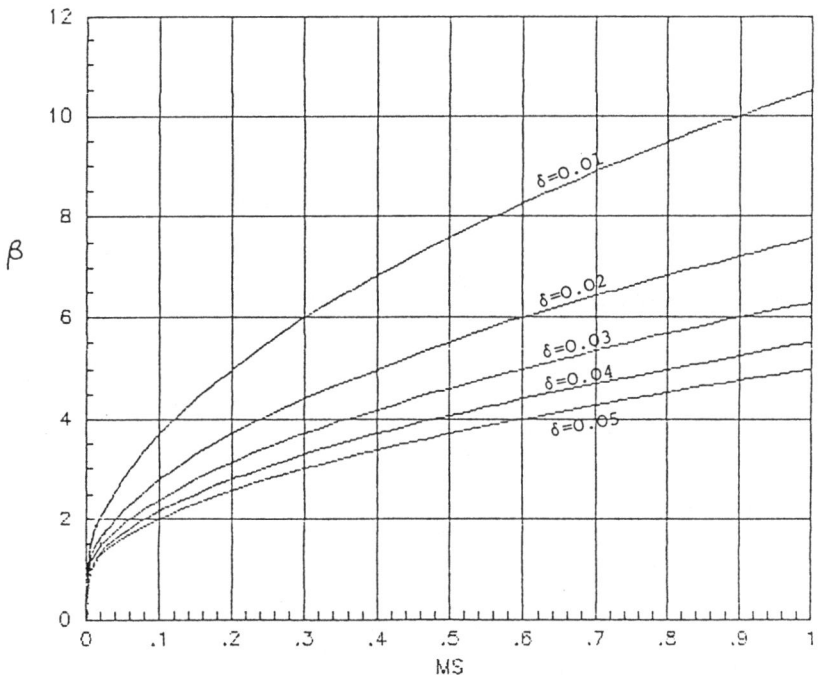

Fig. 3.5 Spectral radius expansion factor (β)
against MS(G) for different amount of
perturbation (β).

CHAPTER FOUR

THE PRIMARY INDICATORS FOR INTERACTIVE DESIGN

4.1 A Complete Set of Indicators for Interactive Design

A complete set of indicators is a set which is sufficient to characterize accurately all relevent aspects of closed-loop behaviour (MacFarlane and Hung, 1985). Any interactive computer-aided control system package must present the designer with a complete set of indicators. These are important in developing an expert system for computer-aided control system design because they are the key to understanding the system and support the interaction between the designer and the computer. Indicators can be used in the following ways:

1. A visual presentation of the indicators to the designer will give him an understanding of the complexity of the system. From the indicators, the characteristics and nature of the system can be evaluated as well as the sources of difficulty in controlling the system.

2. A classification of systems may be made with the information obtained from the indicators.

3. The merits of different designs may be compared.

4. In a complete and systematic design approach, the indicators can contribute to the formulation of suitable design specifications.

5. Design procedures may be developed which involve a systematic manipulation of indicators into a satisfactory form.

4.2 The Primary Indicators

In the frequency response analysis of a multi-input, multi-output

(MIMO) feedback control system, we are provided with a large set of indicators for the system. Different indicators are needed to represent different aspects of the behaviour of the system. Essentially, the large set of indicators arises because of the complexity of the multivariable feedback problem. The designer has the problem of observing changes in a number of indicators at each stage of the design. This can make it difficult for the designer to decide what to do next.

However, with a systematic design approach, the design process may be divided into various stages. At each stage, not all of the available information is essential. In fact, an excess of information may even confuse the designer unless he is highly experienced. Therefore, it would be useful to present him with a set of primary indicators which would specify only the major aspects of the behaviour of the system.

Proposition 4.1:

Characteristic gains and phases

plus

principal gains

are the primary indicators of a feedback control system.

The loci of characteristic values and principal gains may be plotted on the same drawing in Bode form. This is considered to give the minimum set of indicators which enables us to provide a complete description of the system. These indicators are "primary" in the sense that they contain the most crucial information about the system : stability, performance and robustness. Furthermore, the designer should look at the primary indicators initially and use them to guide him through the design process. Thirdly, various other aspects of the system such as the condition number, normality measure, root sensitivity etc. can be deduced from the primary indicators.

Interpretation of the primary indicators

a. The characteristic gains and phases give an indication of the closed-loop stability of the system.

b. The principal gains give an indication of the performance of the system.

c. The divergences between the characteristic and principal gains give an indication of the robustness of the system.

4.3 The Secondary Indicators

All those indicators which are not primary indicators are called secondary indicators. Examples are misalignment angles CS_i , normality indicator $MS(G)$, robust stability indicator, sensitivity indicator, condition numbers ($\kappa(G)$ or cond (g_i)), generalised root locus diagram, Nyquist array, inverse Nyquist array (with Gershgorin or Ostrowski circles), generalised Nyquist diagram (with M circles), closed-loop step responses etc.. The secondary indicators allow the designer to obtain precise values for other measurements that characterize a control system. These measurements include condition numbers, misalignment angle, normality measure, root-sensitivity, amount of interaction, resonance magnitude and frequency, bandwidth of each loop, disturbance rejection, rise times, overshoots etc.. Thus, the relative merits of different compensators can be compared using secondary indicators.

4.4 Manipulation of Gains and Phases

4.4.1 Bode's frequency response design technique

The work of Nyquist (1932) on his stability criterion for feedback

systems laid the corner-stone for feedback control theory. It defined the
conditions for stability of negative feedback systems and made an immense
contribution to the entire field. Bode (1940), however, took a further step
in his famous work on the analysis of the relationship between attenuation and
phase in feedback amplifier design. For a minimum phase system, his formula
is

$$\angle g(j\omega_c) = \frac{1}{\pi} \cdot \int_{\infty}^{\infty} (\frac{d \, \log|g(j\omega)|}{du})(\log \, \coth(\frac{|u|}{2})) \, du \qquad [4.4.1.1]$$

where

$\angle g(j\omega_c)$ is the phase shift at ω_c, and

$u = \log(\omega/\omega_c)$.

The equation shows that phase shift is proportional to the derivative
of the gain. Also, it shows that phase shift depends upon the derivative at
all frequencies. In addition, the contributions to total phase shift at a
particular frequency do not add up equally but in accordance with the
weighting function $\log \coth(|u|/2)$. Computation of phase can be obtained from
the gain characteristic by using [4.4.1.1].

It becomes clear that the design of a feedback system for stability
involves an appropriate manipulation of the gain and phase of the open-loop
transfer function. For example, the rate at which the loop gain is attenuated
is an important consideration because it determines the amount of phase shift
in the cut-off region. Therefore, Bode's design technique is based on the
shaping of the characteristic locus for a feedback system. It has been shown
by Smith (1982) that the Bode gain-phase relationships hold in the
multivariable case with appropriate modifications (Hung and MacFarlane, 1982).

4.4.2 Design philosophy behind the manipulation of gains and phases

In this section, we review the motivation behind the classical
approach. The design problem is posed as follows:

Given a plant and a set of specifications, design controllers such that the resulting overall system will meet the prescribed response characteristics.

As discussed in Chapter One, there are basically three approaches to the problem. These are attribute-centred (i.e. design by synthesis), operation-centred (i.e. design by procedure) and object-centred (i.e. design by search). The first and the third approaches are associated with the optimal control technique and general optimization techniques respectively. The second approach is the approach considered here. It is a classical approach to the problem. The design process consists of combining a series of simple structured units with the plant to meet the specifications. Bode (1945) gave a list of simple structures which can be expected to be appropriate for amplifier design. Each simple structure has a certain type of characteristic which may be suitable for a certain kind of application.

In the feedback design of a control system, we are concerned with the provision of desired closed-loop behaviour. This can be established by an appropriate pattern of relationships among the basic building blocks. This may be carried out by :

(i)   noting the characteristics of various basic building blocks,

(ii)   pointing out their uses, and

(iii)   finding rules for putting the building blocks together to make up a design.

We call these basic building blocks of the controller sub-controllers. The choice of a particular combination of sub-controllers is made by the designer enabling a given set of specifications to be met. A set of general rules can be developed which may be extended to a variety of design situations by the choice of suitable parameters.

The design problem thus becomes a constructive process, involving the combination of individual pieces with simple structures. In a design situation, several combinations of sub-controllers may yield satisfactory results. The selection of any one pattern is carried out using the experience and engineering intuition of the designer.

For certain classes of system, whose characteristics are very similar, a general design pattern may become so well established that systematic procedures can be written out. This allows a set of rules to be written enabling the problem to be tackled systematically.

From the background discussed, it is evident that a characterization of the sub-controllers is most needed. These will enter into the compensated structure in a way which is simple to handle. This is very important in planning any general and systematic design approach which is likely to be of practical value.

4.4.3 Basic types of controller for SISO and MIMO systems

The main types of building blocks for SISO systems are gain adjustment, phase lead, phase lag, and lag-lead controllers. Their use in multivariable systems for the manipulation of gains and phases will be given in Section 6.5. MacFarlane and Belletrutti (1973) have discussed some sub-controllers which are multivariable in nature.

4.4.4 Objective of the manipulation in multivariable systems

In the design of a multivariable feedback control system, we are concerned with a suitable compromise between the conflicting design objectives: stability, performance and robustness. Since the design objectives are mostly concerned with closed-loop phenomena, a set of appropriate open-loop and closed-loop relationships between these quantities is needed.

It can be shown that the stability, performance and robustness of a multivariable feedback system may be discussed in terms of the primary indicators of the open-loop transfer function matrix $G(s)K(s)$. Therefore, it is natural to consider a design technique which is based upon methods of choosing the controller matrix $K(s)$ in such a way as to give the primary indicators a required set of properties. The desired properties of the primary indicators are summarised as follows.

Closed-loop stability

Closed-loop stability can be assessed by the generalised Nyquist diagram of the open-loop transfer function. The characteristic loci must satisfy the Generalised Nyquist Stability Criterion. In the intermediate frequency region, the rate at which the deployed gain can be rolled off without violating the Nyquist Stability Criterion determines the phase shift involved as well as the gain bandwidth. Therefore, appropriate gain-phase trade-offs have to be made for an acceptable compromise between stability and performance.

Closed-loop performance

The closed-loop system performance in terms of input tracking, disturbance rejection and noise rejection depends on the singular values of the return difference and inverse return difference operators. These can be related to the singular values of the open-loop gain matrix. Therefore, the maximum and minimum principal gains of the open-loop gain matrix can be used as indicators of the closed-loop performance.

a. In the low frequency region, the minimum principal gain of the open-loop transfer function has to be suitably large for good tracking and disturbance rejection. This can be achieved by incorporating integral action, which will bring about zero steady-state error.

b. In the high frequency region, however, the maximum principal gain should be small for sensor noise rejection as well as for stability reason. To reduce interaction, one aims to balance up the principal gains at high frequencies and obtain high gains at low frequencies.

Robustness

It has been shown that the robustness of the system can be assessed by the divergences between the corresponding pairs of principal gain and characteristic gain loci of the open-loop transfer function. In the manipulation of gains and phases at both ends of the frequency range, one strives to obtain a minimum degree of divergence between the characteristic and principal gains in the intermediate frequency region.

4.5 Summary

We are now ready to establish a design method which is based on an appropriate manipulation of the primary indicators. This is based entirely on the frequency response behaviour of the open-loop system transfer function. Three design techniques, which represent three hierarchically organized design levels, will be presented. These techniques have been specifically developed with expert system implementation in mind. The characteristic and usage of each kind of sub-controller developed in the design techniques will be noted. In particular, we aim to obtain rules for putting the sub-controllers together to make up a design. The design effort can now be focused on a systematic generation of forward path sub-controllers which, when cascaded with the plant transfer matrix, will give a desired set of characteristic and principal loci.

CHAPTER FIVE

SIMPLE DESIGN TECHNIQUE

5.1 Introduction

In this section, a design technique is presented which is based on the design philosophy of manipulating gains and phases using simple structured sub-controllers for multivariable systems. The design objective is achieved by a manipulation of the primary indicators of the open-loop transfer function into a suitable form.

5.1.1 Design strategy

1. The primary indicators are used for the assessment of stability, performance and robustness of the system.

2. The design of the controller is separated into two distinct stages:

i) Design for good phase properties in the high frequency region.

ii) Design for good gain properties in the low frequency region.

3. The design of the controller is carried out firstly for the high frequency region and then for the low frequency region.

4. Particular attention is paid to the trade-offs in the intermediate frequency region associated with the manipulation of gains and phases at both ends of the frequency scale.

5. The design procedure may be represented as a set of production rules or a structured representation at each design stage. They serve as guidelines for the designer. The design process can be made semi-automatic when a suitable knowledge representation is obtained, which can be extended to deal with more difficult systems.

6. The method is iterative, and one can alternate between the two stages

until the final specifications are met or it is decided that a more complex form of controller is needed.

7. The final controller is a product of the individual sub-controllers and it is the combined effect of these sub-controllers which produces the required overall control action.

5.2 Sub-controllers Based on Singular Value Decomposition (SVD)

(Hung and MacFarlane, 1982)

In this section, the two main types of sub-controller used in the Simple Design Technique (SDT) will be given. This work stems from that of Hung and MacFarlane (1982) on standardization at high and low frequencies. However, a more systematic treatment is given here.

Let $G(s) \in \mathbb{C}^{m \times m}$ have an SVD at one specific frequency s_w

$$G(s_w) = Y_w \, \Sigma_w \, U_w^*$$
[5.2.1]

where Y_w and U_w are unitary

and $\Sigma_w = \text{diag}\{\sigma_i^w\}$ $\qquad i = 1, \ldots, m.$

5.2.1 High frequency sub-controller (HFS)

Asymptotically, as $|s| \longrightarrow \infty$, $G(s)$ takes the form (Hung and MacFarlane, 1981)

$$G(s_\infty) = Y_\infty \, \Sigma_\infty^G \, U_\infty^t$$
[5.2.2]

where Y_∞ and U_∞ are real orthogonal matrices

and $\Sigma_\infty^G = \text{diag}\{\, \sigma_i^\infty / s_i^{r_i}\,\}$

where σ_i^∞ are real and r_i are the orders of infinite zeros of $G(s)$.

We define a high frequency sub-controller for $G(s)$ as

$$K_\infty = U_\infty \, \Sigma_\infty^k \, Y_\infty^t$$
[5.2.3]

where $\Sigma_\infty^k = \text{diag}\{ k_i^\infty \}$

with k_i^∞ real constants.

When this sub-controller is cascaded with the system, and at high frequencies,

$$G(s_\infty)K_\infty = Y_\infty \, \Sigma_\infty^G \, U_\infty^t \, U_\infty \, \Sigma_\infty^k \, Y_\infty^t$$

$$= Y_\infty \, \Sigma_\infty^G \, \Sigma_\infty^k \, Y_\infty^*$$

The product $G(s_\infty)K_\infty$ has the following properties:

1. The input and output gain frames are aligned and thus the system is normal at high frequencies.

2. Since the system is normal, the principal and characteristic gains coincide with no divergence. Therefore, the principal and characteristic loci tend to coincide in a neighbour of $s = \infty$.

3. Since the system is normal as $|s| \longrightarrow \infty$, it is robust in the high frequency region.

4. Since the characteristic gains behave asymptotically as $\sigma_i^\infty \, k_i^\infty \, / \, s^{r_i}$, the phases of the characteristic gain loci will approach $\pm \, r_i \, \pi/2$.

5. The principal gains can be balanced up by a suitable choice of k_i^∞ if the orders of infinite zeros, r_i, are the same.

5.2.2 Low frequency sub-controller (LFS)

Let $G(s)$ be real at $s = 0$ and take the form

$$G(0) = Y_o \, \Sigma_o \, U_o^t \qquad\qquad [5.2.4]$$

where Y_o and U_o are real orthogonal matrices.

Let

$$K_o = U_o \, \Sigma_o^K \, Y_o^t \qquad\qquad [5.2.5]$$

and define a low frequency sub-controller for $G(s)$ as

$$K_L(s) = (\frac{\alpha \, K_o}{s} + I_m) = (K_{LF} \, / \, s + I_m) \qquad [5.2.6]$$

where $\Sigma_o^K = \text{diag}\{ k_i^o \}$ $\qquad i = 1, \ldots, m$

with $k_i^{\,o}$ and α real constants.

At high frequencies, $G(s)K_L(s)$ tends to $G(s)$ and therefore $K_L(s)$ will not affect the system in the high frequency region.

As $|s| \longrightarrow 0$,

$$G(s)K_L(s) \longrightarrow G(0)(\frac{\alpha\ K_o}{s} + I_m)$$

which approximates

$$G(0)\frac{\alpha\ K_0}{s} \hspace{4cm} [5.2.7]$$

and has the following properties:

1. Since

$$G(0)\frac{\alpha\ K_o}{s} = Y_o\ \Sigma_o\ U_o^{\,t}\ U_o\ \Sigma_o^K\ Y_o^{\,t}\ \frac{\alpha}{s}$$

$$= Y_o\ \Sigma_o\ diag\{\ \alpha\ k_i^{\,o}/s\ \}\ Y_o^{\,t}$$

   the input and output gain frames are aligned and the system is normal at very low frequencies.

2. Since the system is normal, the principal and characteristic gain loci tend to coincide in a neighbourhood of $s = 0$.

3. Since the system is normal as $|s| \longrightarrow 0$, it is robust in the low frequency region.

4. The principal gains can be balanced up by a suitable choice of the gain parameters $k_i^{\,o}$.

5. The phases of the characteristic gain loci of $G(0)\frac{\alpha\ K_o}{s}$ will approach $-\pi/2$ because of the effect of integrator $1/s$.

6. The effect of the integral action can be controlled by a suitable choice of the parameter α.

7. Steady-state error and low frequency interaction are eliminated because of the presence of the integrator term.

Another type of low frequency sub-controller is of the form

$$K_L(s) = (\ diag\{\alpha_i\}/s + I_m \ ) = (\ K_{LF} \ / \ s + I_m \ ) \qquad [5.2.8]$$

Phase lag of $-\pi/2$ will be added to the loci at low frequencies. Again, the effect of the integral action can be controlled by a suitable choice of α_i (i=1,...,m). This type of low frequency sub-controller is used for systems which have right-half plane poles, and hence for closed-loop stability the characteristic loci have to encircle the critical point with the correct number of anti-clockwise encirclements (e.g. see Example 5.5.5). Another use for the second type of low frequency sub-controller is for systems whose characteristic loci have different roll-off rates at low frequencies and so one cannot obtain real orthogonal frames at s = 0 (e.g. see Example 5.5.4).

5.3 Basic Design Procedure

5.3.1 Introduction

The Simple Design Technique is based on a cascaded combination of the low and high frequency sub-controllers, possibly modified by the addition of some further basic types of sub-controller for MIMO systems which may be suitable. The design is broken down into two parts which deal with the high frequency behaviour and low frequency behaviour respectively.

5.3.2 High frequency region design

The high frequency sub-controller is used first. We wish to align the input and output gain frames of the system to make it normal. We also want to manipulate the gains and phases in such a way that the characteristic loci satisfy the Nyquist Stability Criterion, and finally we wish to meet the performance specification and reduce interaction.

5.3.3 Low frequency region design

The low frequency sub-controller is the basic tool for performance manipulation. We wish to align the input and output gain frames of the system, balance up the gains, and add integral action for good performance. Particular attention is given to the trade-offs between performance and robustness in the intermediate frequency region resulting from the use of gains in K_o and K_∞.

5.4 Matrix P+I controller

Let the sub-controllers for the high frequency region design be $K_\infty K_i(s)$ and the sub-controllers for the low frequency region design be

$$(\frac{K_{LF}}{s} + I_m) K_j(s).$$

$K_i(s)$ and $K_j(s)$ are basic types of sub-controller which may be needed for further manipulation of the primary indicators. The final controller $K(s)$ is the product

$$K(s) = K_\infty K_i(s) (\frac{K_{LF}}{s} + I_m) K_j(s) \qquad [5.4.1]$$

The effects of the high and low frequency sub-controllers are combined and each comes into proper operation in the appropriate frequency region.

The complete controller becomes a matrix-proportional-plus-integral (P+I) controller. The presence of the intergral term 1/s ensures that

$$K(s) = K_\infty K_i(s) \frac{K_{LF}}{s} K_j(s)$$

at low frequencies, and

$$K(s) = K_\infty K_i(s) K_j(s)$$

at high frequencies.

It also ensures that there will be zero steady-state error in the closed-loop steady state response, since all the gain loci tend to infinity as |s| tends to zero.

5.5 Examples

Examples are now given to illustrate the use of the Simple Design Technique. A discussion on the input and output scaling of a system is given in Appendix F.

5.5.1 Example 1 (from (MacFarlane and Postlethwaite, 1977))

$$G(s) = \begin{bmatrix} s-1 & s \\ -6 & s-2 \end{bmatrix} \frac{1}{(s+1)(s+2)(1.25)}$$

Observations

This system has two poles at -1 and -2. It is a minimum phase system but it is conditionally stable. The primary indicators and the normality indicator MS of the system are shown in Fig.5.1.1 and Fig.5.1.2 respectively. We note that MS is small at ω = 6, which corresponds to small divergences between the characteristic and principal gains at that frequency. The gain divergences are higher at low frequencies than at high frequencies. The indicator MS confirms these observations. The set of generalised Nyquist diagrams is shown in Fig.5.1.3.

High frequency region design

The system has the same roll-off rates at high frequencies. A high frequency sub-controller is obtained as below:

$$K_\infty = \begin{bmatrix} 1.6185 & -1.6187 \\ -0.0003 & 1.6185 \end{bmatrix} .$$

The primary indicators show that the principal and characteristic gain loci coincide in the high frequency region (Fig.5.1.4). MS is small at high frequencies (Fig.5.1.5).

Low frequency region design

A low frequency sub-controller is obtained:

$$K_L(s) =(0.28 \cdot \frac{K_0}{s} + I_2)$$

$$= \frac{1}{s} \begin{bmatrix} s+4.1139 & -1.0284 \\ 6.1690 & s-1.0282 \end{bmatrix}$$

The primary indicators (Fig.5.1.6) show that the principal and characteristic gain loci coincide over the entire frequency spectrum. Thus, we will expect the compensated system to be approximately normal at all frequencies. As expected, MS is small at all frequencies (Fig.5.1.7). A final scalar gain of 8 is added (Fig.5.1.8). The closed-loop step responses of the outputs exhibit no overshoot and are fast with very small interaction (Figs.5.1.9 (a) & (b)).

5.5.2 Example 2 (from Owens (1975))

$$G(s) = \begin{bmatrix} 32.6+16s+2.15s^2 & 9.4+4s+1.1s^2 \\ 6.2+4s+1.05s^2 & 3+4s+s^2 \end{bmatrix} \frac{1}{(s+1)(s+2)(s+3)}$$

The Simple Design Technique is used to design a matrix P+I controller for this system. The suggested bandwidth is 8 rad/s. The system is minimum phase with poles at -1, -2 and -3. The primary indicators of the uncompensated system are shown in Fig.5.2.1.

High frequency region design

We observe that the system has the same roll-off rates at high frequencies. A high frequency sub-controller K_∞ is used to balance up the gains at high frequencies (see Fig.5.2.2):

$$K_\infty = \begin{bmatrix} 2.7992 & -3.0737 \\ -2.9338 & 6.0126 \end{bmatrix}$$

Low frequency region design

A low frequency sub-controller $K_L(s)$ is used to align the input and output gain frames of the system, balance up the gain and add integral action (Fig.5.2.3):

$$K_L(s) = 2(0.5 \cdot K_0/s + I_2)$$

where

$$K_0 = \begin{bmatrix} -0.2559 & 10.9761 \\ -2.1494 & 15.9995 \end{bmatrix}$$

The compensated system is found to have fast closed-loop step responses with small interaction (Figs.5.2.4 (a) & (b)). The normality indicator (MS) and the generalised Nyquist diagram is given in Fig.5.2.5 and Fig.5.2.6 respectively.

5.5.3 Example 3

The system considered here is a pressurised flow-box of a paper-making machine. The details of the model are given in Appendix B. The system, denoted FLOW, has 2 inputs and 2 outputs. The open-loop poles are at -0.3949 and -0.32×10^{-3}. Note that one of the poles is very near the origin and it gives rise to integral action which appears in one of the loci at low frequencies (see Fig.5.3.1). The primary indicators also show that the system has a very large gain imbalance of about 1000 and the divergences between the characteristic and principal gains are very large. Hence, the value of MS is large over all frequencies. (see Fig.5.3.2).

High frequency region design

A high frequency sub-controller is cascaded with the system:

$$K_\infty = \begin{bmatrix} 0.0023 & 999.4759 \\ 1.0005 & -32.3722 \end{bmatrix} .$$

The primary indicators are given in Fig.5.3.3. It can be seen that the characteristic and principal loci all coincide at high and intermediate frequencies. The system has correct gain-phase characteristics at high frequencies. MS is low over all frequencies as the divergences between the two sets of gains are small, which is confirmed by Fig.5.3.4. This illustrates the power of the use of a high frequency sub-controller.

Low frequency region design

An integrator of the form

$$K_L(s) = (\frac{I_2}{s} + I_2)$$

is cascaded with the system. This will give zero steady state error for closed-loop step response. It also injects gain at low frequencies. The primary indicators are shown in Fig.5.3.5. A final scalar gain of 10 is added to give a bandwidth of around 10 rad/s (Fig.5.3.6.). As the system is well-balanced over the intermediate and high frequencies and the low frequency gain is high, the interaction is expected to be small. The closed-loop step responses (Figs.5.3.7 (a) & (b)) can be seen to be fast, non-interactive, accurate in the steady state, with no significant overshoots.

5.5.4 Example 4

The system, GROC, to be considered next is a linearized model of a nuclear rocket engine (see Appendix B for details). GROC has 2 inputs, 2 outputs and 4 states. The open-loop system has poles at -64.8, -3.0×10^{-5}, $-0.603 \pm 0.52j$ and a transmission zero at -0.10. The primary indicators of GROC are given in Fig.5.4.1. Note that one of the loci exhibits some integral action at low frequencies; this is due to a pole which is very near the origin.

High frequency region design

A high frequency sub-controller is obtained:

$$K_\infty = \begin{bmatrix} 1 & 0.00 \\ 0 & 2.49 \end{bmatrix}.$$

The primary indicators are given in Fig.5.4.2. The principal loci are now balanced at around ω = 10. The system is robust because gain divergences between principal and characteristic gains are small.

Low frequency region design

An integrator of the form

$$K_L(s) = (\frac{I_2}{s} + I_2)$$

is cascaded with the system. The primary indicators are given in Fig.5.4.3. We note that the principal loci of the system are now fairly well-balanced. Also, gain divergences are small over all frequencies. A final scalar gain of 10 is used to give a bandwidth of about 10 rad/s (see Fig.5.4.4). The phase margin is then more than 80 degrees. MS is shown in Fig.5.4.5. The closed-loop step responses are shown in Figs.5.4.6 (a) & (b). The compensated system has a fast response and interaction is around 5% .

5.5.5 Example 5

This example is based on a model of an open-loop unstable chemical reactor (Munro, 1972), denoted REAC, with two inputs and two outputs. The state space representation of the system is given in Appendix B.

Observations

There are two right-half plane poles so the system is open-loop unstable. The zeros are both in the left-half plane so there should be no intrinsic difficulty in applying feedback control. The generalised Nyquist stability theorem requires that the characteristic loci must have a net sum of two anti-clockwise encirclements for stability. The primary indicators are given in Fig.5.5.1.

High frequency region design

The system has the same roll-off rates at high frequencies and the gains are balanced. The primary indicators show that the principal and

characteristic gain loci coincide at high frequencies (Fig.5.5.2). The input
and output gain frames are aligned and MS is small at high frequencies. The
high frequency sub-controller is

$$K_\infty = \begin{bmatrix} 0 & 1 \\ -1.77 & 0 \end{bmatrix} .$$

Low frequency region design

A low frequency sub-controller is used as below:

$$K_L(s) = (\frac{1}{s} \cdot I_2 + I_2) .$$

The primary indicators are given in Fig.5.5.3. A scalar gain of 10 is then
injected into the system for performance. The transfer function of the final
controller is

$$K(s) = \begin{bmatrix} 0 & \dfrac{10s+10}{s} \\ -\dfrac{(17.7s+17.7)}{s} & 0 \end{bmatrix} .$$

The primary indicators (Fig.5.5.4) show that the final compensated system now
exhibits good gain and phase margins, has a good bandwidth over which the
principal gains are high, and is robust at the high frequency region. Hence,
the closed-loop step responses are fast, non-oscillatory with fairly low
interaction (see Fig.5.5.5 (a) & (b)). The normality indicator is shown in
Fig.5.5.6.

This design can be compared with the design using the Characteristic
Locus Design Method performed by MacFarlane and Kouvaritakis (1977). The
final controller design given was

$$K(s) = \begin{bmatrix} 0 & \dfrac{10s+9}{s} \\ -\dfrac{(10s+21)}{s} & \dfrac{-8}{s} \end{bmatrix} .$$

The primary indicators and normality indicator of the compensated system are
given in Fig.5.5.7 and Fig.5.5.8 respectively. In comparing the designs from
both techniques, the closed-loop step responses are shown to be very similar

(Fig.5.5.9 (a) & (b)). However, the primary indicators of the design using SDT show that the gain loci are balanced at high frequencies whereas they are not for the Characteristic Locus Design.

The controller design given by Munro (1972) was

$$K(s) = \begin{bmatrix} 0 & \dfrac{10s+10}{s} \\ -\dfrac{(10s+10)}{s} & 0 \end{bmatrix} .$$

The primary indicators and normality indicator of the compensated system are given in Fig.5.5.10 and Fig.5.5.11 respectively. Again, the principal loci are not balanced at high frequencies. The rest is similar to the design using SDT. The closed-loop step responses are given in Figs.5.5.12 (a) & (b).

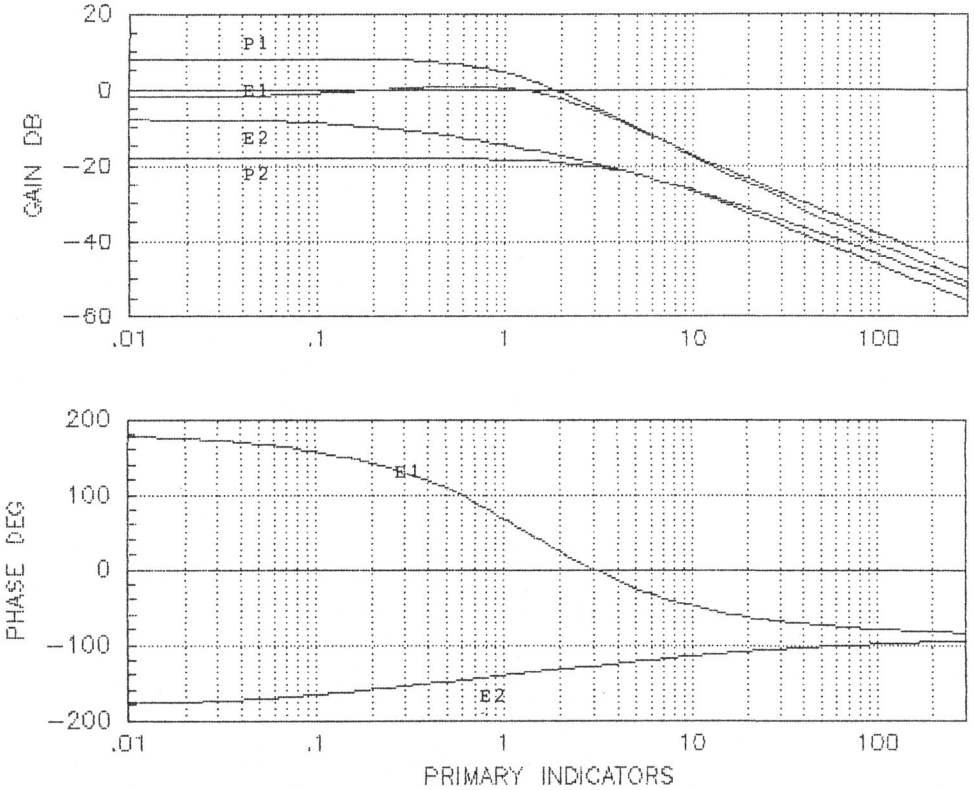

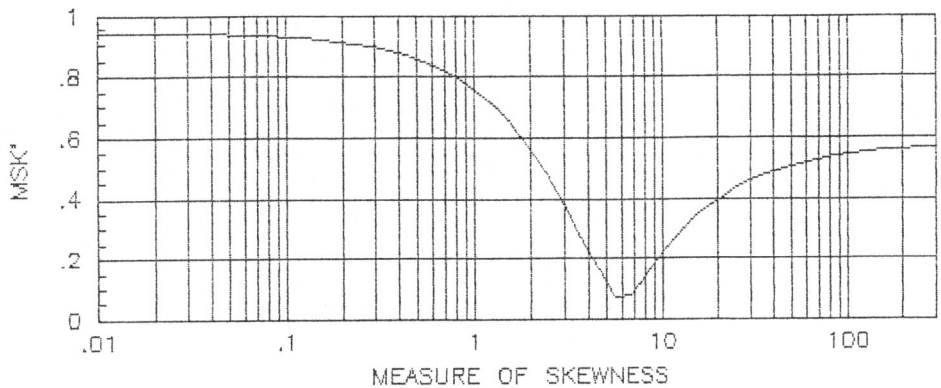

Fig. 5.1.1   **The primary indicators of the
uncompensated system.**

Fig. 5.1.2   **The normality indicator MS of the
system.**

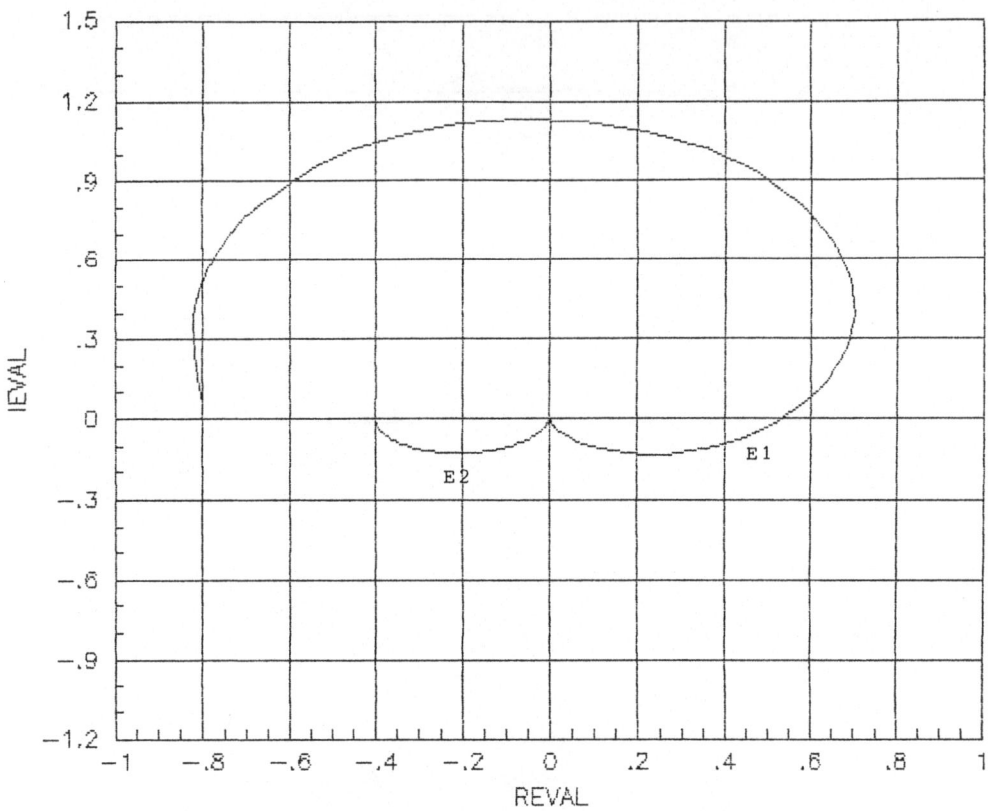

Fig. 5.1.3   Generalised Nyquist diagrams of the
             system.

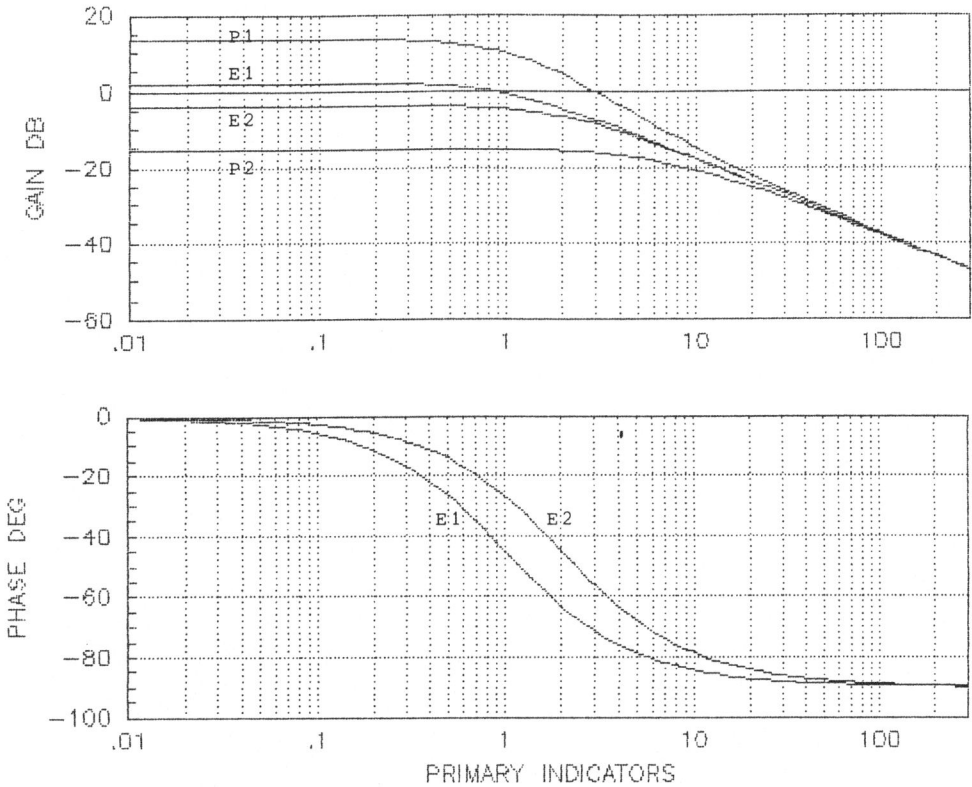

Fig. 5.1.4   The primary indicators after high
frequency region design.

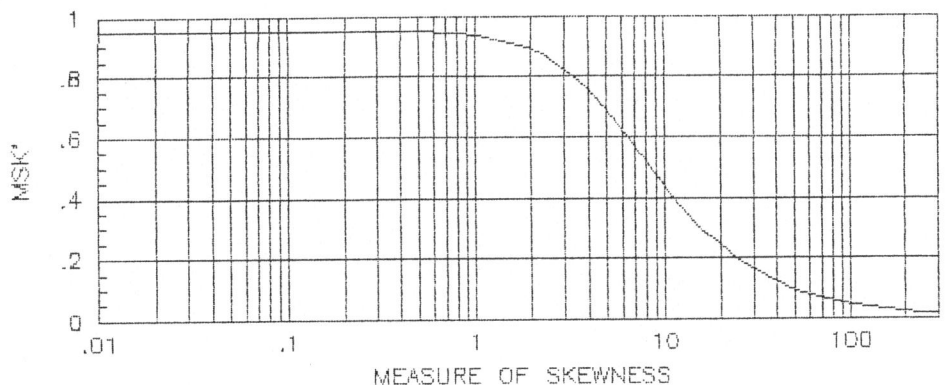

Fig. 5.1.5   The normality indicator MS after
high frequency region design.

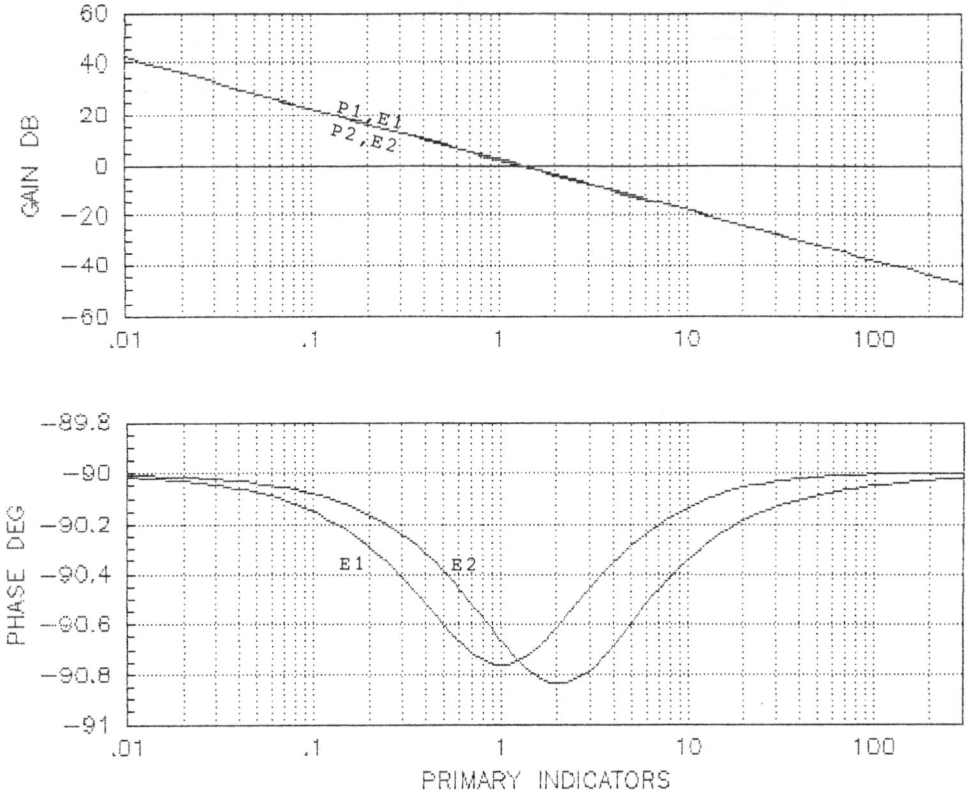

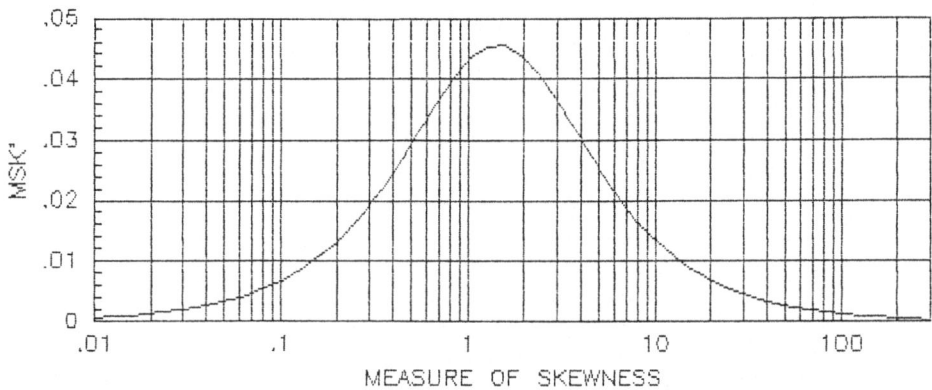

Fig. 5.1.6 The primary indicators after high
and low frequency region design.

Fig. 5.1.7 The normality indicator MS after SDT.

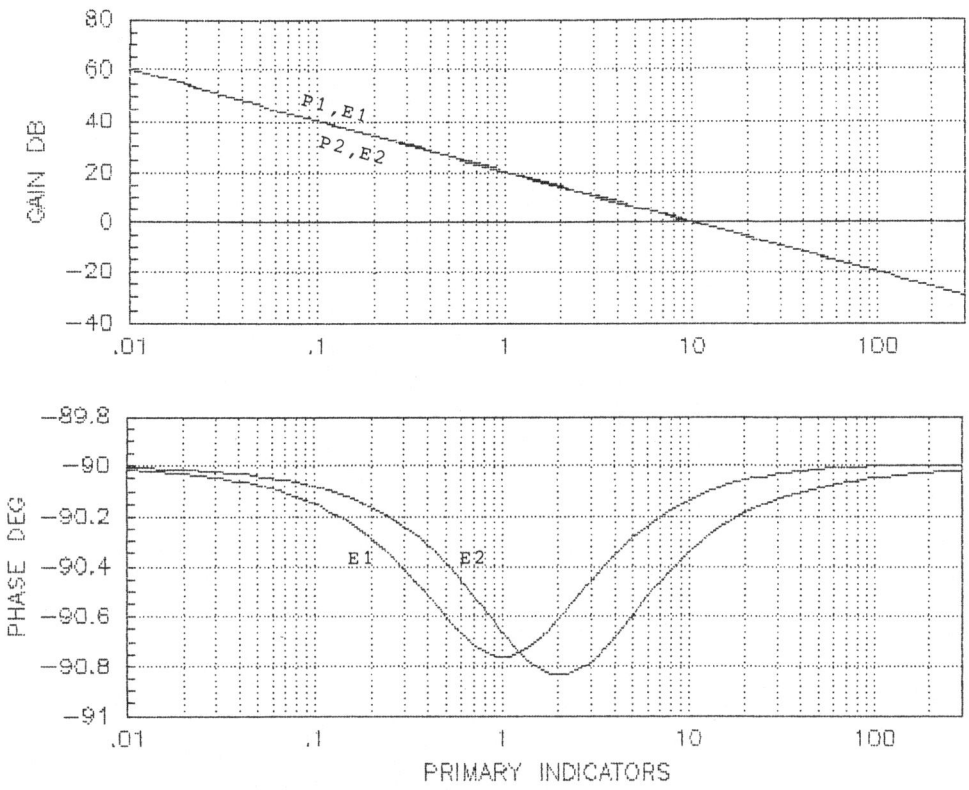

Fig. 5.1.8 The primary indicators after SDT.

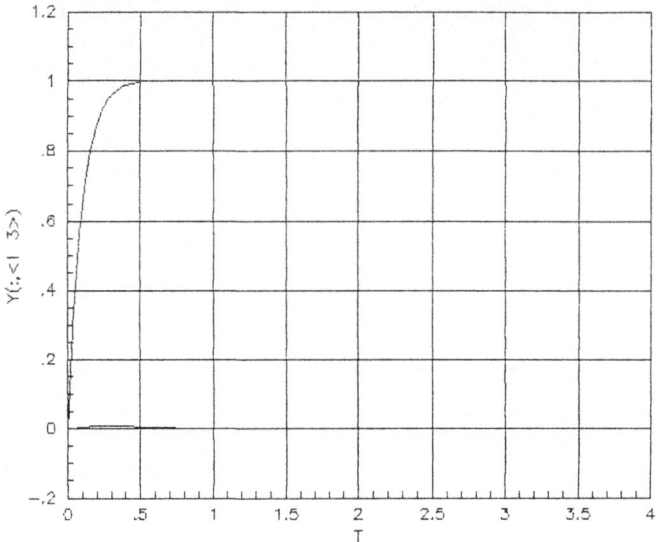

Fig. 5.1.9(a)   Closed-loop step response after
                SDT (step at input 1).

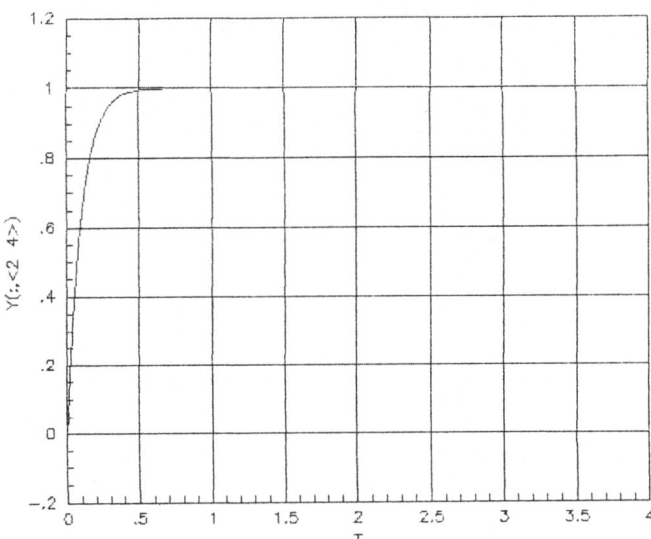

Fig. 5.1.9(b)   Closed-loop step response after
                SDT   (step at input 2).

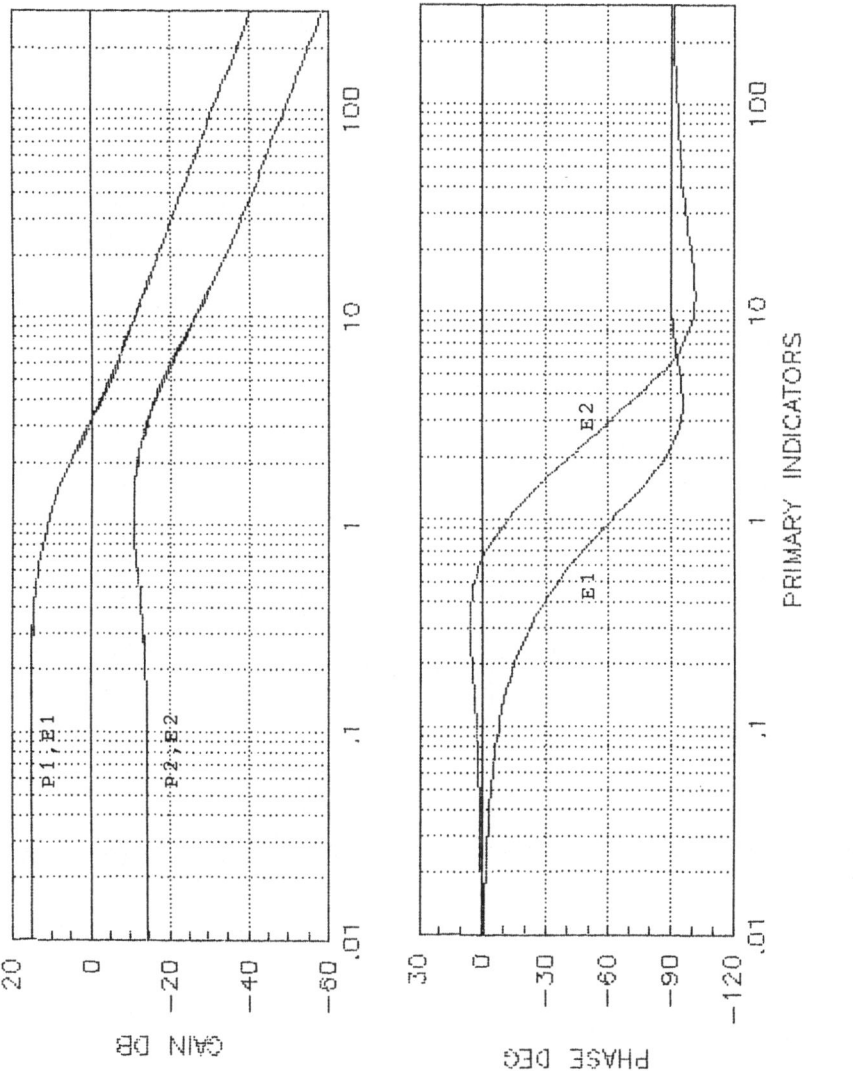

Fig. 5.2.1   The primary indicators of the
uncompensated system.

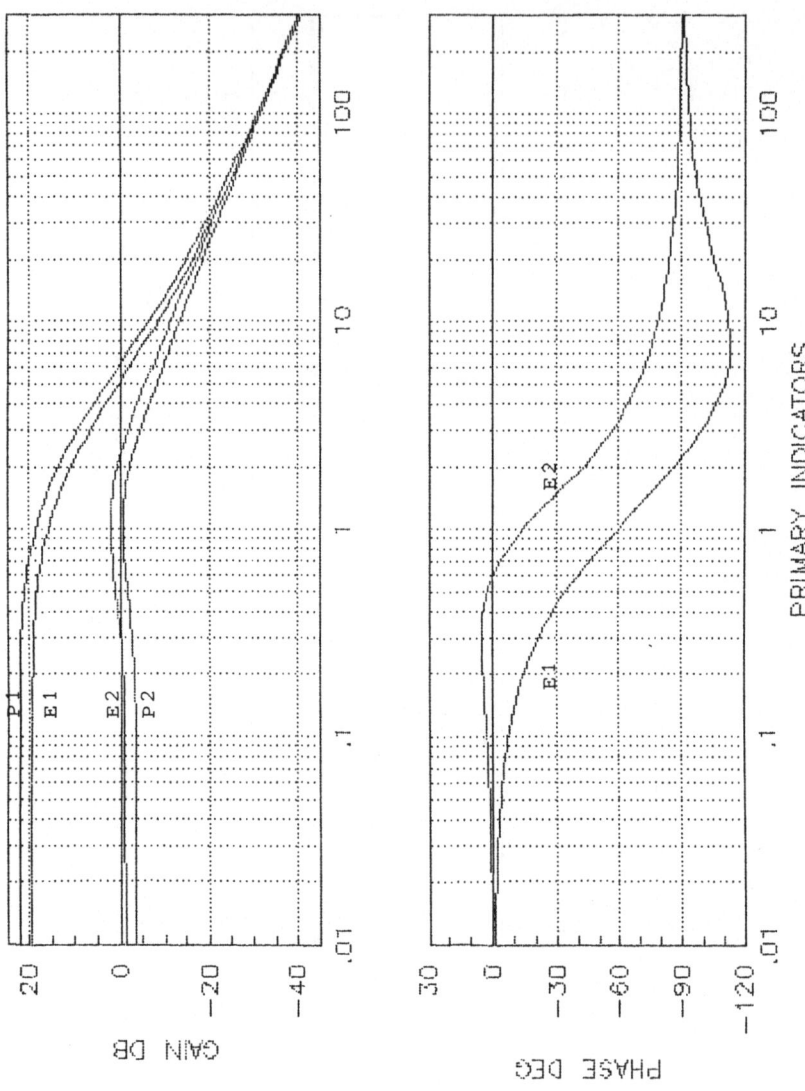

Fig. 5.2.2   The primary indicators after high
             frequency region design.

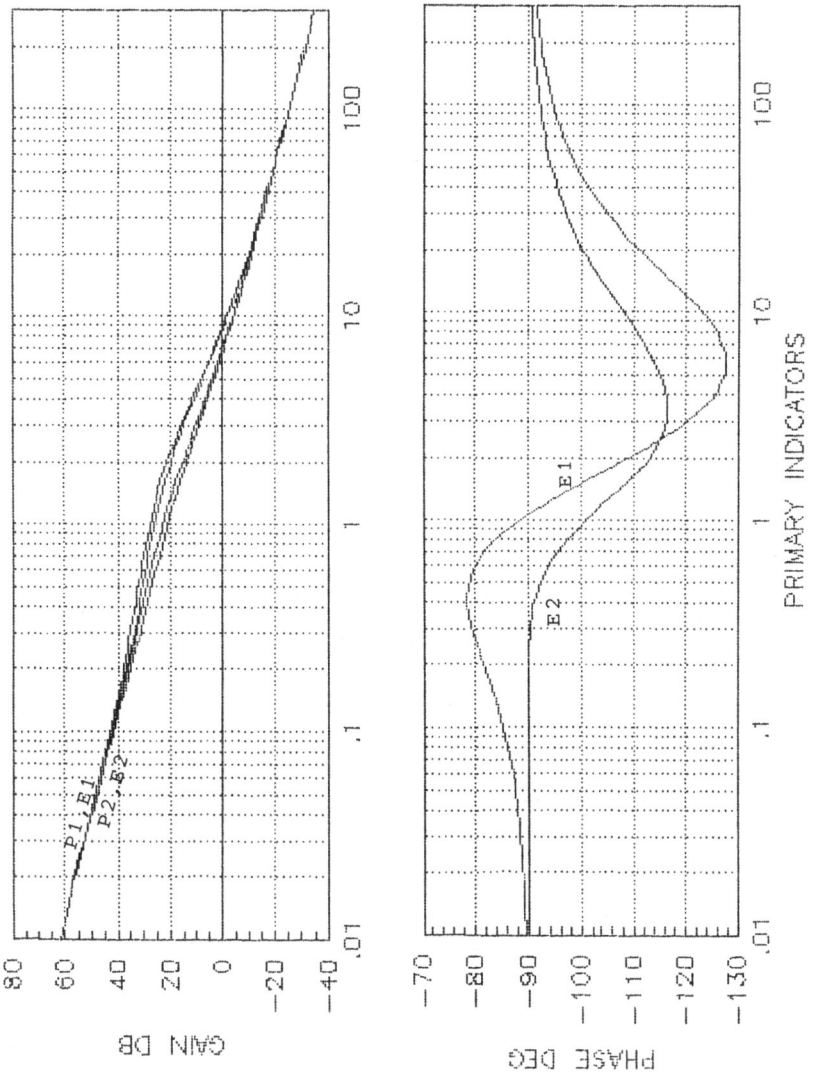

Fig. 5.2.3   The primary indicators after SDT.

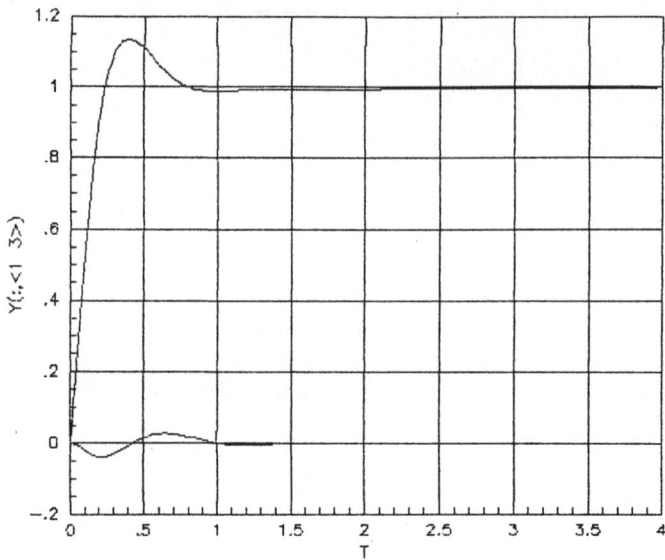

Fig. 5.2.4(a)   Closed-loop step response after SDT
                (step at input 1).

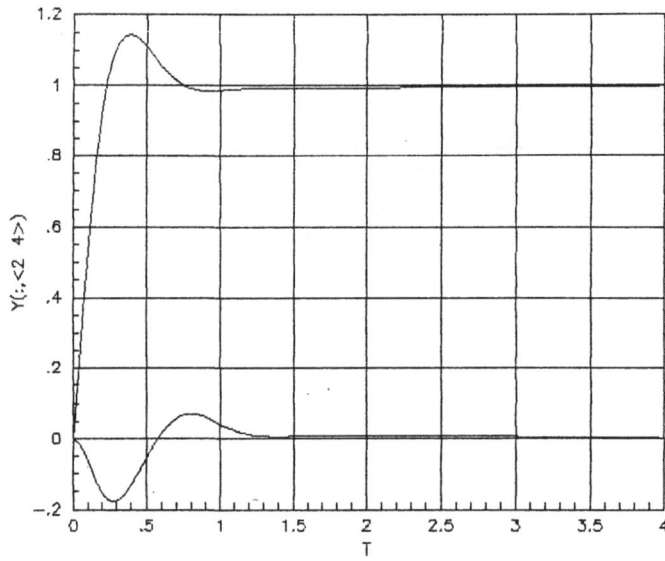

Fig. 5.2.4(b)   Closed-loop step response after SDT
                (step at input 2).

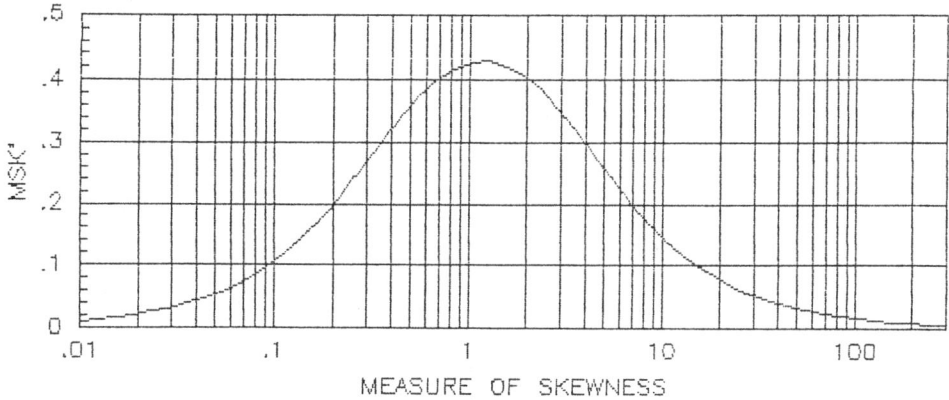

Fig. 5.2.5   The normality indicator MS after SDT

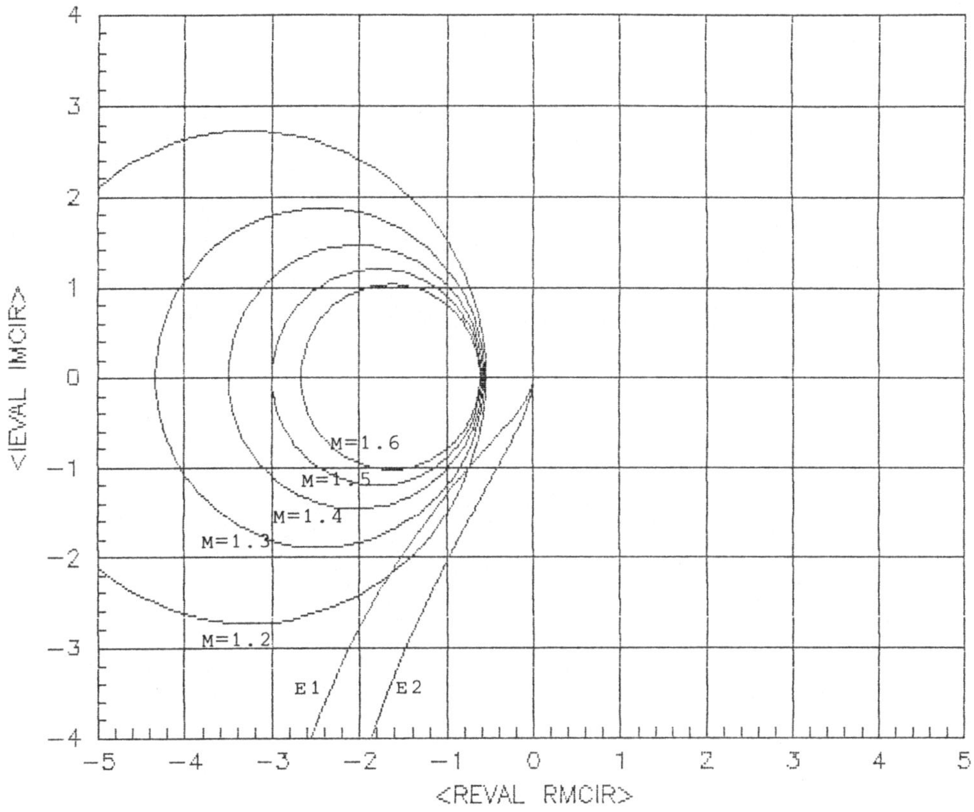

Fig. 5.2.6   Generalised Nyquist diagrams of the compensated system.

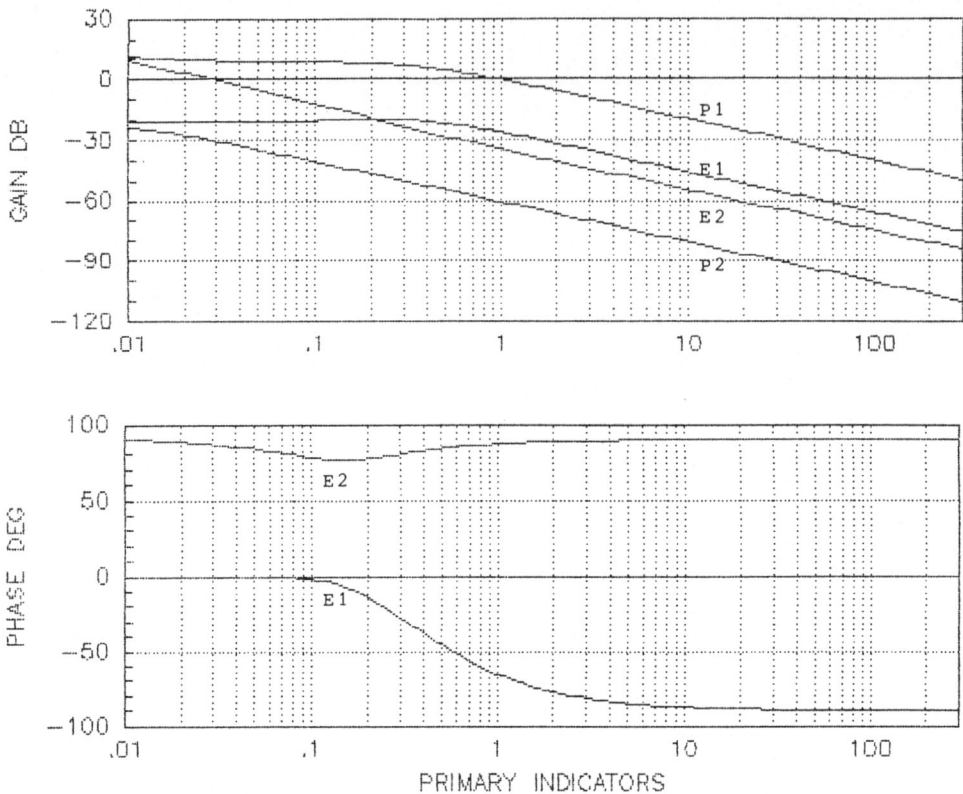

Fig. 5.3.1   The primary indicators of FLOW.

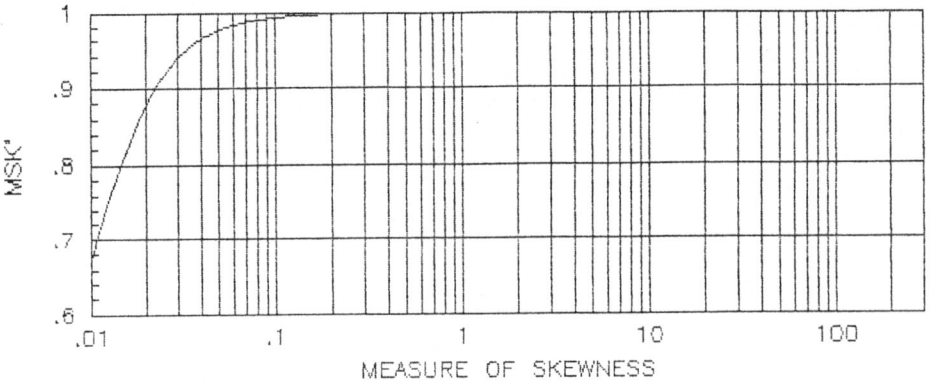

Fig. 5.3.2   The normality indicator MS of FLOW.

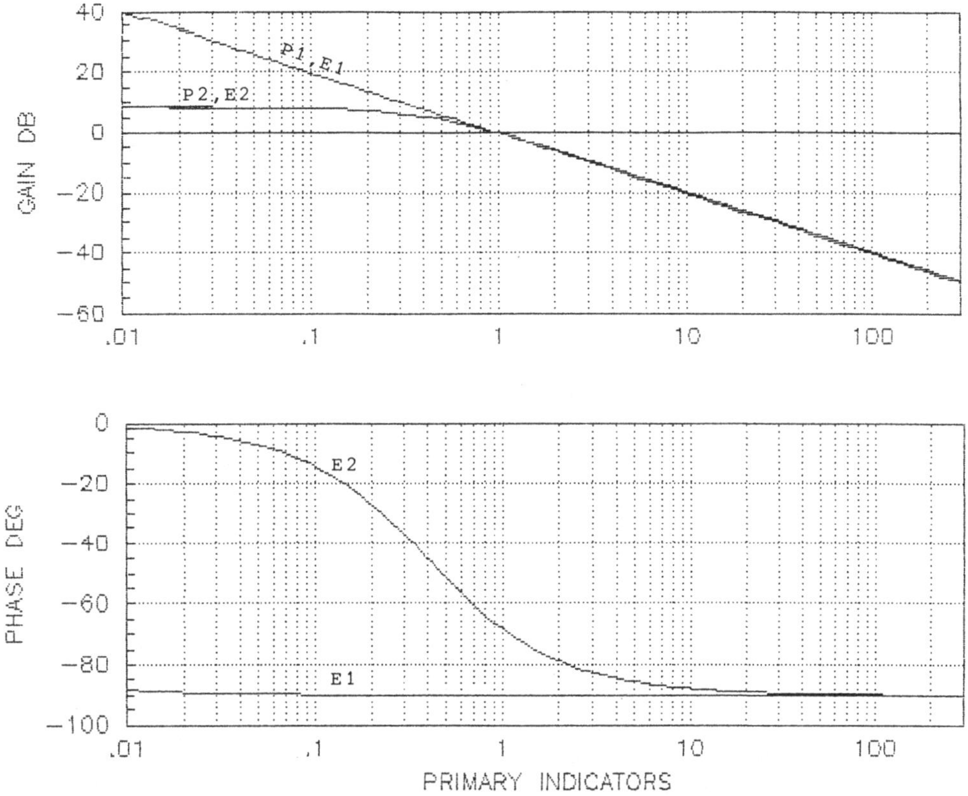

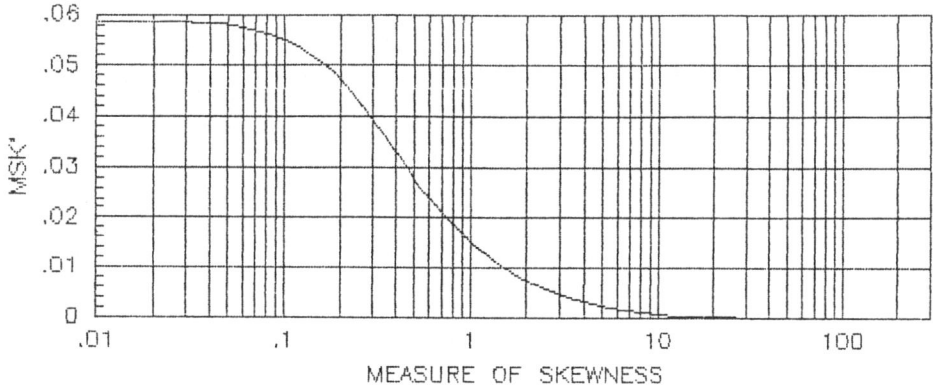

Fig. 5.3.3   The primary indicators after high
frequency region design.

Fig. 5.3.4   The normality indicator MS after high
frequency region design.

74

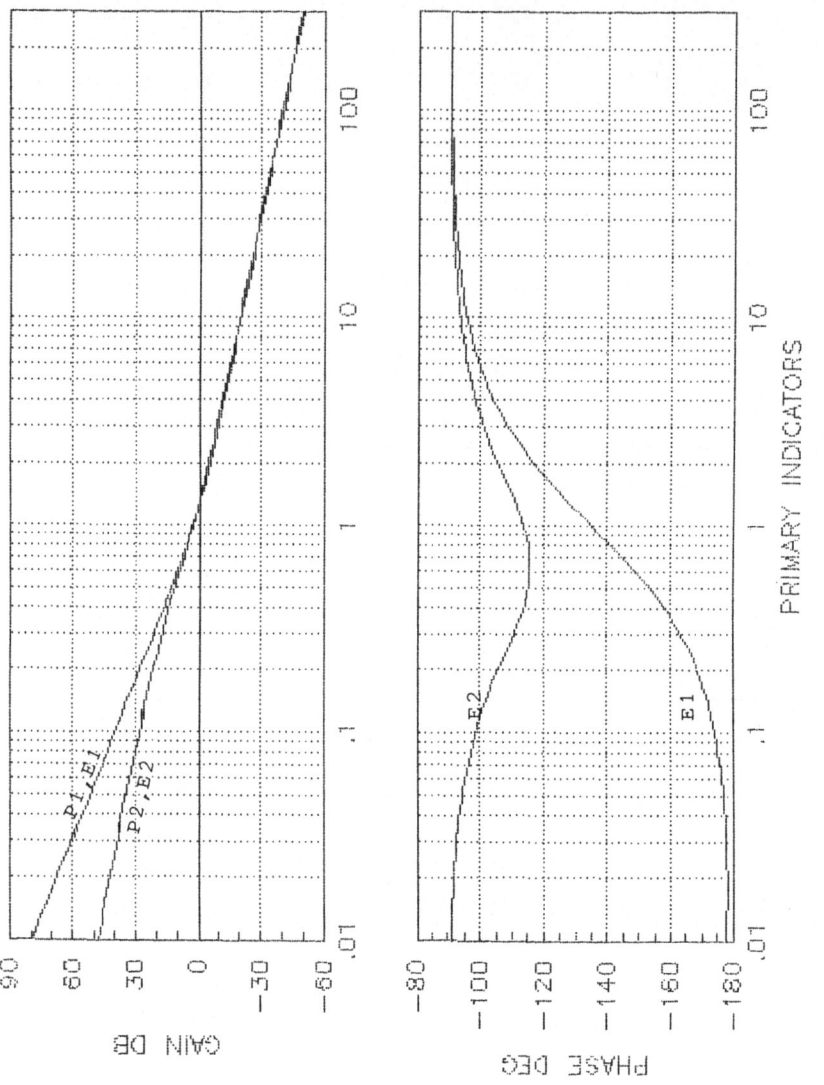

Fig. 5.3.5 **The primary indicators after high and low frequency region design.**

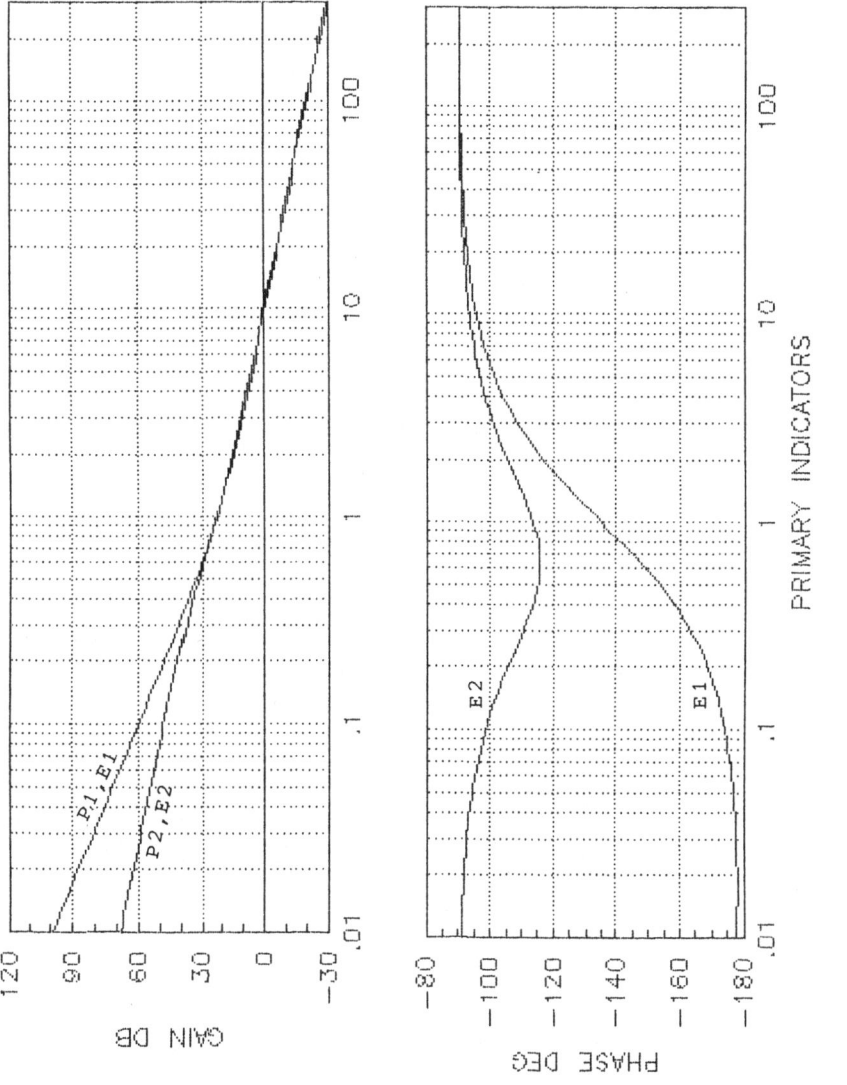

Fig. 5.3.6 The primary indicators after SDT.

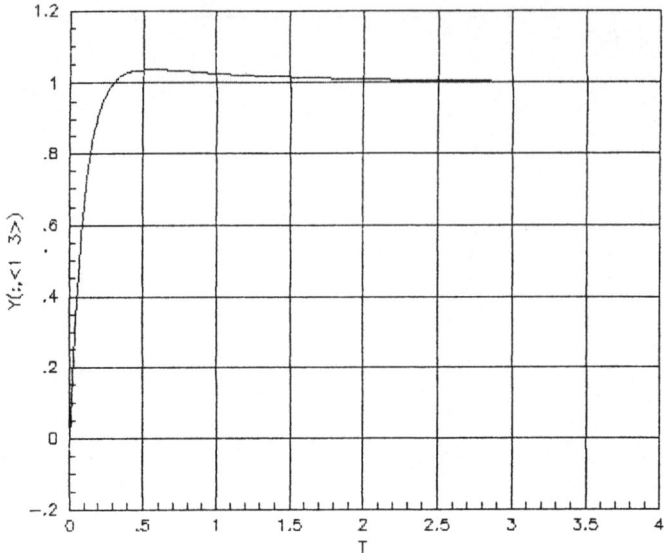

Fig. 5.3.7(a) Closed-loop step response after SDT
(step at input 1).

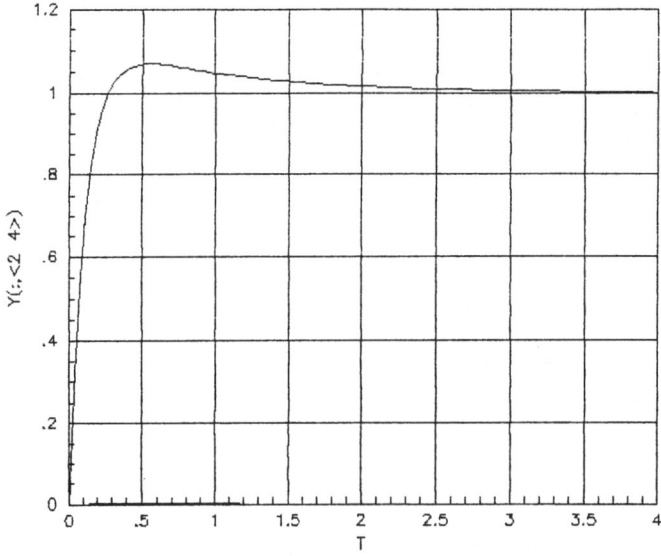

Fig. 5.3.7(b) Closed-loop step response after SDT
(step at input 2).

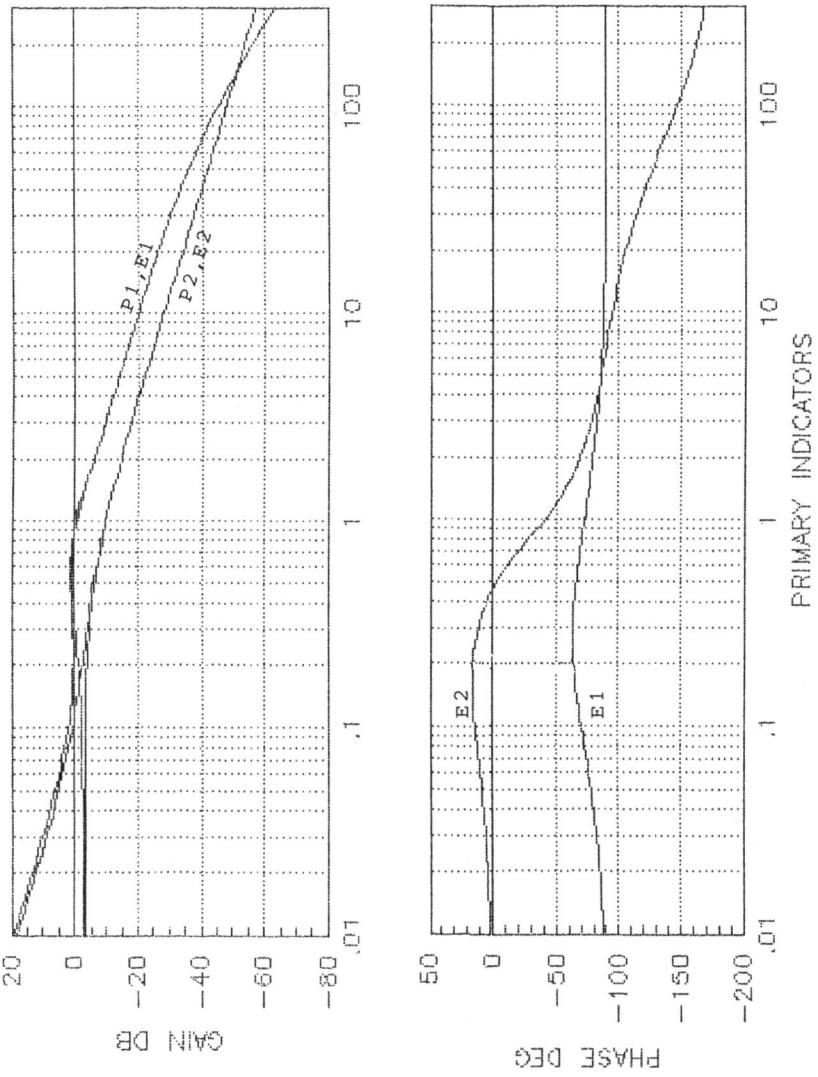

Fig. 5.4.1 The primary indicators of GROC.

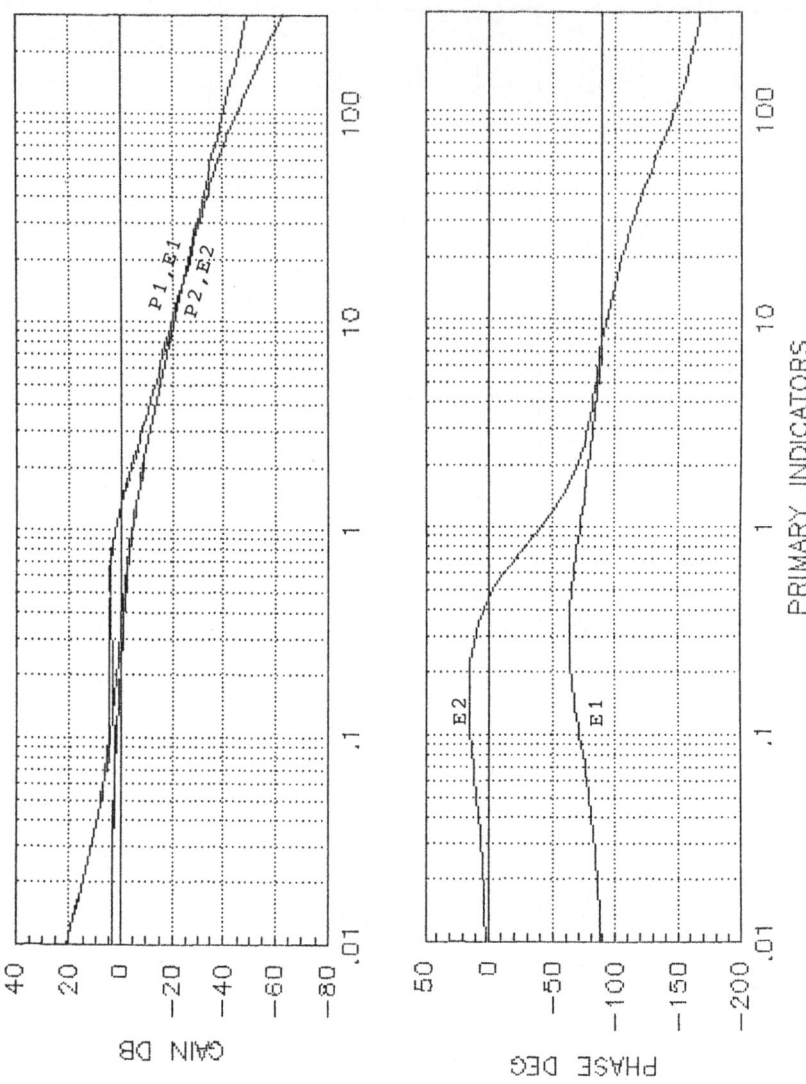

Fig. 5.4.2 The primary indicators after high frequency region design.

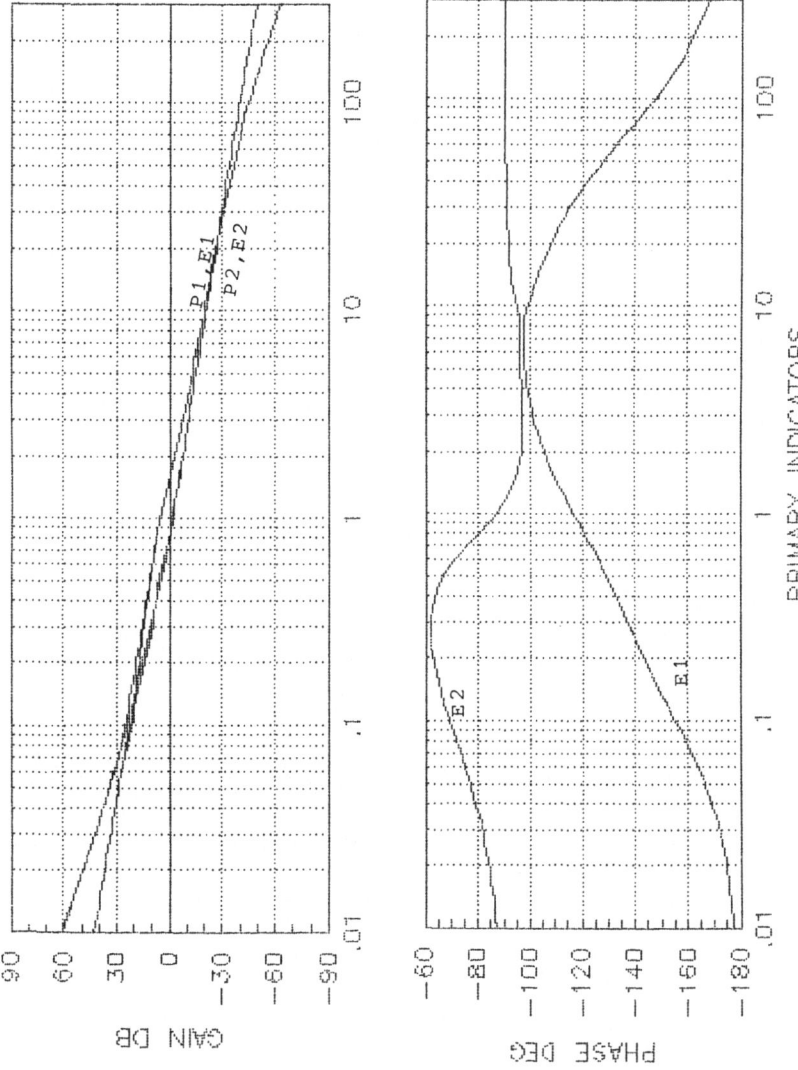

Fig. 5.4.3 The primary indicators after high and low frequency region design.

80

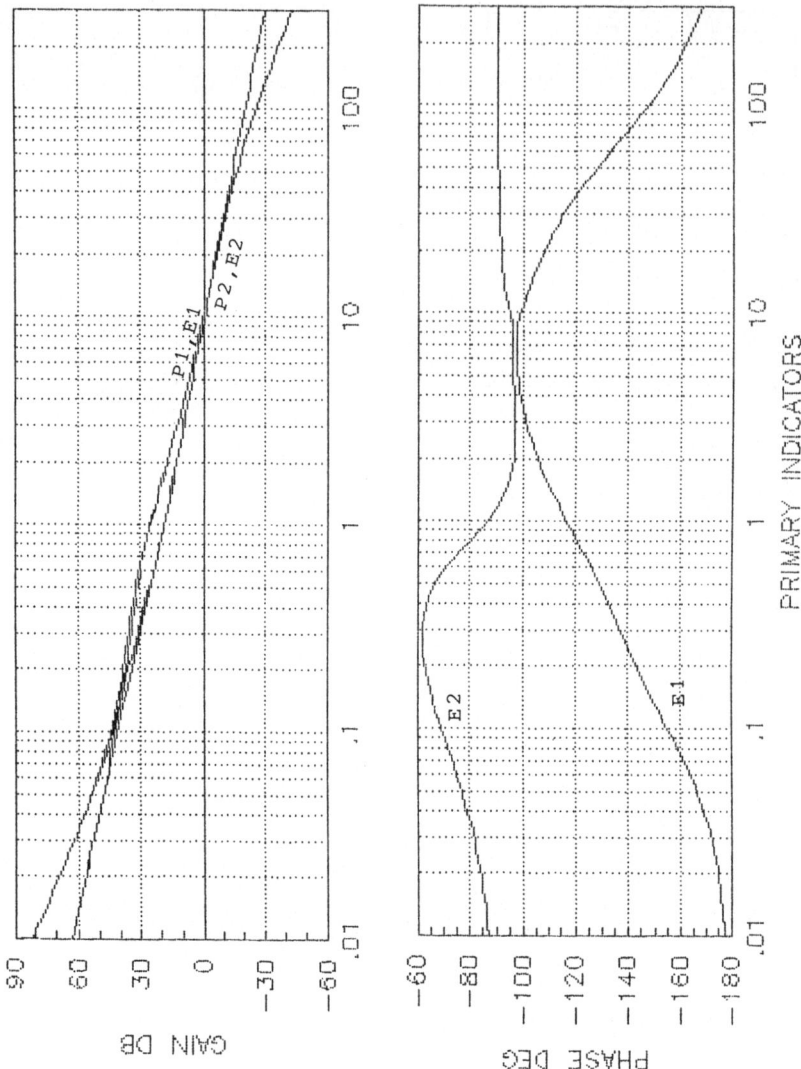

Fig. 5.4.4 The primary indicators after SDT.

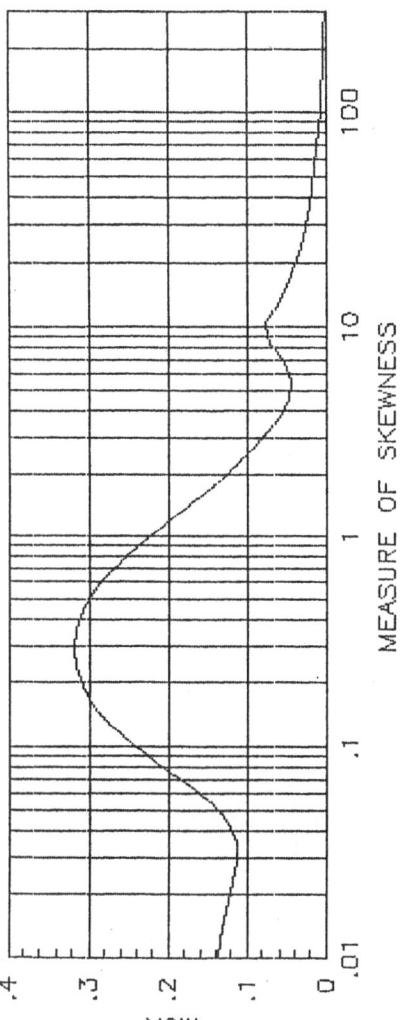

Fig. 5.4.5 The normality indicator MS after SDT.

82

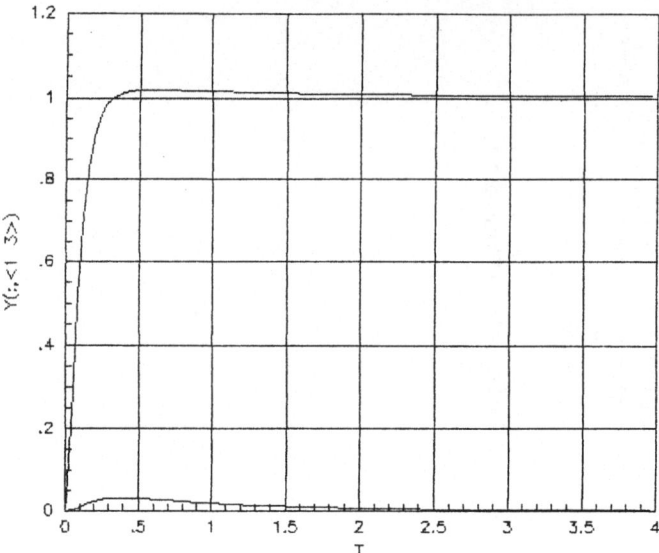

Fig. 5.4.6(a)   Closed-loop step response after SDT
                (step at input 1).

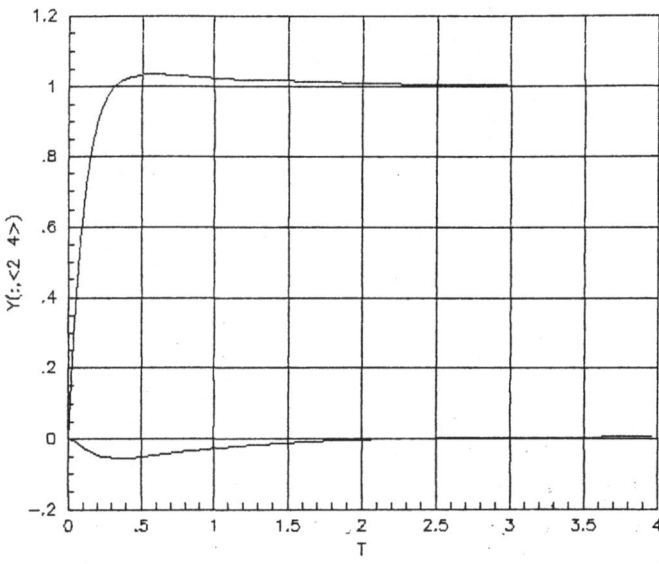

Fig. 5.4.6(b)   Closed-loop step response after SDT
                (step at input 2).

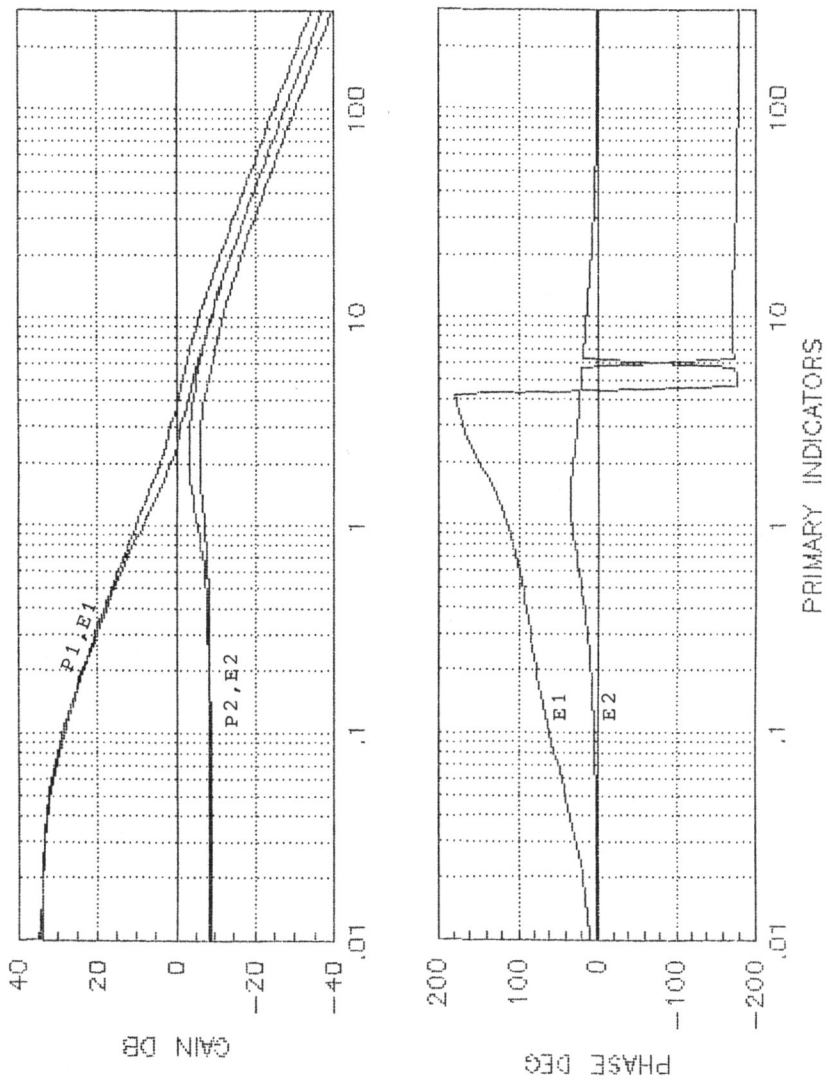

Fig. 5.5.1 The primary indicators of REAC.

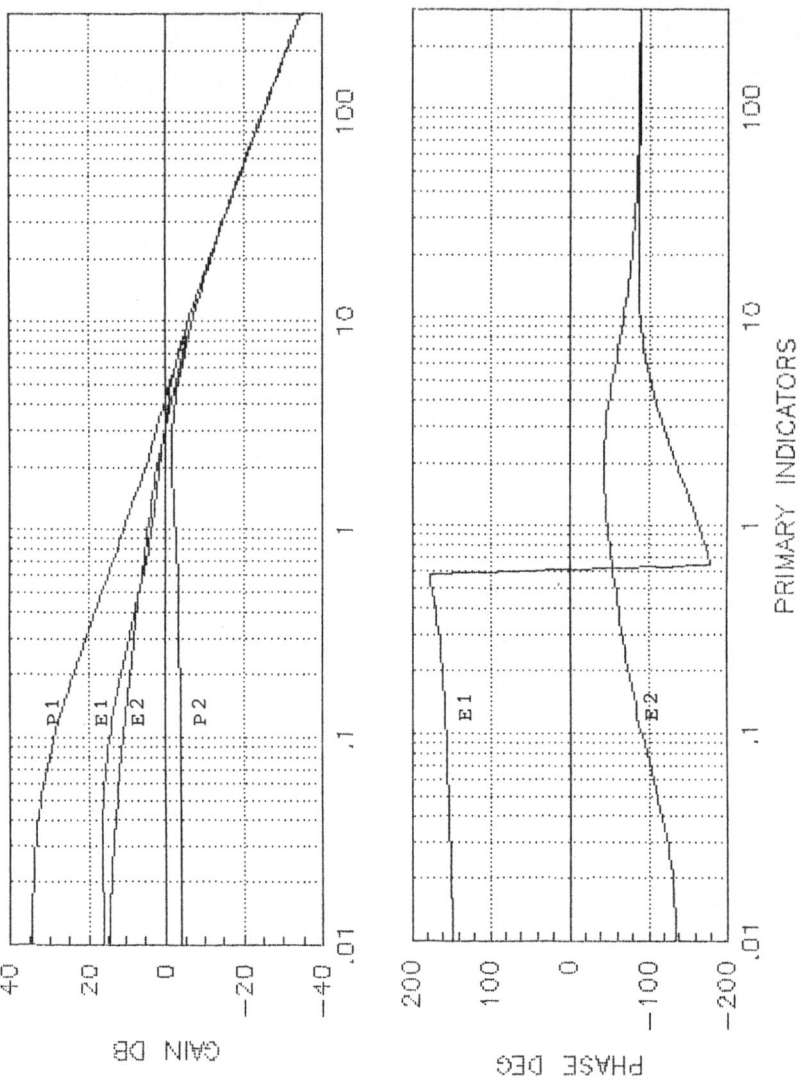

Fig. 5.5.2   The primary indicators after high frequency region design.

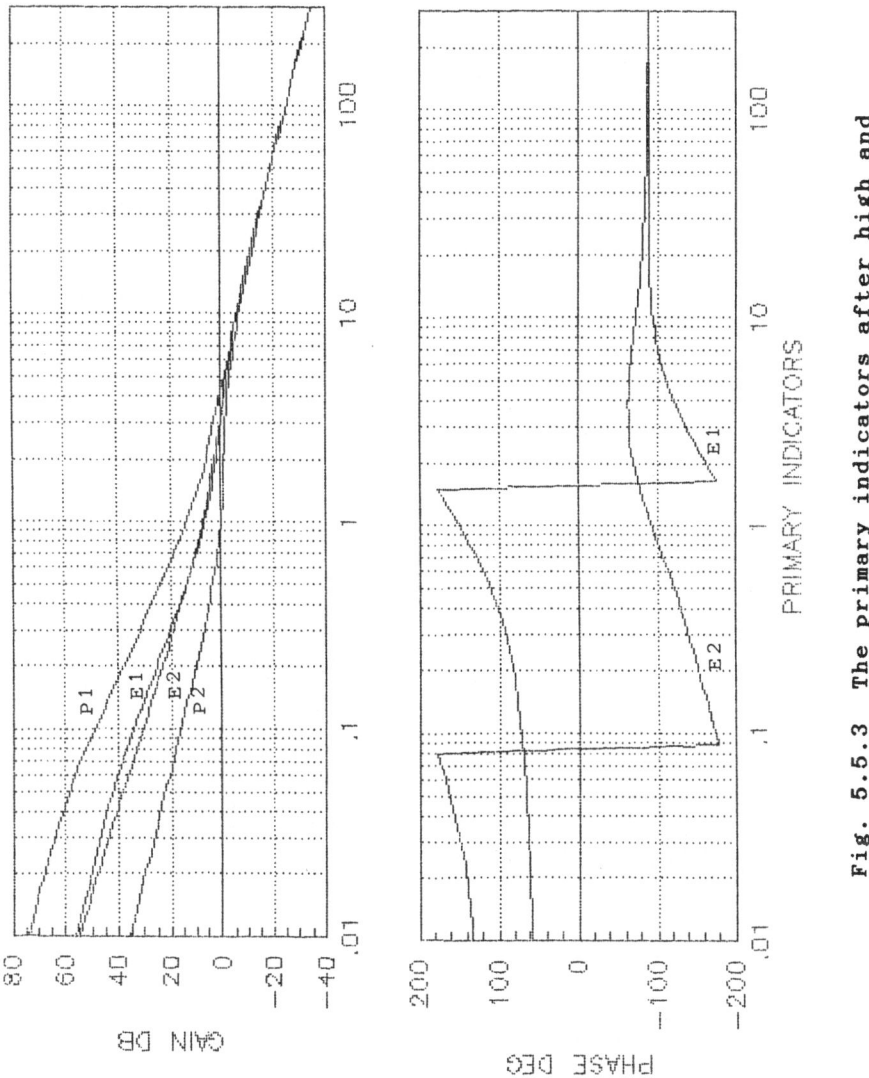

Fig. 5.5.3 The primary indicators after high and low frequency region design.

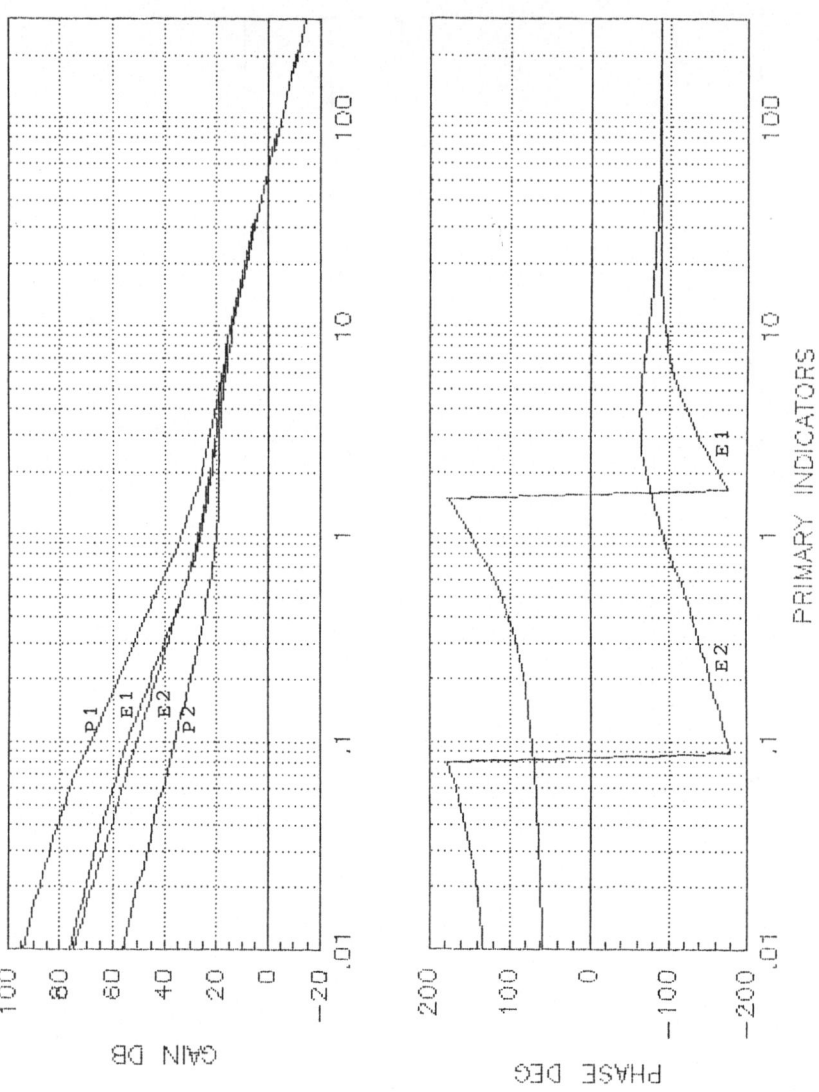

Fig. 5.5.4 The primary indicators after SDT.

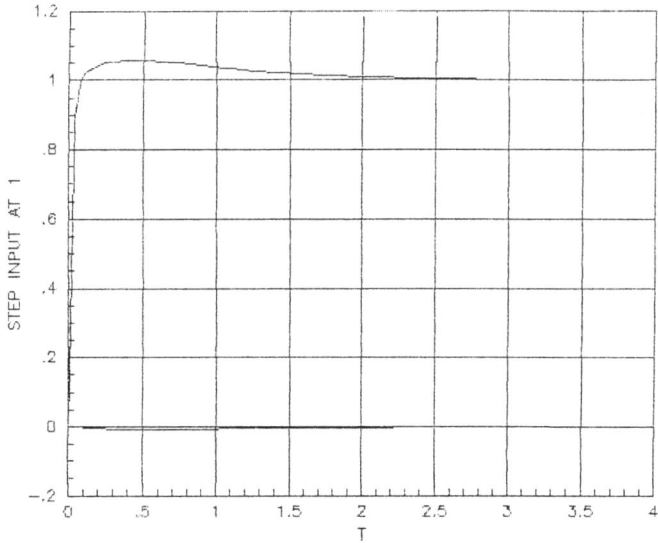

Fig. 5.5.5(a)   Closed-loop step response after SDT
                (step at input 1).

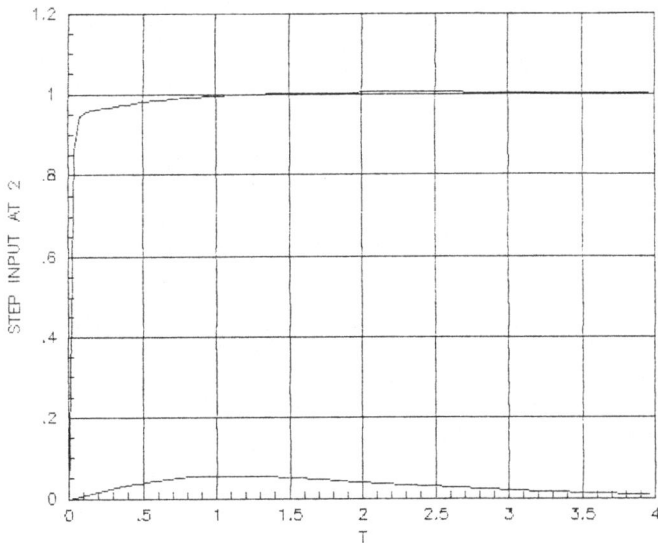

Fig. 5.5.5(b)   Closed-loop step response after SDT
                (step at input 2).

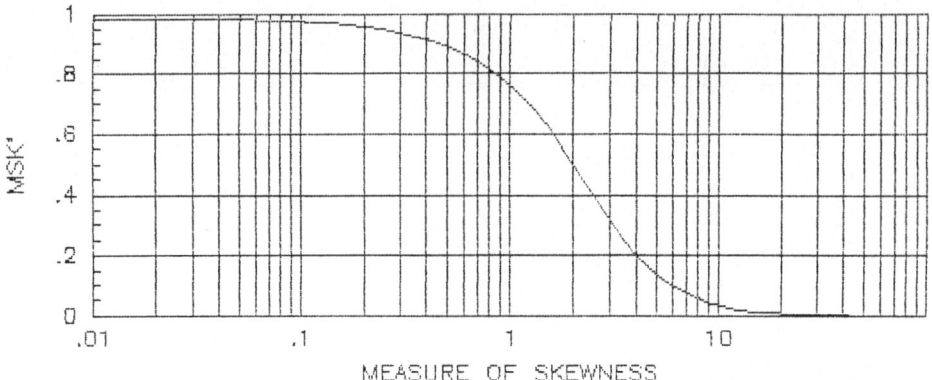

Fig. 5.5.6   The normality indicator MS after SDT.

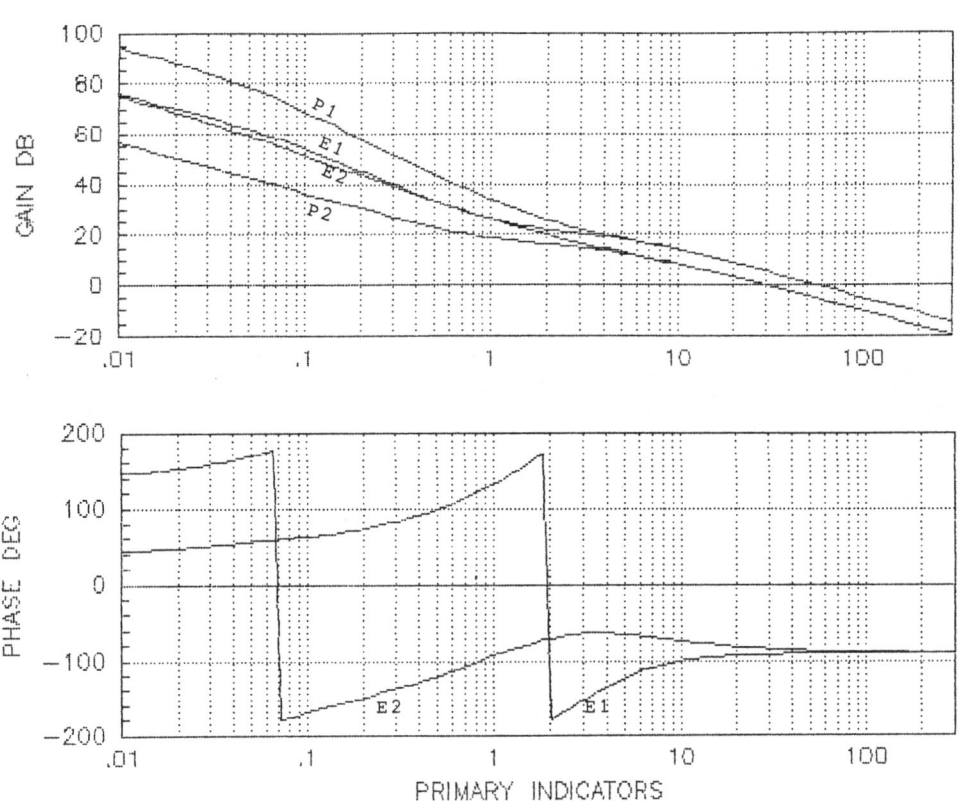

Fig. 5.5.7   The primary indicators of the final
design using the Characteristic Locus
(CL) Design Method.

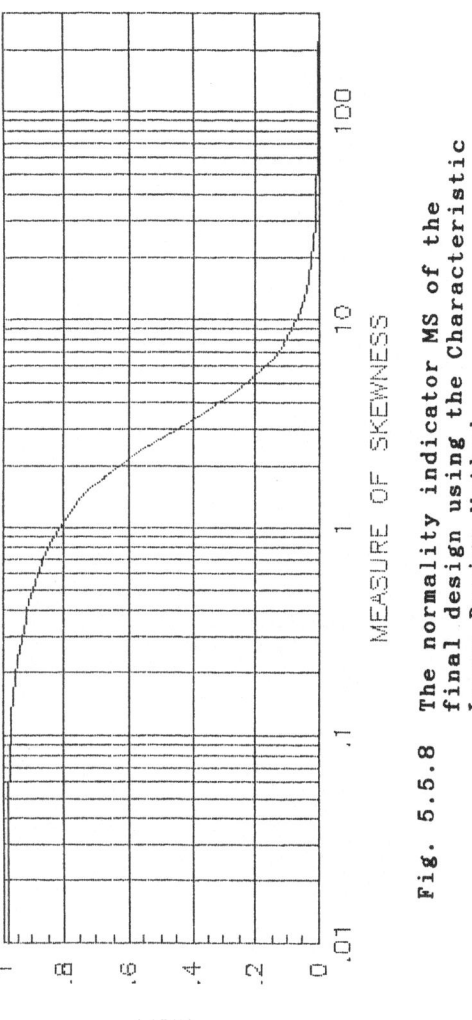

Fig. 5.5.8 The normality indicator MS of the final design using the Characteristic Locus Design Method.

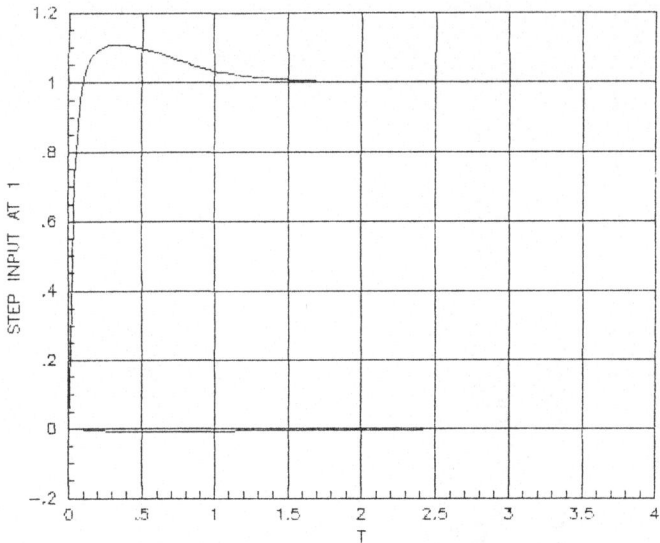

Fig. 5.5.9(a)   Closed-loop step response of the
final design using the CL Design
Method (step at input 1).

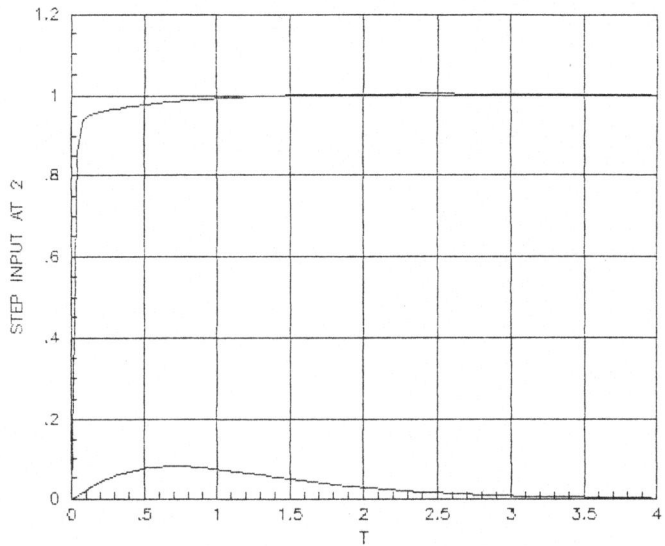

Fig. 5.5.9(b)   Closed-loop step response of the
final design using the CL Design
Method (step at input 2).

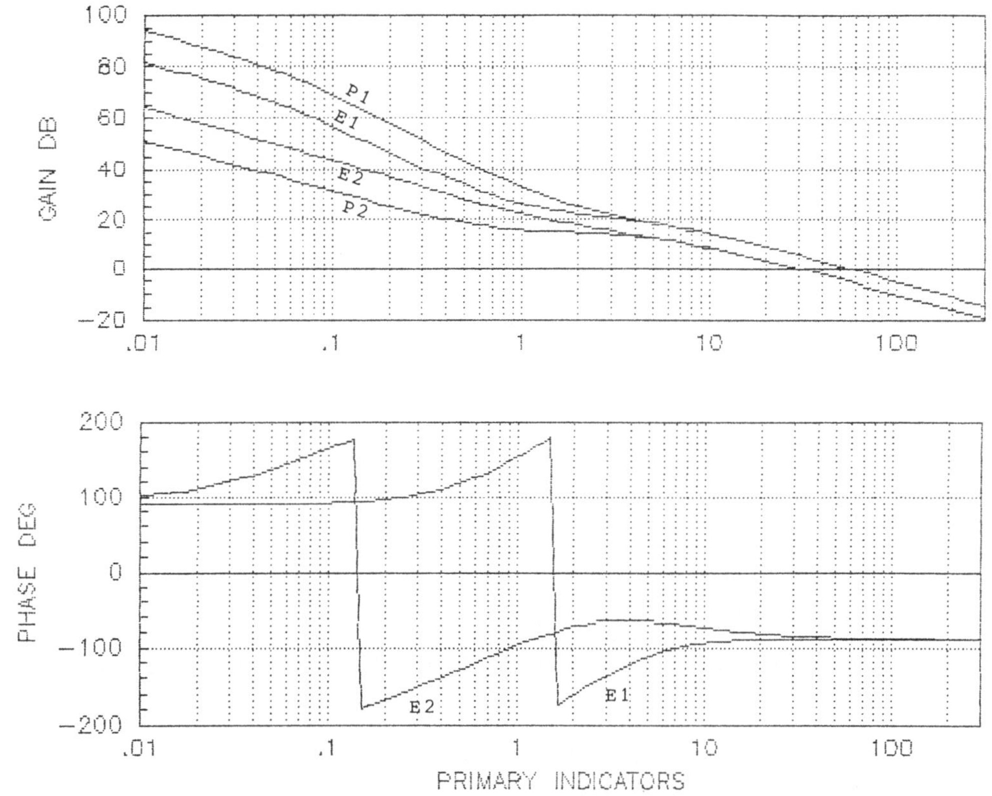

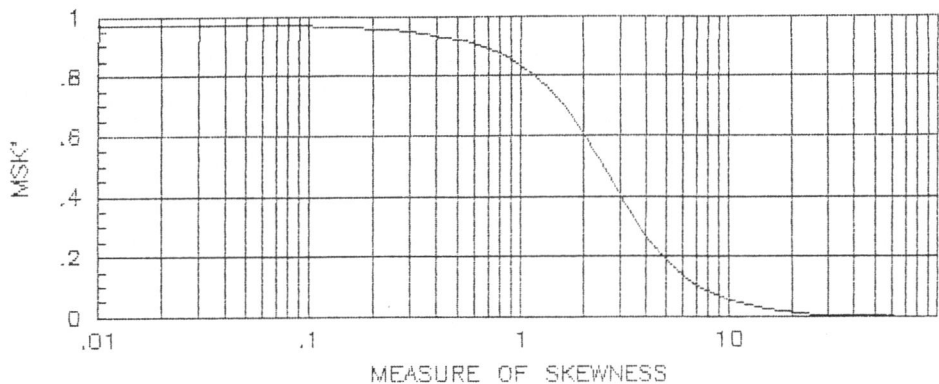

Fig. 5.5.10   The primary indicators of the final
design using INA.

Fig. 5.5.11   The normality indicator MS of the
final design using INA.

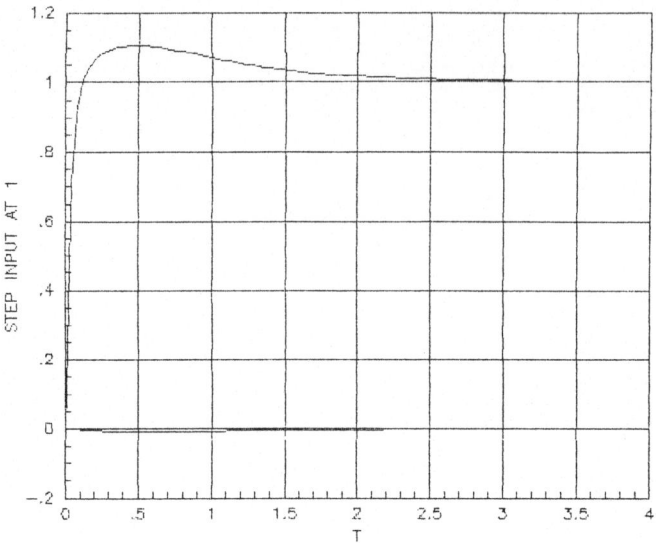

Fig. 5.5.12(a)   Closed-loop step response of the
                  final design using INA
                  (step at input 1).

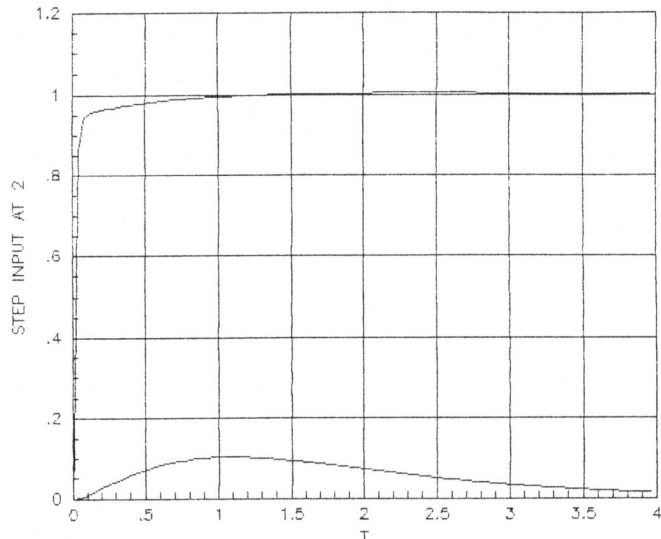

Fig. 5.5.12(b)   Closed-loop step response of the
                  final design using INA
                  (step at input 2).

CHAPTER SIX

REVERSE FRAME ALIGNMENT DESIGN TECHNIQUE

6.1 Introduction

The Simple Design Technique (SDT) was shown to give a simple matrix-proportional-plus-integral controller. In this chapter, this procedure has been extended to deal with systems which require more phase compensation or gain adjustment in an intermediate frequency region than can be provided by the SDT. This is carried out by the use of a dynamic *Reverse Frame Approximation* (RFA) sub-controller. This new design technique, in the extended form presented here, is a redevelopment of the Characteristic Locus Design Method developed by MacFarlane and Kouvaritakis (1977). The Characteristic Locus Design Method is based on the characteristic value decomposition of a matrix whereas the Reverse Frame Alignment technique is based on the singular value decomposition of a matrix. Hence the Reverse Frame Alignment technique can deal with non-square systems and is a more robust design technique.

The key idea behind the technique is the construction of Reverse Frame Approximation sub-controllers at intermediate frequencies. These sub-controllers manipulate the principal gains and phases using simple classical phase lead/lag compensating networks. The Align algorithm (Kouvaritakis, 1974) for approximation of a complex frame by a real frame is used and plays a basic role in the design technique.

6.2 Design Strategy

1. The primary indicators are used for the assessment of stability, perfor-

mance and robustness of the closed-loop system. They are manipulated on the basis of physical reasoning within the guidelines laid down by the experience of the designer.

2. The design of the controller is separated into three distinct stages:

i) Design for good phase properties in the high frequency region.

ii) Design for good gain and phase properties in the intermediate frequency region.

iii) Design for good gain properties in the low frequency region.

3. The design of the sub-controller in the second stage can be repeated at different intermediate frequencies. It should proceed from a high to a low frequency.

4. The design of the controller is performed firstly for the high frequency region, then for the intermediate frequency region, and finally for the low frequency region.

5. The design of the controller is essentially an informed trial-and-error process. One should proceed from a simple compensator to a more complicated one.

6. Particular attention is paid to the selection of compensator parameters in the second stage. An appropriate choice of compensator networks and parameters will give the desired manipulation of gains and phases at intermediate frequencies.

7. The approach is systematic and is based on the classical work of Bode and Nyquist on the frequency response analysis of scalar feedback systems.

8. The design procedure may be represented as a set of production rules or structural representations at each design stage. These serve as guidelines for the designer.

9. The final controller is a product of the individual sub-controllers from the high, intermediate and low frequency regions. It is the combined

effect of the sub-controllers which produce the required overall control action.

6.3 Reverse Frame Approximation Sub-controller

The construction of a sub-controller in an intermediate frequency region is described in this section.

Let $G(s) \in \mathbb{C}^{m \times m}$ have an SVD at one specific intermediate frequency s_w

$$G(s_w) = Y_w \, \Sigma_w \, U_w^* \qquad\qquad [6.3.1]$$

where Y_w and U_w are unitary

and $\Sigma_w = \text{diag}\{ \sigma_i^w \}$ $\qquad i = 1, \ldots, m.$

The RFA sub-controller matrix $K(s)$ has the specific form

$$K(s) = M \, \Sigma^k(s) \, M^{-1} \qquad\qquad [6.3.2]$$

where M is a real matrix with

$$U_w^* M \simeq J \qquad\qquad [6.3.3]$$

where

$$J = \text{diag}\{ e^{jr_i} \} \, . \qquad\qquad (\, i = 1, \ldots, m \,)$$

J is a diagonal matrix with unit modulus elements and r_i are the phases of the diagonal elements. $\Sigma^k(s)$ is a diagonal rational matrix in s.

M is an approximate real inverse of the complex matrix U_w^* and is orthogonal if the approximation is good. M can be calculated as a linear least squares problem, set up as follows: (Edmunds, 1978)

Given $\qquad\qquad\qquad\qquad J = CR + E \qquad\qquad [6.3.4]$

where $C, E \in \mathbb{C}^{m \times m}$

$\qquad R \in \mathbb{R}^{m \times m}$

$\qquad J = \text{diag}\{ j_i \} \quad i = 1, \ldots, m$

with $\quad j_i \in \mathbb{C}$

and $\quad |j_i| = 1$

The objective is to

$$\text{minimize } \| \, E \, \|_F$$

where $\| \cdot \|_F$ denotes the Frobenius norm of a matrix.

For a solution to the problem, the interested reader may refer to Edmunds (1978).

$\Sigma^k(s)$ is a diagonal rational matrix of the following form:

$$\Sigma^k(s) = \text{diag}\left\{ \frac{(s+a_i)(s+c_i)}{(s+b_i)(s+d_i)} \right\} \qquad (i = 1,\ldots,m)$$

The diagonal elements are simple classical compensators such as phase lead, phase lag or lag-lead compensators. The appropriate choice of parameters a_i, b_i, c_i and d_i will manipulate the principal gains and phases of the loci.

Asymptotically, as $| \, s \, | \longrightarrow \infty$,

$$K(s_\infty) = M \, \Sigma^k(s_\infty) \, M^{-1}$$
$$= M \, \text{diag}\{1,\ldots,1\} \, M^{-1}$$
$$= I \ .$$

Therefore, when this sub-controller is cascaded with the system, at high frequencies, it will not affect the properties of the system.

6.4 Gain and Phase Adjustment using the RFA sub-controllers

In this section, we show how the gains and phases of the open-loop transfer function can be manipulated using the RFA sub-controllers.

Let $G(s) \in \mathbb{C}^{m \times m}$ have an SVD at one specific intermediate frequency s_w:

$$G(s_w) = Y_w \, \Sigma_w \, U_w^*$$
$$= (Y_w U_w^*)(U_w \Sigma_w U_w^*)$$
$$= P_w \cdot \text{diag}\{e^{j\theta_i}\} \cdot P_w^* \cdot U_w \cdot \text{diag}\{\sigma_i\} \cdot U_w^* \qquad [6.4.1]$$

The uncompensated system has principal gains σ_i and principal phases θ_i,

$(i = 1, \ldots, m)$.

An RFA sub-controller is

$$K(s_w) = M \, \Sigma^k(s_w) \, M^{-1}$$

$$= M \cdot \text{diag}\left\{\frac{(jw + a_i)(jw + c_i)}{(jw + b_i)(jw + d_i)}\right\} \cdot M^{-1} \qquad (i = 1, \ldots, m)$$

$$= M \cdot \text{diag}\{k_i e^{jp_i}\} \cdot M^{-1} \qquad\qquad\qquad [6.4.2]$$

where $k_i = \left|\dfrac{(jw + a_i)(jw + c_i)}{(jw + b_i)(jw + d_i)}\right|$

and $\quad p_i = \angle \dfrac{(jw + a_i)(jw + c_i)}{(jw + b_i)(jw + d_i)}$

Hence,

$$G(s_w)K(s_w) = Y_w \cdot \Sigma_w \cdot U_w^* \cdot M \cdot \text{diag}\{k_i e^{jp_i}\} \cdot M^{-1}$$

$$= Y_w \cdot \text{diag}\{\sigma_i\} \cdot U_w^* \cdot M \cdot \text{diag}\{k_i e^{jp_i}\} \cdot M^{-1}$$

Using [6.3.3],

$$G(s_w)K(s_w) \simeq Y_w \cdot \text{diag}\{\sigma_i\} \cdot \text{diag}\{e^{jr_i}\} \cdot \text{diag}\{k_i e^{jp_i}\} \cdot M^{-1}$$

$$\simeq Y_w \cdot \text{diag}\{\sigma_i k_i e^{j(r_i + p_i)}\} \cdot M^{-1}$$

$$\simeq Y_w \cdot \text{diag}\{\sigma_i k_i\} \cdot \hat{U}_w^* \qquad\qquad\qquad [6.4.3]$$

where

$$\hat{U}_w^* = \text{diag}\{e^{j(r_i + p_i)}\} \cdot M^{-1}$$

Also, let

$$Y_w \hat{U}_w^* = \hat{P}_w \cdot \text{diag}\{e^{j\phi_i}\} \cdot \hat{P}_w^* \, . \qquad\qquad [6.4.4]$$

Hence, the compensated system at $s = jw$ has principal gains $\sigma_i k_i$ and principal

phases ϕ_i approximately. $(i = 1, \ldots, m)$

6.5 The Use of Classical Compensators in the RFA sub-controllers

6.5.1 Introduction

The objective in using an RFA sub-controller is to obtain gain/phase adjustment in the intermediate frequency region. This is done by putting classical dynamic compensating networks on the diagonal of an RFA sub-controller. However, their selection with an appropriate choice of parameters is not obvious. A practical guide to their use is now given. This will cover phase lead, phase lag, lag-lead, lead-lag lead-lead, lag-lag and conjugate pole-pair cancellation compensating networks. The aim is to provide a systematic way of manipulating the primary indicators at intermediate frequencies. The technique is to treat the compensators as graphical objects through a *normalization* process.

6.5.2 Normalization of the classical compensating networks

While using classical compensating networks on the diagonal of an RFA sub-controller, the choice of parameters for the classical compensators is an immediate task which a designer has to face. In order to provide a systematic approach to the problem of selecting parameters, we standardize all the classical compensators by a normalization procedure. The objective is to transform the selection of pole/zero positions into a selection of appropriate gain/phase characteristics at a particular frequency. Below is a description of the normalization procedure for each basic classical compensator.

Phase lead compensation

A phase lead compensating network transfer function is

$$G(s) = s+a/s+b \qquad b > a$$

$$= s+a/s+\alpha a \qquad \alpha > 1$$

We define $n = \sqrt{\alpha} \cdot a$ as the transformation parameter. Therefore,

$$G(s) = (s + n/\sqrt{\alpha})/(s + n \cdot \sqrt{\alpha}) \ .$$

Figure 6.1 shows a series of normalized phase lead networks with $n = 1$ as α varies from 1 to 10. The physical meaning of n is that it is the <u>frequency</u> at which maximum phase is introduced into the compensated system, whereas α is the parameter controlling the **maximum amount of phase** that is added. The graph with n = 5 as α varies from 1 to 10 is given in Fig.6.2. A graph is obtained by plotting the maximum phase advance and gain attenuation against α at s = nj (Fig.6.3). This graph is useful in practice for the selection of α. Phase lead compensation is used when we need to introduce phase advance at high or intermediate frequencies. This is done at the expense of low frequency gain attenuation.

Phase lag compensation

Phase lag compensation is the dual of phase lead in many ways. A phase lead compensating network transfer function is

$$G(s) = s+b/s+a \qquad b > a$$

$$= s+\alpha a/s+a \qquad \alpha > 1$$

Again, n = $\sqrt{\alpha} \cdot a$ is defined as the transformation parameter and

$$G(s) = (s + n \cdot \sqrt{\alpha})/(s + n/\sqrt{\alpha}) .$$

Figure 6.4 shows a series of phase lag networks with n = 1 as α varies from 1 to 10. n and α have the same physical meaning as in phase lead compensation. A graph is obtained by plotting the maximum phase lag and gain increase against α at s = nj (see Fig.6.3). This graph can be used for the selection of α. Phase lag compensation is used when we need to adjust the shape of a gain locus (i.e. to increase its slope) at intermediate frequencies. This is done at the expense of phase lag introduced at that region.

Lag-lead compensation

A particular type of lag-lead compensation is introduced with the following transfer function:

$$G(s) = (s+a)(s+a)/(s+a \cdot \sqrt{\alpha})(s+a/\sqrt{\alpha}) .$$

In this case, the normalization parameter n is the same as a.   Figure 6.5 shows a series of lag-lead networks with n = 1 as α varies from 1 to 10.   The parameter α controls the amount of gain attenuation at that frequency.   Figure 6.6 gives a plot of α against gain attenuation.   Lag-lead compensators may be used to deal with the occurence of a resonance peak on the gain loci.

Lead-lag compensation

A particular type of lead-lag compensation is introduced with the following transfer function:

$$G(s) = (s+a\cdot\sqrt{\alpha})(s+a/\sqrt{\alpha})/(s+a)(s+a)$$

In this case, the normalization parameter n is the same as a.   Figure 6.7 shows a series of lead-lag networks with n = 1 as α varies from 1 to 10.   The parameter α controls the amount of gain increase at that frequency (see Fig.6.6).   Lead-lag compensators may be used to deal with the occurence of a resonance peak on the gain loci.

Lead-lead compensation

The use of lead-lead compensation should be broken down into two phase lead compensations.   This will introduce excessive phase advance which may not be possible or efficient if performed by one phase lead compensator.

Lag-lag compensation

Similar to lead-lead compensation, the use of lag-lag compensation should be broken down into two phase lag compensations.   This will introduce a large gain increase which may not be possible or efficient if performed by one phase lag compensator.

Conjugate pole-pair cancellation compensation

This is introduced as a special second order compensating network. The objective is to cancel the pair of complex conjugate poles of the system which usually cause a resonance peak and sharp phase changes.

Let p and $\bar{p}$ be a pair of complex conjugate poles of the system.

$$G(s) = (s+p)(s+\bar{p})/(s+z1)(s+z2)$$

where z1 and z2 are poles of the compensator in the left-half plane which are usually assigned far to the left on the real axis.

6.6 An Optimizer for Parameter Tuning

The choice of parameters when forming the RFA sub-controllers depends on the intuition and experience of the designer. The normalization procedure helps the designer to find an initial set of parameters. However, it would be useful to have an optimizer to find an optimum solution. The optimizer will involve tuning the parameters of an RFA sub-controller to minimize an objective function J.

Let $G(s) \in \mathbb{R}(s)^{mxm}$ be the transfer function of a plant after an RFA sub-controller has been cascaded. $G(j\omega_f) \in \mathbb{C}^{mxm}$ should therefore be a function of the pole/zero positions of the RFA sub-controller as well as a function of the frequency at which the RFA sub-controller is formed. Let these parameters be a_1, ..., a_k and the vector of parameters $(a_1 ... a_k)^t$ be denoted by a, where $a \in E^k$. It is assumed that ω_f lies in a frequency interval of interest $[\omega_{min} , \omega_{max}]$ ($f = 1,..., \gamma$). Also, let

$\{\sigma_i^f\}$ denote the set of singular values for $G(j\omega_f)$,

$\{g_i^f\}$ denote the set of eigenvalues for $G(j\omega_f)$,

phaspec denote a specified phase margin,

$\text{cond}\{\sigma^f\}$ denote the condition number of $G(j\omega_f)$ in 2-norm,

$\angle g_i{}^f$ denote the phase margin of $g_i{}^f$

$MS(\omega_f)$ denote the measure of skewness for $G(j\omega_f)$.

$(i = 1,\ldots,m)$

The problem may be stated as follows:

$$\text{minimize } \{ J(a) = \frac{w_1}{\gamma}\cdot\sum_{f=1}^{\gamma} MS(\omega_f) + \frac{w_2}{h}\cdot\sum_{f=1}^{\gamma} (\sum_{i=1}^{m} \alpha_i{}^f\theta_i{}^f)$$
$$+ w_3(1 - 1/(\max_f \{cond\{\sigma^f\}\})) \}$$

$$[6.6.1]$$

where

w_i are non-negative weighting factors, $(i = 1, 2, 3)$

$\theta_i{}^f = (\text{phaspec} - \angle g_i{}^f) / \text{phaspec}$,

$$\alpha_i{}^f = \begin{cases} 0 , & \theta_i{}^f \leq 0 \\ 1 , & \theta_i{}^f > 0 \end{cases}$$

h is the number of occurences when $\alpha_i{}^f = 1$.

The objective function $J(a)$ is a scalar function and we seek the parameter vector a that minimizes $J(a)$. The first term in $J(a)$ corresponds to the average measure of skewness over the frequency interval $[\omega_{min}, \omega_{max}]$. It has a range from 0 to 1. The second term corresponds to the violation of a specified phase margin. The average amount of phase which exceeds the specified phase margin is calculated. It is then divided by the specified phase margin. The minimum value will be zero and the value is 1 if the system has zero phase margin. The third term represents the gain balancing of the loci. It is calculated as one minus the inverse of the maximum condition number in the interval $[\omega_{min}, \omega_{max}]$ to give a range from 0 to 1. The weightings w_1, w_2 and w_3 are given by the designer at the start of the optimization procedure. They represent the relative importance of each term in the objective function. The following table is suggested for the weightings:

Value	Description
0	Not important
1	Weakly important
2	Mildly important
3	Moderately important
4	Very important
5	Extremely important

Each value expresses a degree of importance about that part of the function that needs to be minimized. It has been found that further refinements of the scale suggested above are not necessary. In fact, a wider scale may cause the function value to change only slightly in successive evaluations.

An optimizer for use in the Reverse Frame Alignment technique has been implemented in the commerical analysis and design package MATRIX$_x$ (Integrated Systems, 1984). The objective function to be minimized is a function of two variables, which are the pole and zero positions in the RFA sub-controller. There are several considerations when choosing the optimization method to be used in our problem. Firstly, the objective function J(a) expressed in [6.6.1] cannot be differentiated easily. This is because the function is dependent on obtaining the eigenvalues and singular values of a general complex matrix. Hence, methods that rely on obtaining the derivatives of the function are not applicable. Therefore, direct search methods are used for the solution of our problem.

Secondly, it is found that each function evaluation takes around 20s to complete. The reason is that we are using 'functions' in MATRIX$_x$ and the instructions have to go through an interpreter. Hence, we would like to use methods that require a minimum of function evaluations as they take up a lot of computing time.

Thirdly, we would like to use a method that can easily be extended for use in problems of higher dimensions (i.e. when the number of variables is more than two). In that case, we can tune the parameters of more than one locus at one time or we can tune the frequency at which the RFA sub-controller is formed as well.

Finally, we have considered whether to put any constraints on the values of the variables. If constraints are used, then the designer needs to specify the regions where the pole and zero lie. However, if unconstrained optimization is desired, the designer only inputs a starting point for the search. In either case, the designer can make use of his findings in the previous trial designs. In order to explore the problem more thoroughly, two direct search methods are used.

The first method, alternating variable method with Fibonacci search, is based on constrained optimization. The second method is the simplex method of Nelder and Mead. A good description of these methods can be found in Murray (1972). The alternating variable method is simple to implement. However, when comparing it with the simplex method, it is less efficient and generally requires more function evaluations. Also, the method has not made use of all the information of the magnitude of the function values whereas the simplex method uses a number of heuristics involving the magnitude of the function values.

The simplex method is found to be very efficient in practice and it is faster when compared with the alternating variable method. A comparison of the two methods in an example is given in Table 6.1(a). Although the simplex method is for unconstrained optimization, constraints can be considered by expressing them as penalty terms in the objective function. In the end, the optimizer using the simplex minimization algorithm was extended to include an additional variable which is the frequency at which an RFA sub-controller is formed. Some results obtained from an example are given in Section 6.9.3.

In the paper by Nelder and Mead (1965), a stopping criterion based on the variation in the function values over the simplex is proposed. A pre-set value is suggested and when improvement on successive iterations is less than that value, the search will stop. It is then assumed that the minimum is located. However, it is difficult to decide on a pre-set value because we know very little about the function before optimization. The specified tolerance for termination may take a very long time to reach. Hence, we would like the designer to have complete control over the optimizer and let him decide whether he is satisfied with the result after every few iterations.

6.7 Design Procedure

6.7.1 Introduction

The Reverse Frame Alignment design technique is based on a cascaded combination of the high frequency, RFA and the low frequency sub-controllers plus any further basic types of sub-controllers for MIMO systems which may be suitable. The design is broken down into three parts, each of which is concerned with a particular frequency region of interest.

6.7.2 High frequency region design

The high frequency sub-controller used in the Simple Design Technique can be kept for use in the Reverse Frame Alignment design technique. We wish to align the input and output frames of the system to make it normal. We also want to manipulate the gains and phases in such a way that the characteristic loci satisfy the Nyquist Stability Criterion. Also, we wish to meet the performance specification and reduce interaction.

6.7.3 Intermediate frequency region design

The RFA sub-controller is the basic tool. We want to manipulate the principal gains and phases of the system near a chosen intermediate frequency. The manipulation of gains will affect the performance of the system and the manipulation of phases will affect the stability or stability margin of the system. A number of RFA sub-controllers may be cascaded with the system at different intermediate frequencies so as to obtain a fine manipulation of the gains and phases to satisfy the performance and stability criteria. However, they will not interfere with the high frequency region design.

Each locus is specifically chosen to be manipulated by the use of a simple classical compensating network at the corresponding position in the diagonal rational matrix $\Sigma^k(s)$ of an RFA sub-controller. The selection of the network is determined by the objective of the manipulation. For example, a phase lead network will give phase advance to the corresponding locus but it will lower the gain of that locus at low frequencies. On the other hand, a phase lag network can adjust the shape of a gain locus at the expense of an increase of phase at a range of chosen frequencies.

6.7.4 Low frequency region design

The low frequency sub-controller is the basic tool. We wish to align the input and output gain frames of the system, balance up the gains and add integral action for performance.

6.8 The Final Controller

Let the sub-controllers for the high frequency region design be $K_\infty K_i(s)$. Also, let the sub-controllers for the intermediate frequency region be $\quad\prod\limits_{j=1}^{n} M_j \Sigma^k(s_j)M_j^{-1}$ and the sub-controller for the low frequency region

design be $(K_{LF}/s + I_m)K_r(s)$. The final controller $K(s)$ is the product

$$K(s) = K_\infty K_i(s)\left\{ \prod\limits_{i=1}^{n} M_j \cdot \Sigma^k(s_j) \cdot M_j^{-1} \right\} (K_{LF}/s + I_m)K_r(s) \qquad [6.7.1]$$

The effects of the different sub-controllers are combined and each comes into proper operation in an appropriate frequency region. $K_i(s)$ and $K_r(s)$ are basic types of sub-controllers which may be needed for further manipulation of the primary indicators. The complete controller is a dynamic controller with integral action. At high frequencies,

$$K(s) = K_\infty K_i(s)K_r(s) \quad .$$

At low frequencies,

$$K(s) = K_\infty K_i(s)\left\{ \prod\limits_{j=1}^{n} M_j \cdot \Sigma^k(s_j) \cdot M_j^{-1} \right\}(K_{LF}/s)K_r(s) \quad .$$

The presence of the integral term will ensure zero steady-state error in the closed-loop steady state response.

6.9 Examples

6.9.1 Example 1

This example is a 2-input, 12-state, 2-output gas turbine model which will be referred to as AUTO. The details of the system are given in Appendix B. The system has been studied before using a closed-loop design method by Bloch & Postlethwaite (1981), an inverse Nyquist array method by Patel and Munro (1982), and a Characteristic Locus method by Edmunds, Jeanes and Maciejowski (1983). Here, the Simple Design Technique is used first to design a controller for the system. The Reverse Frame Alignment technique is used next. It is found that the Reverse Frame Alignment technique provides a systematic method for the design of the controller.

1. Simple Design Technique

High frequency region design

The primary indicators of AUTO are shown in Fig.6.8.1. The system has different roll-off rates at high frequencies and the gains are balanced at around the bandwidth frequency. A high frequency gain of 1.5 is chosen. The primary indicators (Fig.6.8.2) show that the gains are balanced at around s = 10j.

Low frequency region design

A low frequency sub-controller is added together with a scalar gain of 4 (Fig.6.8.3). The indicator MS is given in Fig.6.8.4. The low frequency sub-controller is

$$4(\ I_2 + K_0/s\ ) \qquad .$$

The closed-loop step responses of the compensated system are given in Figs.6.8.5 (a) & (b). The overshoot is 30% and interaction is around 18%. One of the outputs has a rather sluggish response (see Fig. 6.8.5 (b)).

2. Reverse Frame Alignment technique

DESIGN ONE

High frequency region design

The high frequency sub-controller used in the SDT is kept. The primary indicators show that one of the loci is rather flat near the bandwidth frequency. Therefore, we need to obtain gain adjustment using an RFA sub-controller.

Intermediate frequency region design

An RFA sub-controller is used at s = 3j. Figure 6.8.6 gives the accuracy of approximation over the intermediate frequency region by finding the condition number of M. This provides an additional secondary indicator for the designer. A phase lag network is used to manipulate the gain of one of the loci.   The primary indicators show that the RFA sub-controller has achieved its objective (Fig.6.8.7).   The RFA sub-controller is

$$K_1(s) = M_1 \cdot diag\{ \; 1 \; ,(s+10)/(s+1)\} \cdot M_1^{-1} \; .$$

Low frequency region design

A low frequency sub-controller is added and a scalar gain of 4 is used to obtain a good performance for the system (Fig.6.8.8). MS is given in Fig.6.8.9. The low frequency sub-controller is

$$4(\; I_2 + 0.1K_0/s \;) \; .$$

The closed-loop step responses of the design are given in Figs.6.8.10 (a) & (b). Both loci show a fast response to a step input. The response which had an overshoot of 30% before remains the same.

DESIGN TWO

High Frequency region design

The high frequency sub-controller used in SDT is kept.

Intermediate frequency region design

The RFA sub-controller used in DESIGN ONE is kept. A second RFA sub-controller is used at s = 7j to introduce phase advance into the second locus.

$$K_2(s) = M_2 \cdot diag\{ \; 1 \; ,(s+4)/(s+12)\} \cdot M_2^{-1} \; .$$

However, the gain of that locus has been lowered (Fig.6.8.11) and further gain balancing is needed. The primary indicators (Fig.6.8.12) now show that the system has good gain balancing and acceptable phase margins for both loci at high and intermediate frequencies. What remains is the low frequency design.

Low frequency region design

The low frequency sub-controller is

$$4(\; I_2 + 0.1K_0/s \;) \; .$$

The primary indicators and MS are given in Fig.6.8.13 and Fig.6.8.14. The closed-loop step responses of the final design are given in Figs.6.8.15 (a) & (b). Because of the extra phase introduced at the bandwidth frequency, the 30% overshoot of one of the responses has been brought down to around 13%.

6.9.2 Example 2

The second example we consider is a CH-47 tandem rotor helicopter. The system, denoted GHEL, is a 2-input, 2-output, 4-state system. The details are given in Appendix B. The open-loop poles are at -2.2597, -0.0938 $\pm$ 0.2899j and 0.3873. Hence the system is open-loop unstable. There is one finite zero at -0.018. LQG design using full-state loop transfer recovery technique has been done with satisfactory results (Doyle and Stein, 1981; Bloch and Postlethwaite, 1981). Moreover, a closed-loop design and a Characteristic Locus design were given in Bloch and Postlethwaite (1981). In this section, the system is designed using the SDT and the Reverse Frame Alignment technique. The design objective is to control two measured outputs, which are vertical velocity and pitch attitude, by manipulating collective and differential collective rotor thrust commands. A bandwidth of about 10 rad/s is suggested by Doyle and Stein (1981).

High frequency region design

Figure 6.9.1 gives the primary indicators of the system without any compensation. The primary indicators of the system after cascading a high frequency sub-controller are shown in Fig.6.9.2.

$$K_\infty = \begin{bmatrix} 0.0418 & 4.3932 \\ -0.9991 & 0.1839 \end{bmatrix}$$

The gains are balanced at $\omega = 10$ and the divergences between the characteristic and principal gains are small at intermediate frequencies. The primary indicators, after including an integrator, are shown in Fig.6.9.3. One locus has a very small phase margin at the crossover frequency. The closed-loop step response is found to be highly oscillatory and hence SDT cannot give a satisfactory design. Therefore, keeping the high frequency sub-controller, we proceed to the next stage using the Reverse Frame Alignment technique.

Intermediate frequency region design

An RFA sub-controller is used to introduce phase advance of around 55 degrees (pole-zero ratio = 10) to one locus at ω = 8.

$$K_1(s) = M_1 \cdot \text{diag}\{ (s + 2.5298)/(s + 25.2982),\ 1\} \cdot M_1^{-1}$$
$$M_1 = \begin{bmatrix} 0.1443 & 0.9447 \\ 4.5111 & -0.7823 \end{bmatrix}$$

The resulting primary indicators are shown in Fig.6.9.4. The side-effect of putting in phase lead is that the gain of that locus is lowered. A better gain balancing of the loci at around ω = 10 is obtained using a second high frequency sub-controller (Fig.6.9.5).

$$K_2 = \begin{bmatrix} 1 & 0 \\ 0 & 4 \end{bmatrix}$$

Another RFA sub-controller is then used to bring up the lower locus at ω = 2 with a pole-zero ratio of 6. (Fig.6.9.6).

$$K_3(s) = M_2 \cdot \text{diag}\{1,\ (s + 4.899)/(s + 0.8165)\} \cdot M_2^{-1}$$
$$M_2 = \begin{bmatrix} 0.7399 & -0.5916 \\ -1.0174 & -1.2520 \end{bmatrix}$$

We note that the gain divergences are very small at high and intermediate frequencies and hence the system is robust.

Low frequency region design

Finally, an integrator is added to remove steady state error (Fig.6.9.7).

$$K_L(s) = (0.3/s + 1) I_2$$

The normality indicator (MS) is shown in Fig.6.9.8. The value of MS is small over the high and intermediate frequencies as expected. Figures 6.9.9 (a) & (b) show the time responses of the resulting closed-loop outputs to a step change in both reference inputs. The interaction obtained in both outputs is good and the steady state error is zero. Hence, the behaviour of both outputs is quite acceptable. Finally, the set of generalised Nyquist diagrams is shown in Fig.6.9.10.

Parameter optimization

As an example of the use of the parameter optimizer, we shall return to the design of the sub-controller $K_3(s)$. $K_3(s)$ was formed at $\omega = 2$ with a pole-zero ratio of 6, which results in a zero and pole at -4.899 and -0.8165 respectively. Here, we tune the two parameters by minimizing an objective function (see [6.6.1]) and examine the results when different weightings are used in the function. The optimizer is set up as follows:

(i)   The second locus is chosen for manipulation.

(ii)   A logarithmic equally spaced frequency list $\{\omega_1,...\omega_8\}$ of 8 points is used where $3 \le \omega_i \le 11$.

(iii)   The desired phase margin over the relevant frequency interval is chosen to be 60 degrees.

(iv)   The frequency at which the RFA sub-controller is formed is 2 rad/s.

The objective function to be minimized is a linear combination of four terms. The first one is the average value of MS over the frequency interval [3,11]. The second term penalizes any violation of the specified phase margin. The third term is a function of the condition number over the interval [3,11]. The fourth one is a penalty term which constrains the pole and zero to be in the left-half plane. Different weightings on the first three terms of the objective function have been tried.

To compare the alternative variable method and the simplex method, the procedures were repeated, again with an initial guess of zero at −4.899 and pole at −0.8165. In the simplex method, the side length of the initial simplex is 0.3. In the alternative variable method, the zero is constrained between −10 and −1 and the pole is constrained between −5 and zero. The results are summarized in Table 6.1(a). It is found that the simplex method is generally faster and requires less function evaluations. With the simplex method, the average value of MS, minimum phase margin and maximum condition number of the matrix in 2-norm over the frequency interval [3,11] are given in Table 6.1(b).

From Table 6.1(a), it can be seen that from case 1 to 4, the optimiser suggested a zero at below −6 whereas the initial guess was at −4.899. In case 4, where equal weights are put on all three terms, the average MS is less than the case of initial guess although the phase margin and condition number are similar (from Table 6.1(b)). The primary indicators with the parameters in case 4 are shown in Fig.6.9.11. With the same low frequency sub-controller as before, the primary indicators shown in Fig.6.9.12 are similar to Fig.6.9.7. An inspection of the normality indicator (Fig.6.9.13) shows that MS is smaller than before in the interval [3,11]. Although there is no appreciable difference in the overshoot of the closed-loop step responses (Figs.6.9.14 (a) & (b)), interaction has been reduced to less than 5% in both channels.

In case 5, where the phase margin is weighted heavily against the other two terms, the use of a phase lead network is suggested. However, this is undesirable because the average MS will be 0.5017, which is quite high, and the condition number is much larger than before (from Table 6.1(b)).

In case 6, where phase margin is weighted heavily against MS, a better phase margin of at least 48.8 degrees is obtained. The primary indicators are shown in Fig.6.9.15. Again, the same low frequency sub-controller is cascaded with the system (see Fig.6.9.16). As shown in Figs.6.9.17 (a) & (b), the closed-loop step responses have a slight drop in overshoot but interaction is more than before. The value of MS is higher than before as expected (see Fig.6.9.18).

6.9.3 Example 3

The design of a controller for a nuclear powered turbo-generator TGEN (see Appendix B for details) is considered. The plant inputs are the throttle valve position and the excitation control. The controlled outputs are the generator terminal voltage and the generator load angles. The objective of the design is a fast non-interacting response for the closed-loop system.

An analysis of the plant model shows that all the ten poles of the system are all in the left-half plane. Hence, no anticlockwise encirclements of the -1 point are required for closed-loop stability. All the finite zeros of the system are on the left-half plane as well and the system is of minimum phase. The Simple Design Technique is used first to design the system. The Reverse Frame Alignment technique is then used.

1. Simple Design Technique

High frequency region design

The primary indicators of TGEN are shown in Fig.6.10.1. The system has different roll-off rates at high frequencies and the gains are balanced at around the bandwidth frequency (s = 10j). A high frequency gain of 100 is used (see Fig.6.10.2) The high frequency sub-controller is

$$K_\infty = \begin{bmatrix} -100 & 0 \\ 0 & 1 \end{bmatrix}$$

Low frequency region design

A low frequency sub-controller is added. The primary indicators and the indicator MS are shown in Fig.6.10.3 and Fig.6.10.4 respectively. The loci have to be brought down with a scalar gain of 0.001 because of the very bad phase behaviour associated with a pair of complex conjugate poles of the system (see Fig.6.10.5). A lightly damped resonance can be observed from the gain plot. The closed-loop step response (Figs 6.10.6 (a) & (b)) shows that the overshoot is 22% and interaction is around 30% with the second input. Both outputs give a rather sluggish response because of small loop gains. The design is not satisfactory and we use Reverse Frame Alignment technique next.

2. Reverse Frame Alignment technique

High frequency region design

The high frequency sub-controller used in the SDT is kept.

Intermediate frequency region design

A sub-controller with a conjugate pole-pair cancellation network on its diagonal is used (Fig.6.10.7).

$$K_1(s) = \begin{bmatrix} \dfrac{(s + \alpha_1)\,(s + \alpha_2)}{(s + z_1)\,(s + z_2)} & 0 \\ 0 & 1 \end{bmatrix}$$

where

$\alpha_1 = 0.36 + 6.34j$, $\alpha_2 = 0.36 - 6.34j$, $z_1 = 20$ and $z_2 = 25$.

It can be seen that the sudden phase change at around $\omega = 6.34$ has been reduced. Next, the gains are balanced again at high frequencies (Fig.6.10.8).

$$K_2 = \begin{bmatrix} 15 & 0 \\ 0 & 1 \end{bmatrix}$$

The indicator on the accuracy of approximation when using an RFA sub-controller is given in Fig.6.10.9. An RFA sub-controller is formed to introduce phase lead into the second locus at $\omega = 11$ with a pole-zero ratio of 7.

$$K_3(s) = M_3 \cdot \text{diag}\{1, (s + 4.1576)/ (s+29.1033)\} \cdot M_3^{-1}$$

where
$$M_3 = \begin{bmatrix} 0.8555 & -0.5164 \\ 0.5240 & 0.8681 \end{bmatrix}$$

It can be seen that the second locus has a better phase margin than before (Fig.6.10.10). Also, the gain divergences between principal and characteristic gains are smaller than before. Next, another RFA sub-controller is cascaded to introduce phase advance into the first locus at $\omega = 11$ with a pole-zero ratio of 6.

$$K_4(s) = M_4 \cdot \text{diag}\{ (s +4.4907)/(s + 26.944), 1\} \cdot M_4^{-1}$$

where
$$M_4 = \begin{bmatrix} 0.8560 & -0.5169 \\ 0.5169 & 0.8560 \end{bmatrix}$$

The primary indicators (Fig.6.10.11) show that the system now has a much better phase margin for both loci at intermediate frequencies.

Low frequency region design

A low frequency sub-controller

$$(I_2 + 0.3K_0/s)$$

with
$$K_0 = \begin{bmatrix} 6.9931 & 6.2752 \\ -0.0357 & 1.3171 \end{bmatrix}$$

is used. The primary indicators and MS are given in Fig.6.10.12 and Fig.6.10.13. The loci are lowered with a scalar gain of 0.35 (see Fig.6.10.14). The closed-loop step responses of the design are given in Figs.6.10.15 (a) & (b). The overshoot is 10% and interaction is around 6%. for the first input, which is acceptable. As for the second input, the overshoot and interaction are less than 3% & 25% respectively. The very large gain divergence at 6.34 rad/s is very difficult to remove at this stage. At the next level of design using an observer-based controller, we will be able to obtain a much better result.

Parameter optimization

In the final stage of the intermediate frequency region design above, an RFA sub-controller at ω = 11 with a pole-zero ratio of 6 is used. This results in a zero at −4.4907 and a pole at −26.944. Here, we would like to tune the parameters of pole, zero and the frequency at which the RFA sub-controller is formed. The set up of the optimizer is given below:

(i)   The first locus is chosen for manipulation.

(ii)   A logarithmic equally spaced frequency list $\{\omega_1, \ldots, \omega_8\}$ of 8 points is used where $4 \leq \omega_i \leq 11$.

(iii)   The desired phase margin over the frequency interval [4,11] is 60 degrees.

The objective function to be minimized is described in Section 6.6. An additional penalty term is added to force the pole and zero into the left-half plane. The results of putting different weights on the first three terms of the objective function are summarised in Table 6.2(a). More details are given in Table 6.2(b). It is interesting to note from Table 6.2(a) that although different weights are used in cases 1, 3 and 4, they all give the same result. From Table 6.2(b), we note that there are improvements on MS and the condition number over the frequency interval [4,11]. The specified phase margin is satisfied. Therefore, it seems that an "optimum" set of parameters would be zero at −6.4942, pole at −27.2504 with the RFA sub-controller formed at ω = 7.5587. Next, we conclude the design using the above set of parameters. The primary indicators are given in Fig.6.10.16. A low frequency sub-controller

$$K_L(s) = (0.3\ K_0/s + I_2)$$

is added where

$$K_0 = \begin{bmatrix} 4.5019 & 3.9576 \\ -0.8084 & 0.6216 \end{bmatrix}.$$

The primary indicators of the design as shown in Fig. 6.10.17 are similar to the previous design (see Fig.6.10.12). The gain in each loop is reduced by 0.35 to give a good phase margin at the crossover frequency. The closed-loop step responses given in Figs.6.10.18 (a) & (b) are again similar to the previous design except that there is a slight reduction on interaction in the second loop. The peak value of MS is around 0.83 (Fig.6.10.19) which is less than before.

The set of parameters given in case 2 is found unacceptable as seen from its phase margin and condition number given in Table 6.2(b).

6.10 Non-Square Systems

6.10.1 The squaring-up problem

A plant with transfer function matrix $G(s) \in R(s)^{m \times \ell}$ is non-square if $m \neq \ell$. If there are more inputs than outputs, then we have the problem of finding a precompensator $K(s) \in R(s)^{\ell \times m}$ for the plant $G(s)$ such that the closed-loop system is stable, has good performance and good robust stability properties. The first step is the formation of a high frequency sub-controller, K_∞, which is non-square and has dimension $(\ell \times m)$. We recall from [5.2.3] that

$$K_\infty = U_\infty \Sigma_\infty^k Y_\infty^t$$

where U_∞ and Y_∞ are real orthogonal matrices of dimension $(\ell \times \ell)$ and $(m \times m)$ respectively. Σ_∞^k is of dimension $(\ell \times m)$ with the last $(\ell - m)$ rows all zeros. The squared-up system is square and the design techniques for square multivariable feedback systems described so far apply with no modifications.

6.10.1.1. Example

Consider a plant with transfer function matrix

$$G(s) = \begin{bmatrix} 2(s+1) & 1 & (s+3) \\ (s+3) & 4(s+1) & 2 \end{bmatrix} \frac{1}{(s+2)\,(2s+3)}$$

The system is non-square with 3 inputs and 2 outputs. The set of principal loci of the system is shown in Fig.6.11.1.

High frequency region design

A high frequency sub-controller is formed with

$$K_\infty = U_\infty \, \Sigma_\infty^k \, Y_\infty^t$$

where

$$U_\infty = \begin{bmatrix} 0.3141 & 0.8389 & 0.4455 \\ 0.9486 & -0.2963 & -0.1111 \\ 0.0385 & 0.4565 & -0.8889 \end{bmatrix}$$

$$Y_\infty = \begin{bmatrix} 0.1602 & 0.9870 \\ 0.9870 & -0.1602 \end{bmatrix}$$

$$\Sigma_\infty^k = \begin{bmatrix} 1.0 & 0.0000 \\ 0.0 & 0.1917 \\ 0.0 & 0.0000 \end{bmatrix}$$

Hence, $$K_\infty = \begin{bmatrix} 1.6377 & 0.0524 \\ -0.4088 & 1.0273 \\ 0.8700 & -0.1022 \end{bmatrix}$$

The primary indicators of G(s) K_∞ are shown in Fig.6.11.2. It can be seen that the characteristic and principal gains at high frequencies have been balanced.   The cascaded system is square and exhibits simple frequency response characteristics.

Low frequency region design

A low frequency sub-controller is formed as follows:

$$K_L(s) = (\frac{0.6\ K_o}{s} + I_2)$$

$$K_o = \begin{bmatrix} 1.8434 & -0.3746 \\ -2.277 & 2.4852 \end{bmatrix}$$

The primary indicators (Fig.6.11.3) show that the characteristic and principal gains coincide over the entire frequency spectrum. Thus, MS is expected to be small over all frequencies (see Fig.6.11.4). A scalar gain of 6 is added and the final compensated system has a bandwidth of about 10 rad/s (see Fig.6.11.5). The closed-loop step responses are found acceptable as shown in Figs.6.11.6 (a) & (b).

6.10.2 The squaring-down problem

We shall now consider the case where a plant has more outputs than inputs (m>ℓ) which is often encountered in practice. It is well-known that we can only have independent control over at most ℓ linear combinations of the m outputs. Therefore, we have an excess of measurement information and would like to make use of the degrees of freedom to our advantage. Kourvaritakis (1974) has presented a careful study of the problem in his Ph.D. thesis. In his approach, he aims to find a post-compensator, L, such that the squared-down system has zeros placed at desired locations. The new system should have good phase properties by the creation of zeros. As discussed by Kourvaritakis, there are severe restrictions on the structure of L since each row of the desired output matrix must be a combination of the rows of the original C matrix. In actual practice, the structure of L may be further restricted by the design specifications. In spite of the above limitation on the degree of freedom of L, we aim to obtain a post-compensator L which can create finite left-half plane zeros in the squaring-down process.

Here, we propose a compensation scheme for designing non-square systems which makes use of the output information. Fig.6.12 gives the diagram of the proposed scheme.

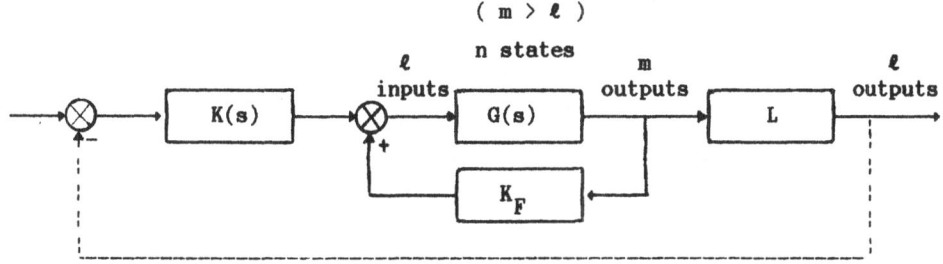

Fig.6.12 Compensation scheme for the squaring-down problem

L is a squaring-down post-compensator of dimension (ℓxm). K_F is a constant or dynamic output feedback matrix of dimension (ℓxm). K(s) is a combination of pre-compensators for shaping the primary indicators of the open-loop system. The purpose of K_F is to improve the phase properties of the system by pole placement. Conditions for the existence of K_F have been investigated by some researchers. Davison and Wang (1975) have shown that if m+ℓ-1 ≥ n, then a constant K_F exists which allows the closed-loop poles to be arbitrarily close to a specified set of poles. A further result due to Munro and Novin-Hirbod (1979) states that if m+ℓ-1 < n, then a full-rank dynamic compensator K_F(s) with degree r can be constructed which will allow arbitrary pole placement,

where $r \geq \dfrac{n-(m+\ell-1)}{\max\{m,\ell\}}$.

For a robust design, it is desirable for the assigned poles to be insensitive to perturbations in the coefficient matrices of the closed-loop system. A robust pole placement method by output feedback is proposed by Chu and Nichols (1984). Some necessary and sufficient conditions for closed-loop eigenstructure assignment by output feedback are given in Fletcher et al. (1985).

The arrangements of Fig.6.12 can also be adopted for the design of square systems (i.e. when m=ℓ, L=I_m). However, we propose it mainly for non-square systems for two reasons. Firstly, for non-square systems (m>ℓ), we can make use of the extra information from output for feedback. If m+ℓ−1 ≥ n, then only a constant matrix for output feedback is required. After pole placement in the inner loop, the design of K(s) using either the Simple Design Technique or the Reverse Frame Alignment technique should be straight-forward.

Secondly, for square systems, we propose a more systematic design method which is based on the use of observers and state feedback. In fact, the observer-based approach can be viewed as another form of output feedback. The approach, which is general and well-established, will be discussed in detail in the next section.

6.10.2.1. Example

This example to be considered is the original non-square model of the chemical reactor REAC discussed in Section 5.5.5. The plant, denoted NSRE, is an open-loop unstable, 2-input, 4-state, 3-output system (see Appendix B). The specification required that output y_1 (or a linear combination of y_1 and y_3) and output y_2 are to be controlled using input u_1 and u_2 (Munro, 1972). Hence, the first step is to find a squaring-down matrix L.

Let $\tilde{y}_1$, $\tilde{y}_2$ be the new set of outputs. L should be of dimension (2x3). Therefore,

$$\begin{bmatrix} \tilde{y}_1 \\ \tilde{y}_2 \end{bmatrix} = \begin{bmatrix} \alpha & 0 & \beta \\ 0 & 1 & 0 \end{bmatrix} \begin{bmatrix} y_1 \\ y_2 \\ y_3 \end{bmatrix} = \begin{bmatrix} \alpha y_1 + \beta y_3 \\ y_2 \end{bmatrix}$$

where α, β are constants.

The original C matrix of the plant is

$$\begin{bmatrix} 1 & 0 & 0 & 0 \\ 0 & 1 & 0 & 0 \\ 0 & 0 & 1 & -1 \end{bmatrix}$$

and let x_i (i = 1,...,4) be the states of the system.

We have

$$\begin{bmatrix} y_1 \\ y_2 \\ y_3 \end{bmatrix} = C \cdot \begin{bmatrix} x_1 \\ x_2 \\ x_3 \\ x_4 \end{bmatrix}$$

Therefore
$$\begin{bmatrix} \tilde{y}_1 \\ \tilde{y}_2 \end{bmatrix} = \begin{bmatrix} \alpha x_1 + \beta(x_3 - x_4) \\ x_2 \end{bmatrix}$$

$$= \begin{bmatrix} \alpha & 0 & \beta & -\beta \\ 0 & 1 & 0 & 0 \end{bmatrix} \cdot \begin{bmatrix} x_1 \\ x_2 \\ x_3 \\ x_4 \end{bmatrix}$$

For simplicity, α and β were chosen as 1 as Munro did in his design.

Hence,

$$L = \begin{bmatrix} 1 & 0 & 1 \\ 0 & 1 & 0 \end{bmatrix}$$

The squared-down system has finite zeros at -1.193 and -5.012.

The next step is to obtain an output feedback matrix K_F. Since the system has 2 inputs ($\ell=2$), 3 outputs ($m=3$) and 4 states ($n=4$), the inequality $m+\ell-1 \geq n$ is satisfied. A full-rank output feedback matrix

$$K_F = \begin{bmatrix} 1 & -3 & 2 \\ 2.5 & 0 & 1 \end{bmatrix}$$

is found which places the poles at -1.6417, $-4.476 \pm 1.156j$ and -21.2372. The primary indicators of the system after squaring down and output feedback are given in Fig.6.13.1, which show a simple frequency response characteristic. The final step is to design pre-compensators for the square system using the Simple Design Technique.

High frequency region design

A high frequency sub-controller is obtained as follows:

$$K_\infty = \begin{bmatrix} 0 & 1 \\ -1.7763 & 0 \end{bmatrix} \cdot$$

From the primary indicators (see Fig.6.13.2), it can be seen that the system now has correct gain-phase characteristics as well as good gain balancing at high frequencies.

Low frequency region design

A low frequency sub-controller

$$K_L(s) = (\frac{3.5 \, K_o}{s} + I_2)$$

is obtained where

$$K_o = \begin{bmatrix} 1.1381 & -0.7217 \\ -0.677 & 4.5394 \end{bmatrix} .$$

The primary indicators (Fig.6.13.3) show that the gain loci coincide at both frequency ends and the principal loci are fairly well balanced at intermediate frequencies. The divergances between the characteristic and principal gains are small as well. The normality indicator MS (Fig.6.13.4) shows that the maximum value is 0.26 which is acceptable. Comparing with the design of REAC in Section 5.5.5., the value of MS is much smaller at low frequencies. Finally, a scalar gain of 6 is introduced for the final compensated system to have a bandwidth of around 12 rad/s (see Fig.6.13.5). The closed-loop step responses of the outputs are fast and the interaction is small (see Figs.6.13.6 (a) & (b)).

125

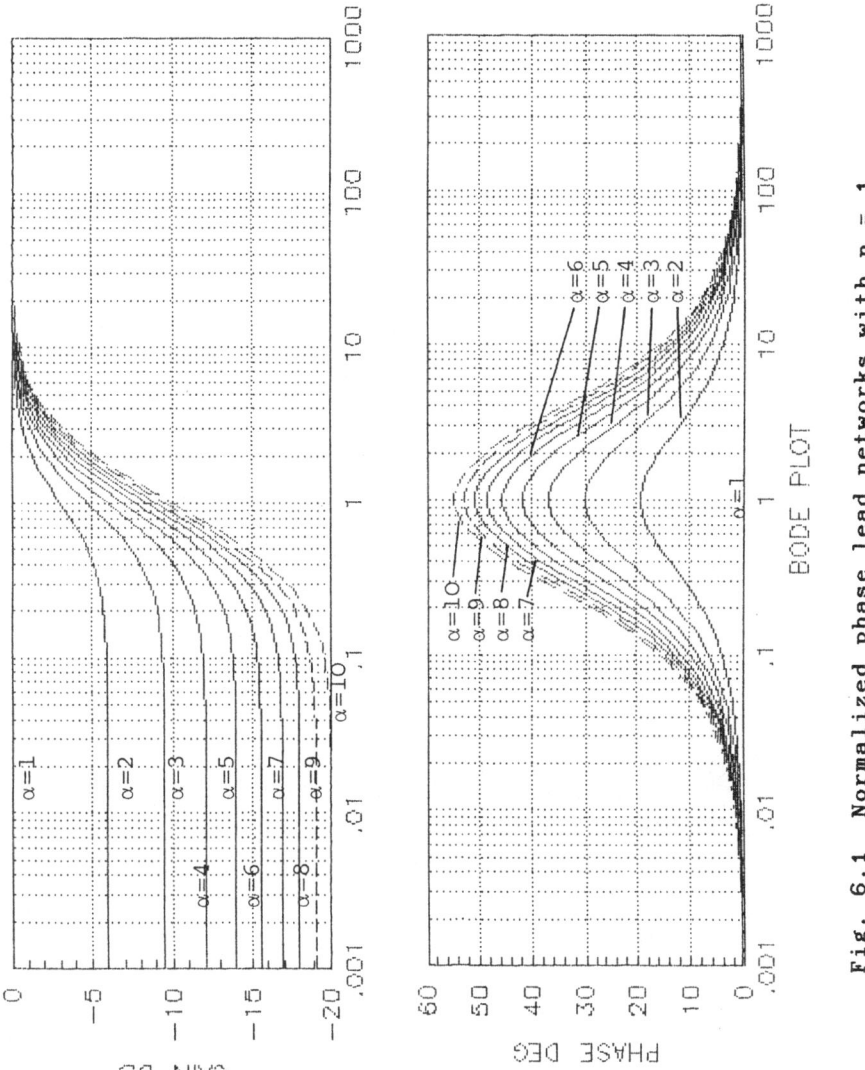

Fig. 6.1 Normalized phase lead networks with n = 1 and α varies from 1 to 10.

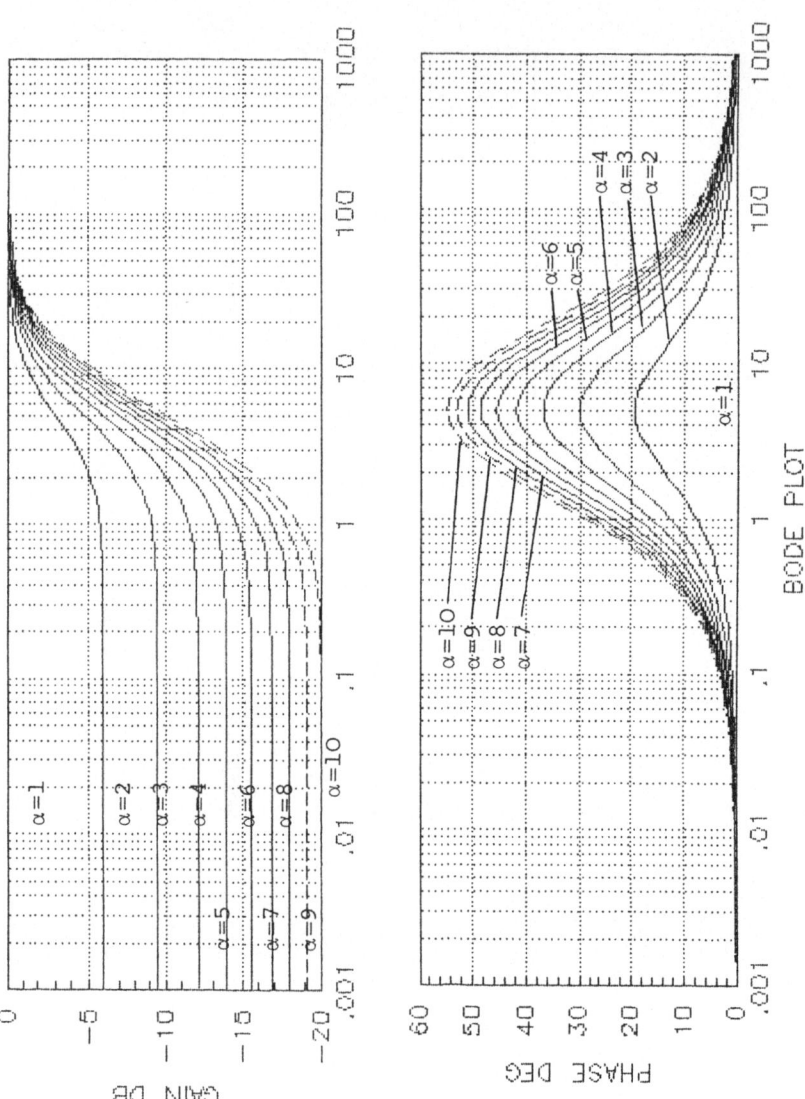

Fig. 6.2 Normalized phase lead networks with n = 5 and α varies from 1 to 10.

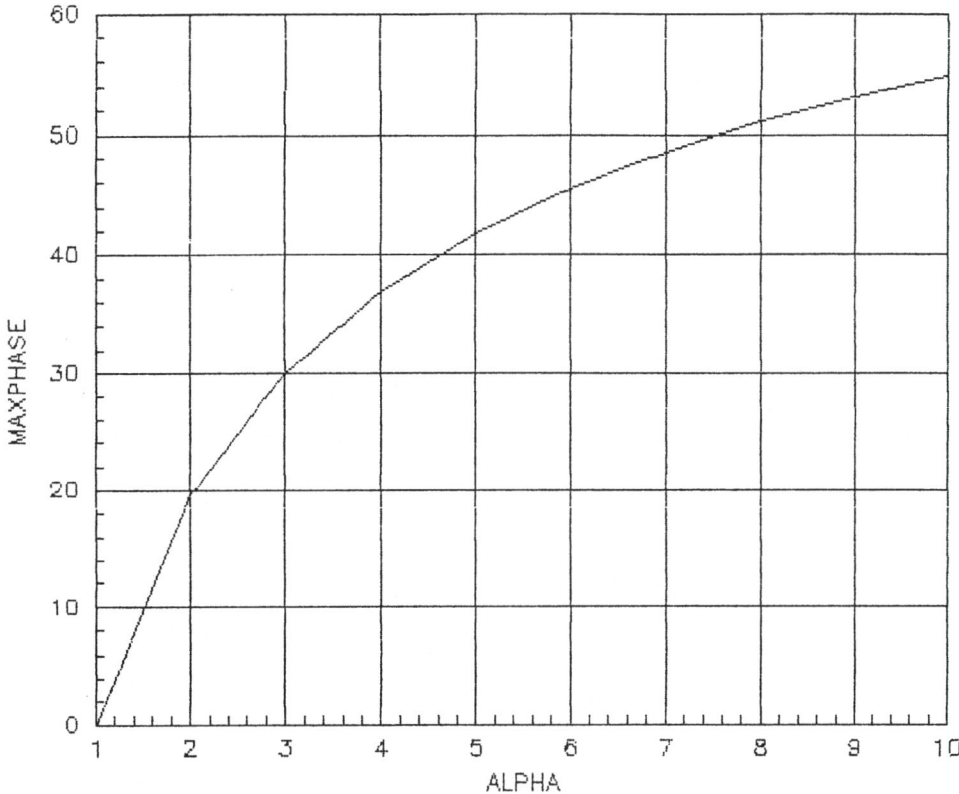

Fig. 6.3(a)   Maximum phase advance (lag) against α for
phase lead (lag) network.

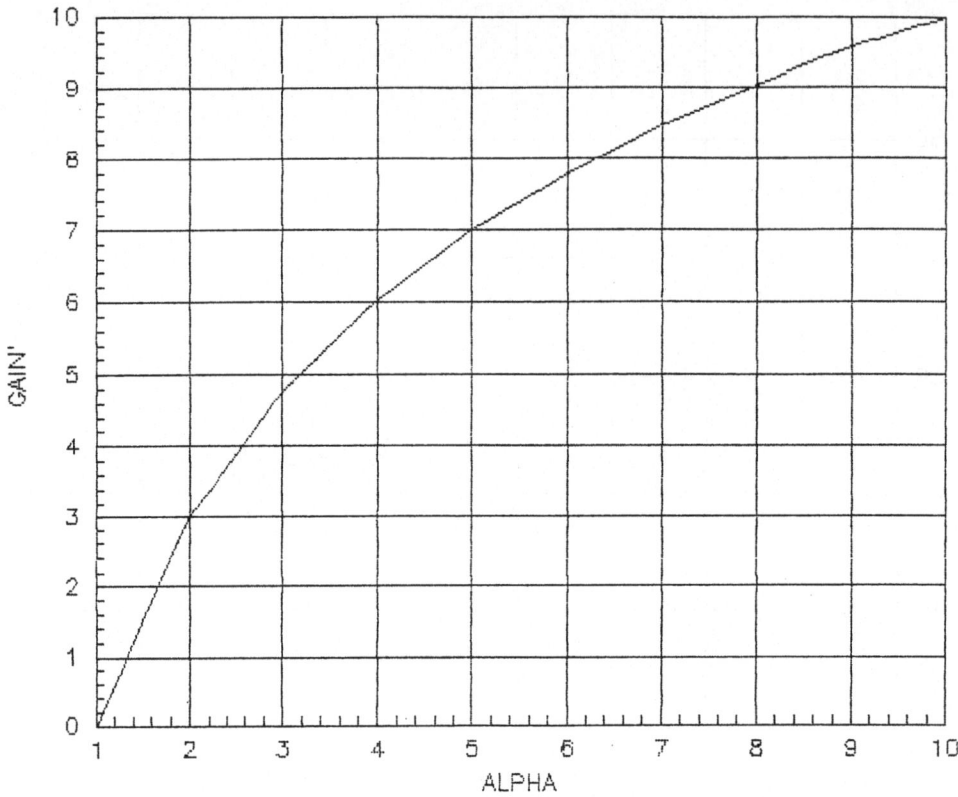

Fig. 6.3(b)   Maximum gain attenuation (increase) against α
for phase lead (lag) network.

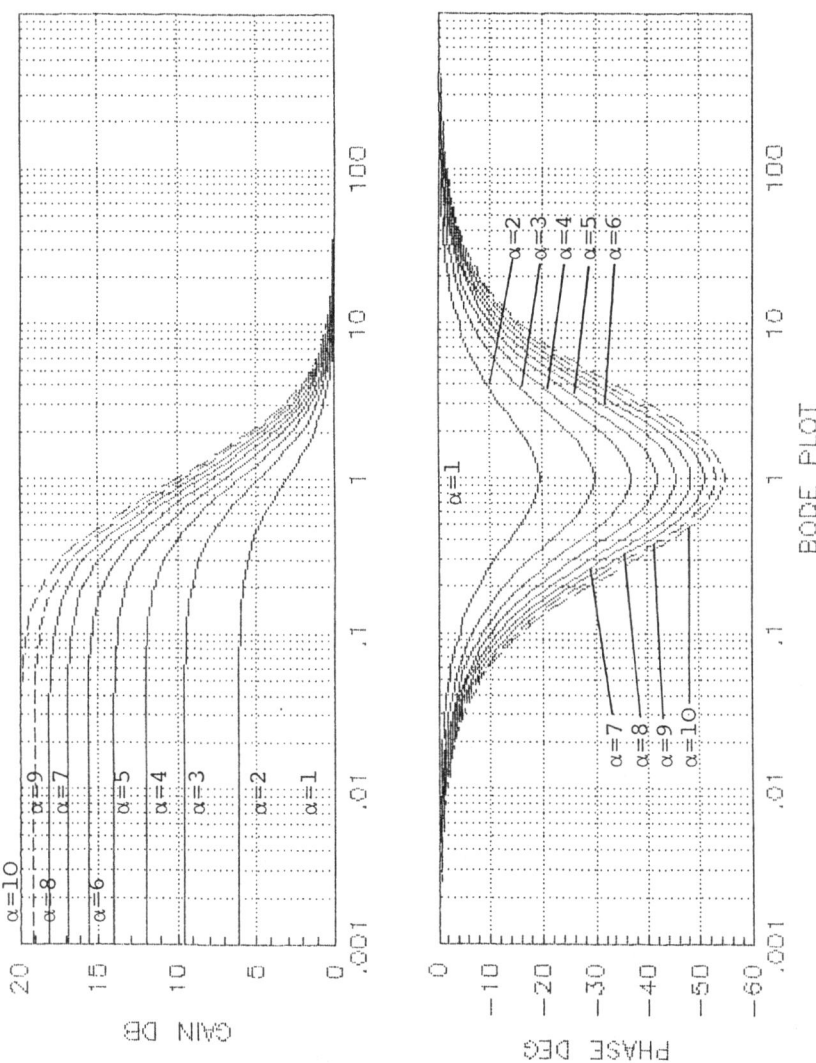

Fig. 6.4 Normalized phase lag networks with n = 1 and α varies from 1 to 10.

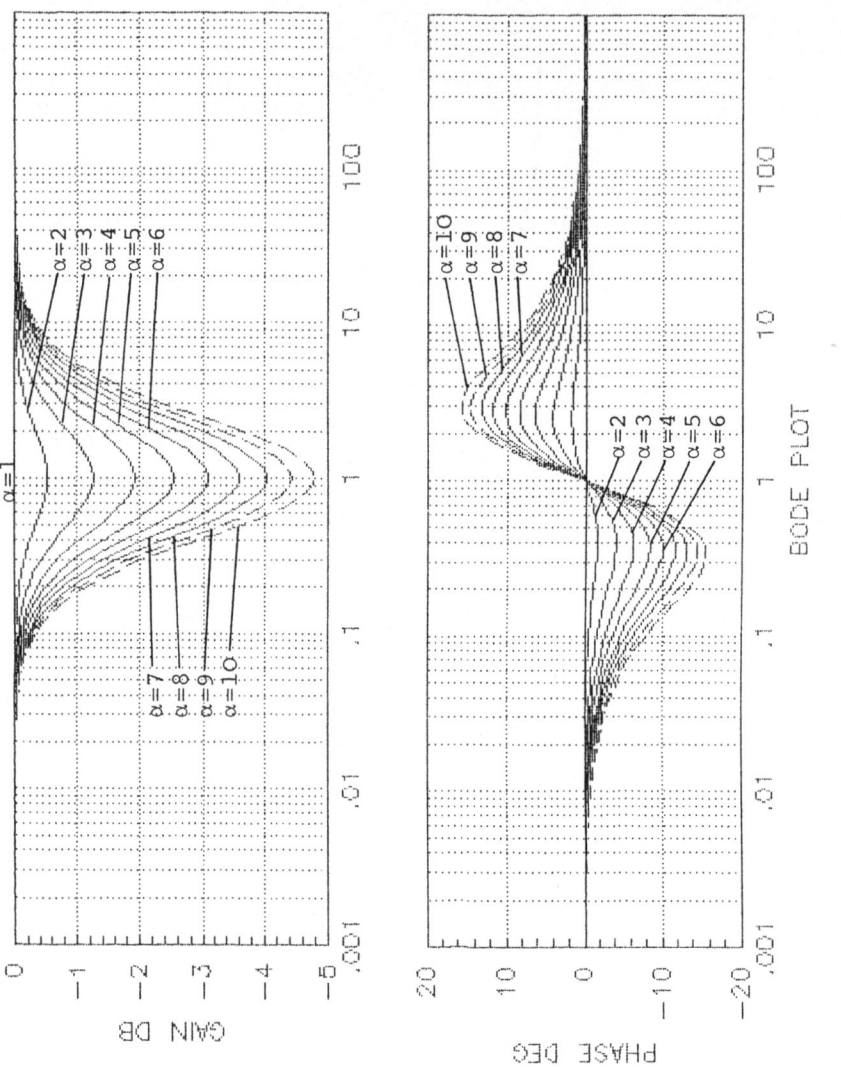

Fig. 6.5 Normalized lag-lead network with n = 1 and α varies from 1 to 10.

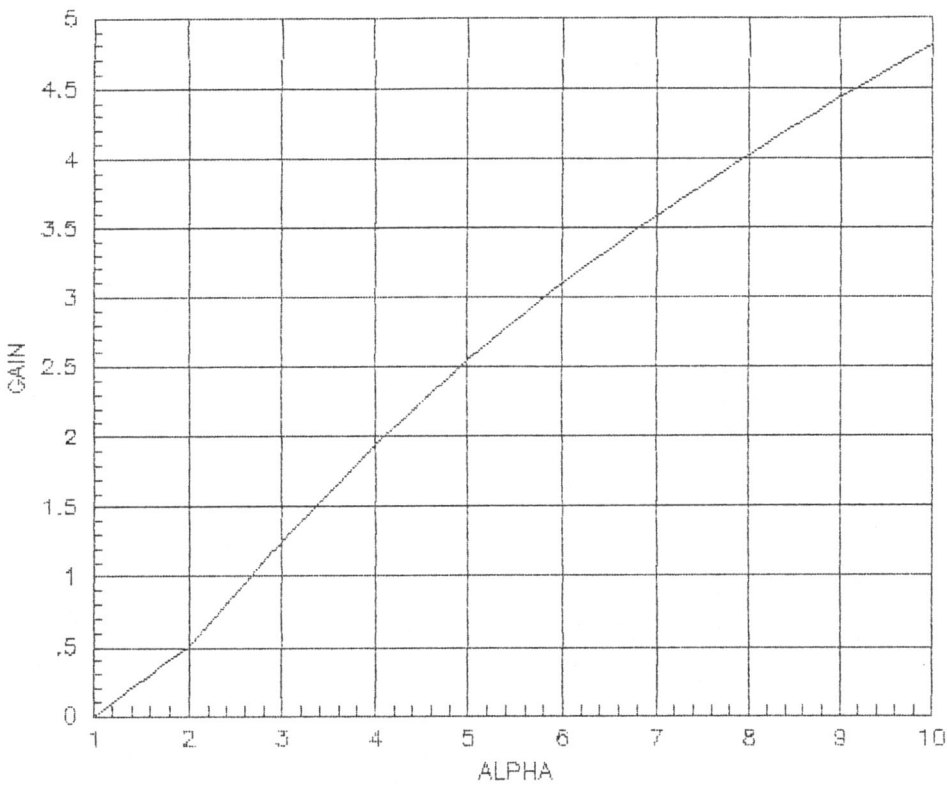

Fig. 6.6 Gain attenuation (increase) against α for
lag-lead (lead-lag) network.

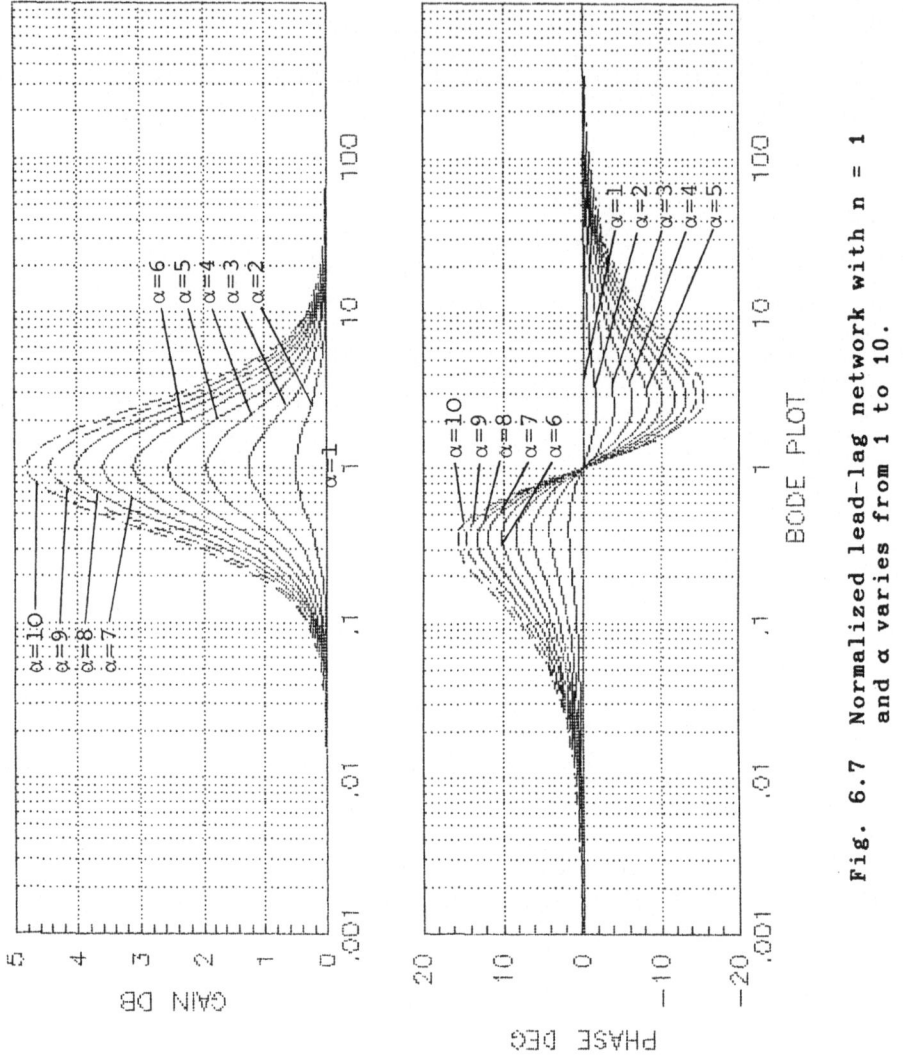

Fig. 6.7 Normalized lead-lag network with n = 1 and α varies from 1 to 10.

133

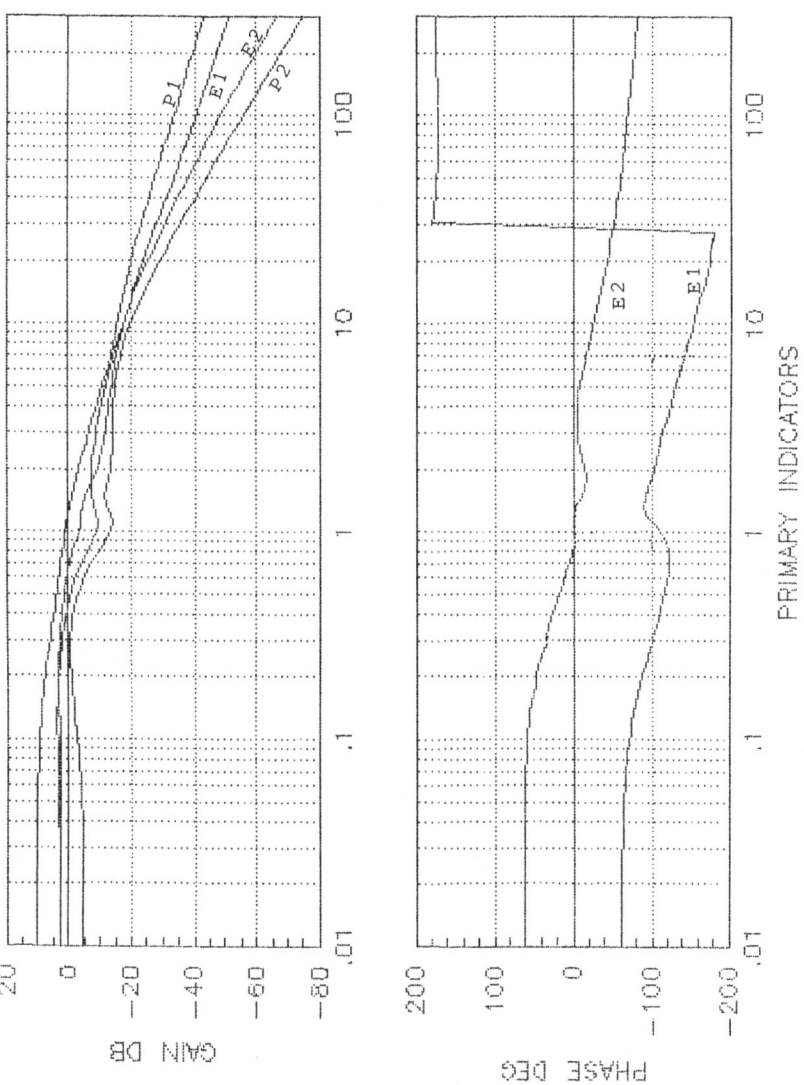

Fig. 6.8.1 The primary indicators of AUTO.

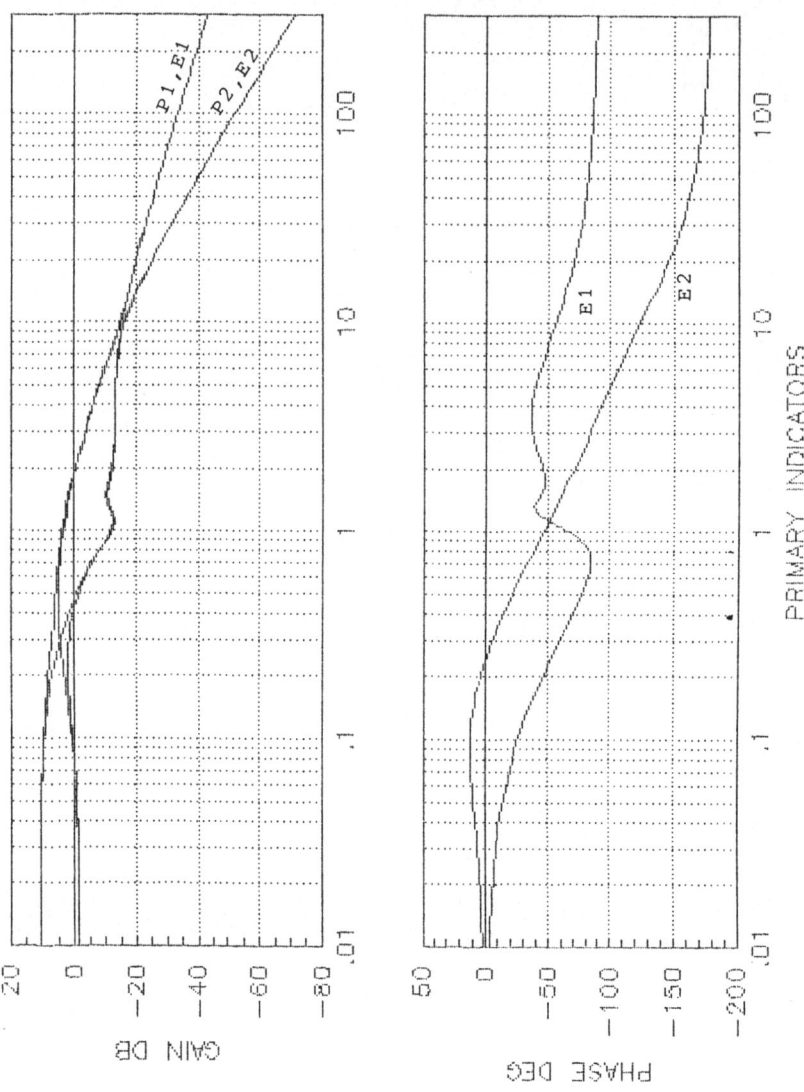

**Fig. 6.8.2 The primary indicators after high
frequency region design.**

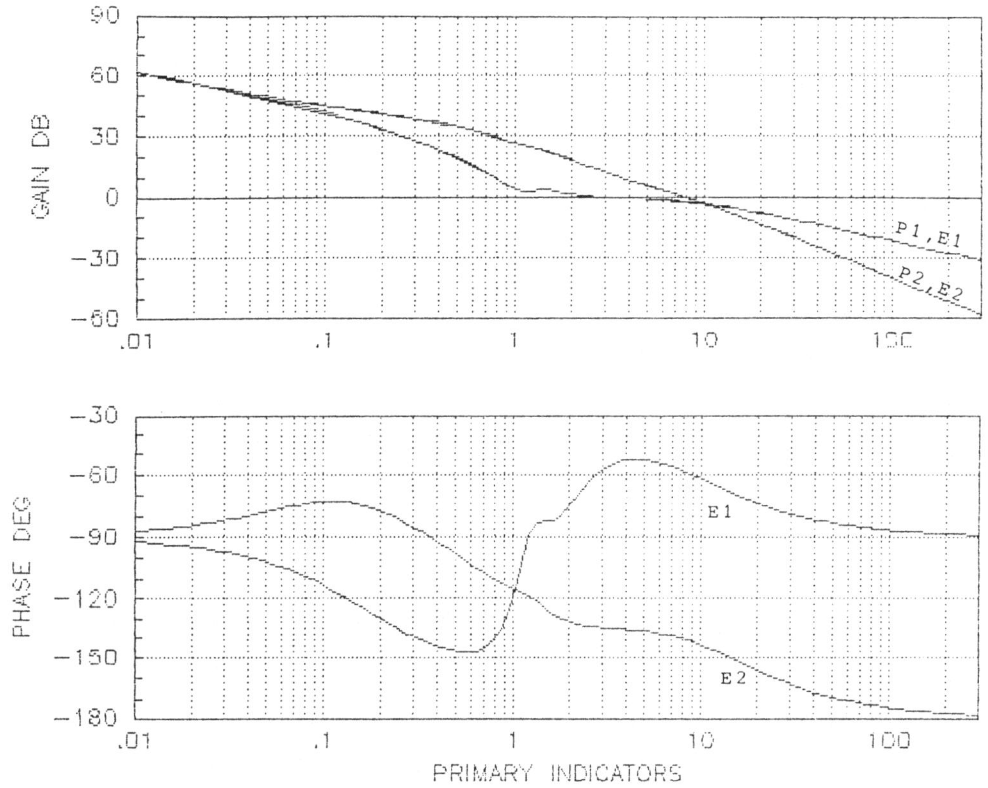

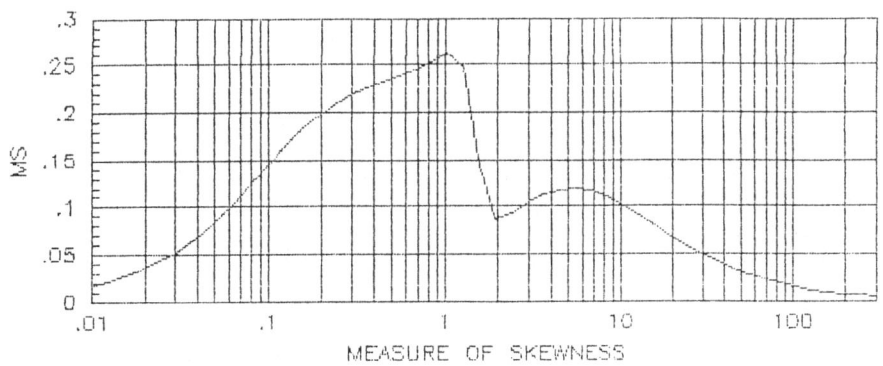

Fig. 6.8.3   The primary indicators after SDT.

Fig. 6.8.4   The normality indicator MS after SDT.

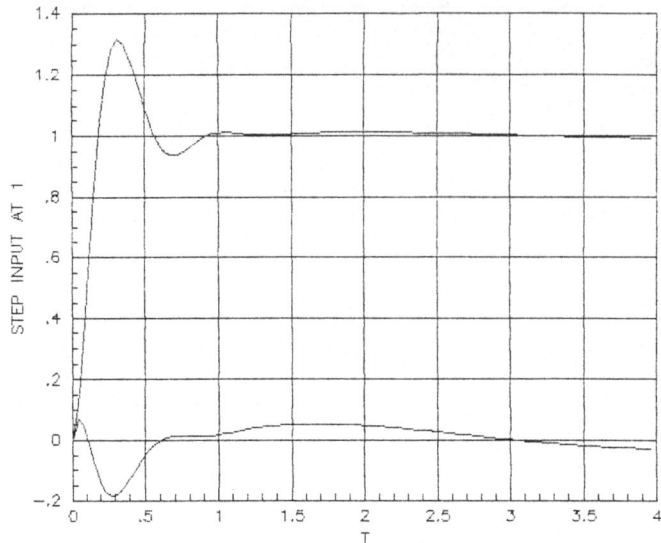

Fig. 6.8.5(a)   Closed-loop step response after SDT
(step at input 1).

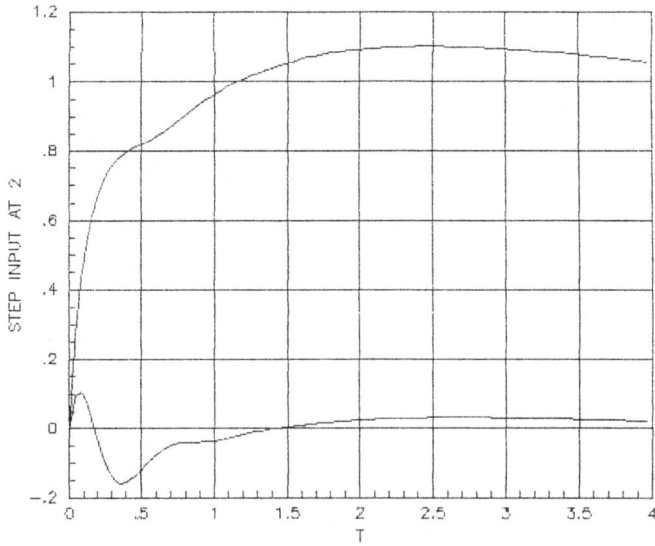

Fig. 6.8.5(b)   Closed-loop step response after SDT
(step at input 2).

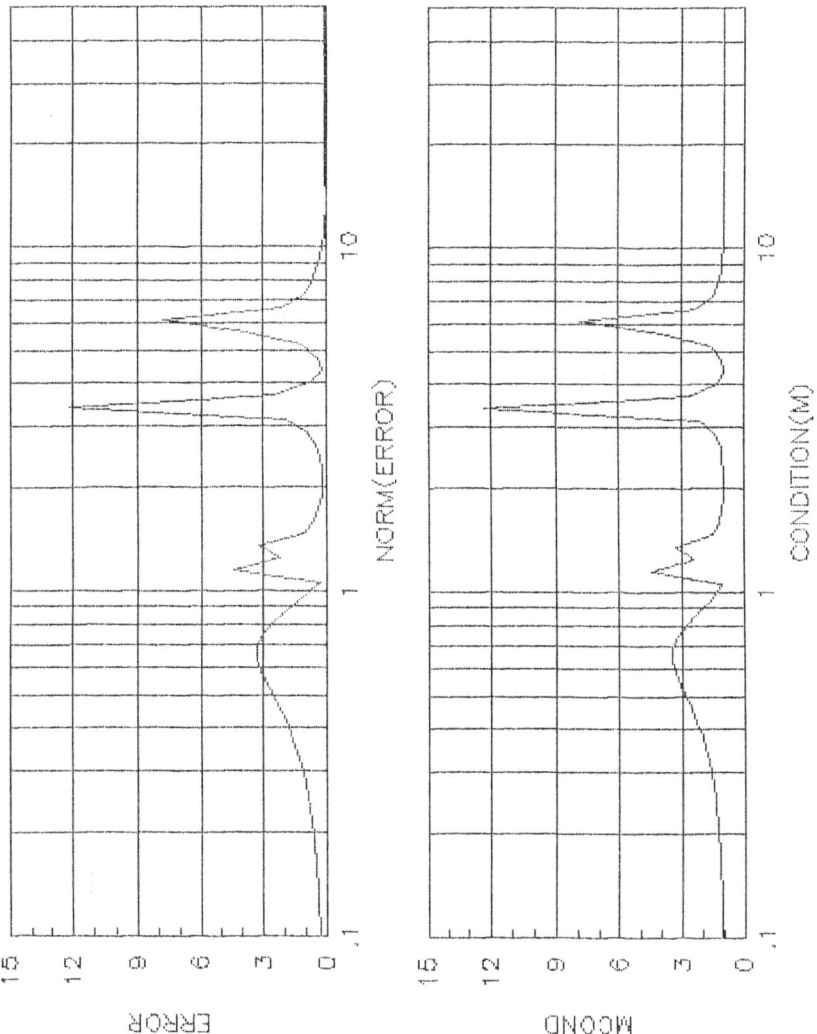

Fig. 6.8.6   The accuracy of approximation using
an RFA sub-controler.

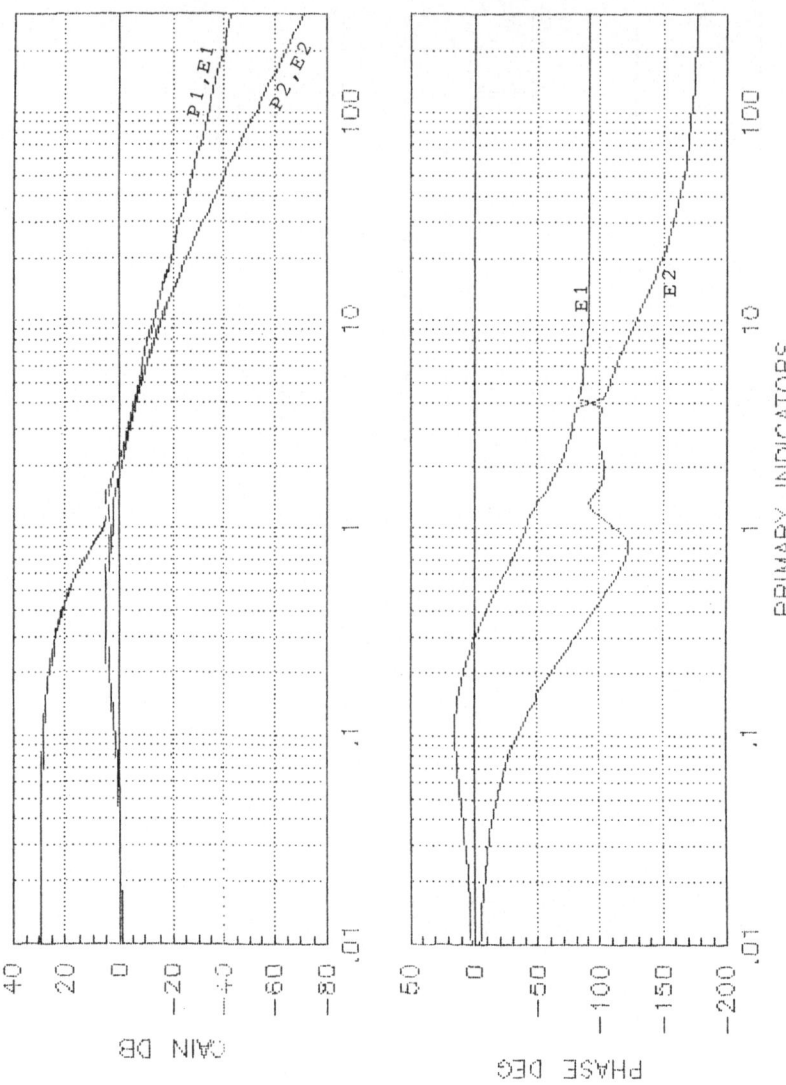

Fig. 6.8.7 The primary indicators after an RFA sub-controller.

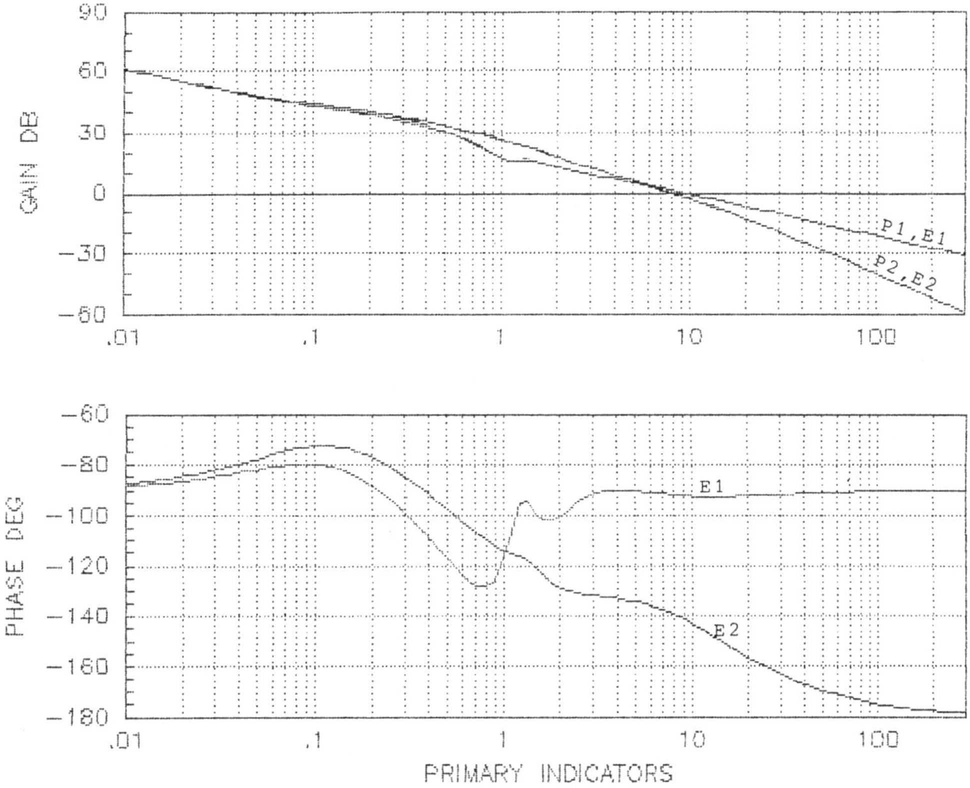

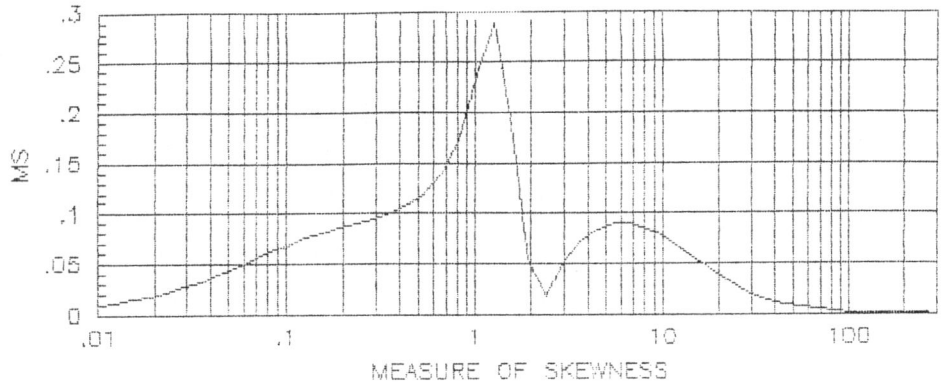

Fig. 6.8.8   The primary indicators after DESIGN ONE.

Fig. 6.8.9   The normality indicator MS after DESIGN ONE.

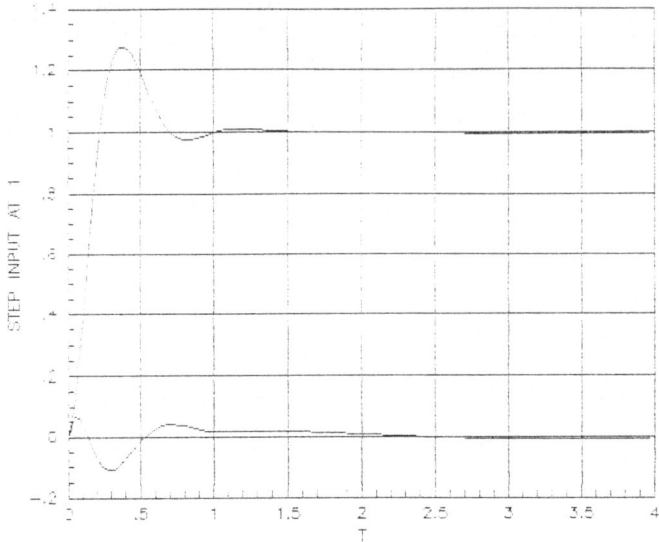

Fig. 6.8.10(a)   Closed-loop step response after
DESIGN ONE (step at input 1).

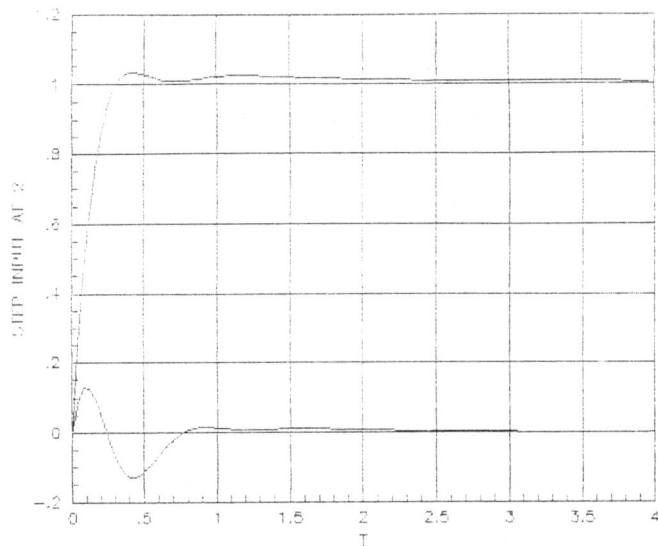

Fig. 6.8.10(b)   Closed-loop step response after
DESIGN ONE (step at input 2).

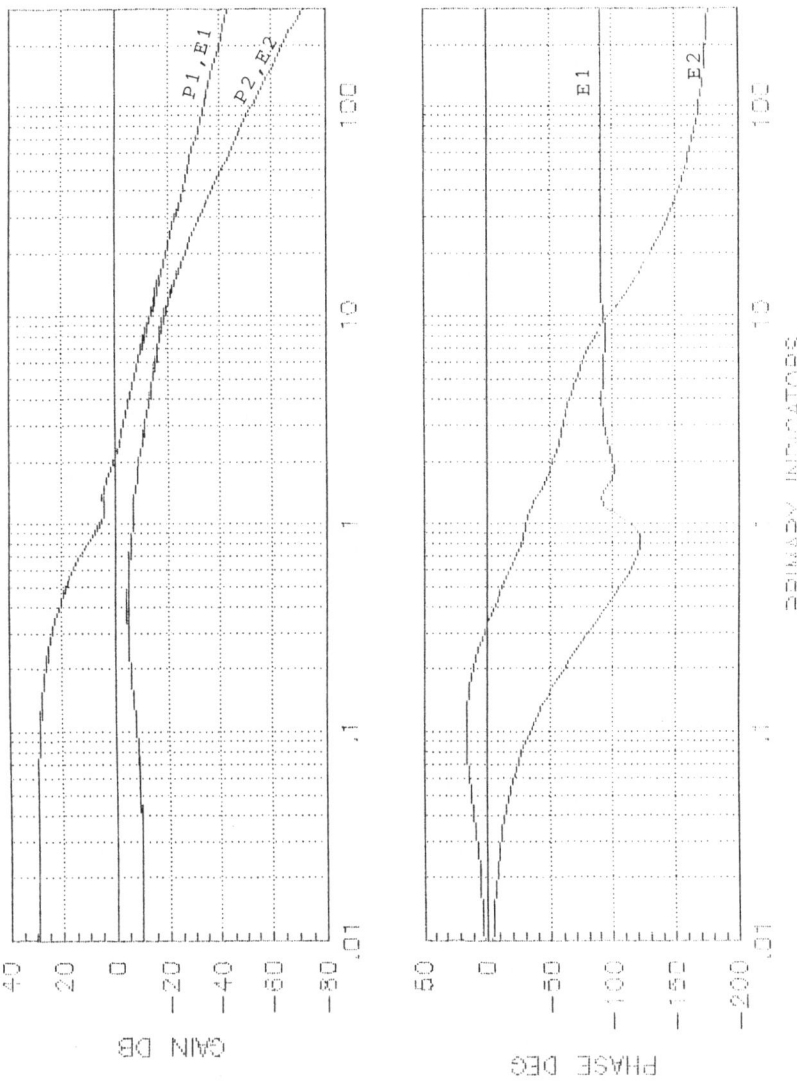

Fig. 6.8.11   The primary indicators after a
second RFA sub-controller.

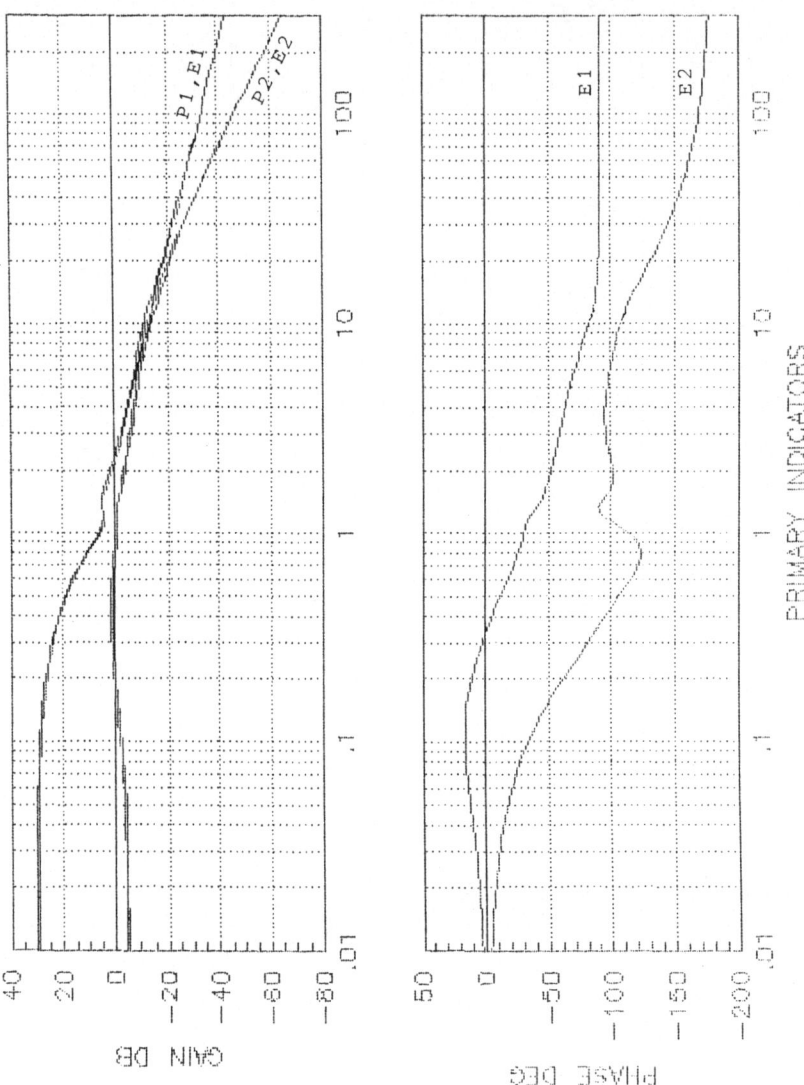

Fig. 6.8.12 The primary indicators after gain
balancing at intermediate frequencies.

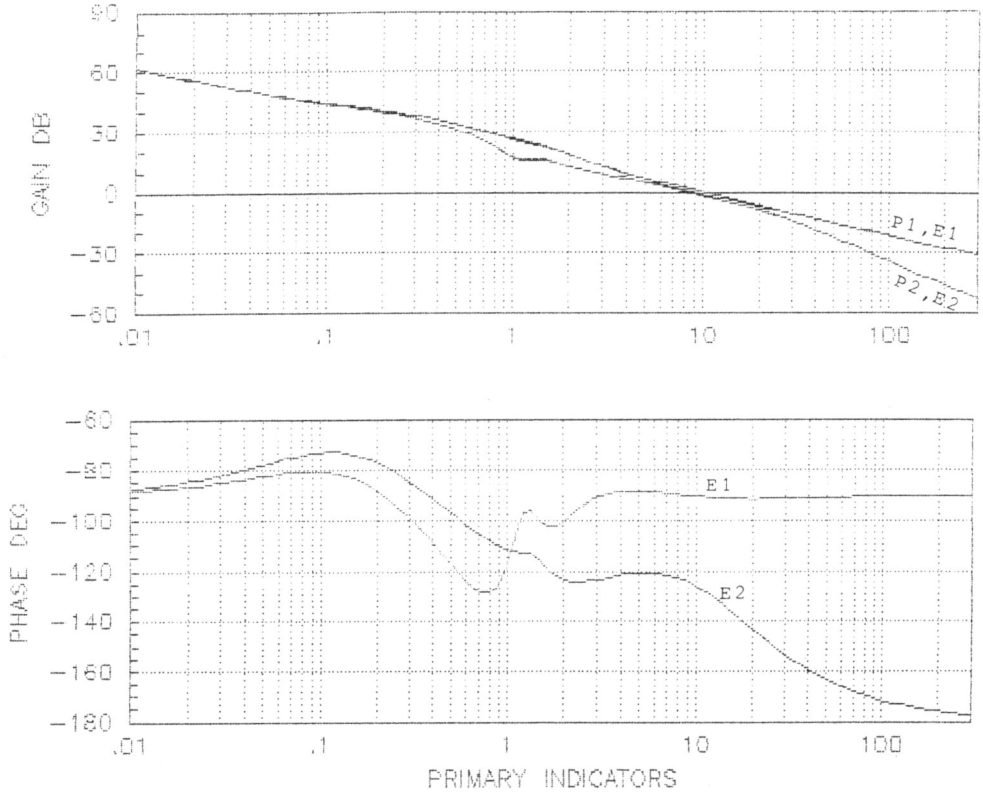

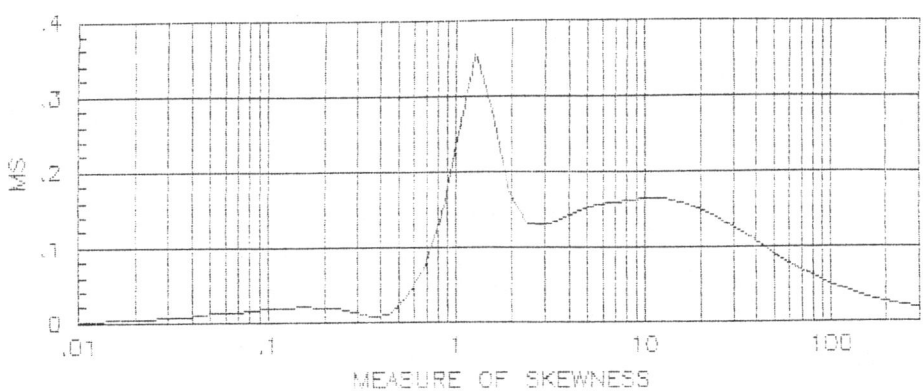

Fig. 6.8.13   The primary indicators after DESIGN
TWO.

Fig. 6.8.14   The normality indicator MS after DESIGN
TWO.

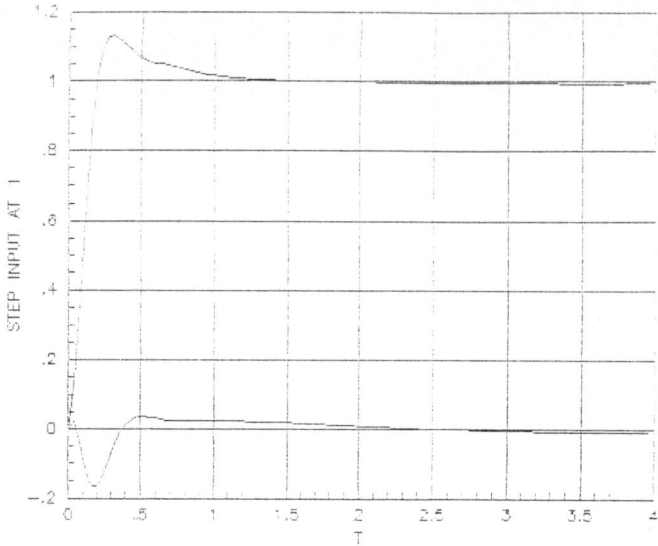

Fig. 6.8.15(a)   Closed-loop step response after
                 DESIGN TWO (step at input 1).

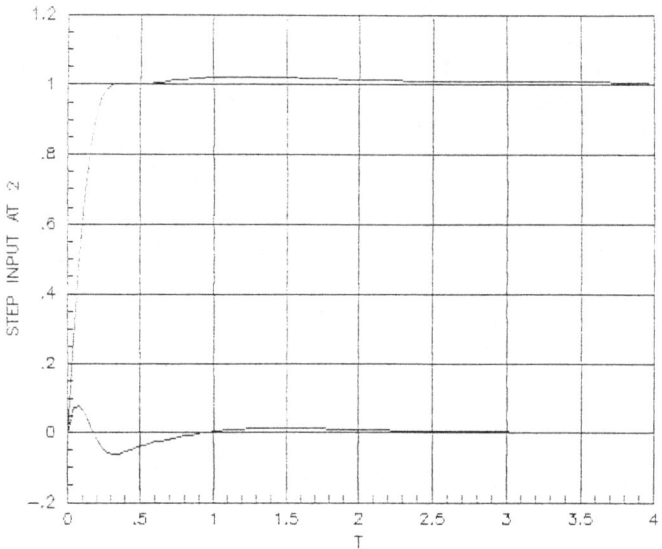

Fig. 6.8.15(b)   Closed-loop step response after
                 DESIGN TWO (step at input 2).

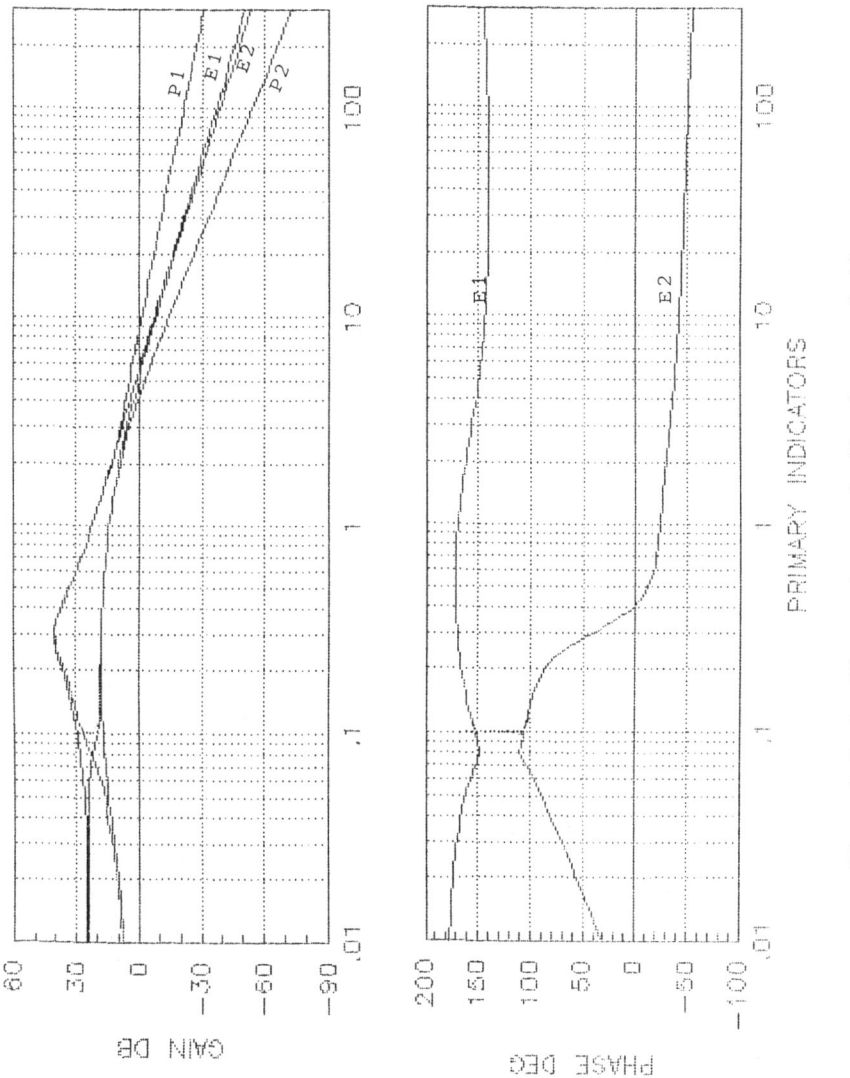

Fig. 6.9.1   The primary indicators of the
uncompensated system.

146

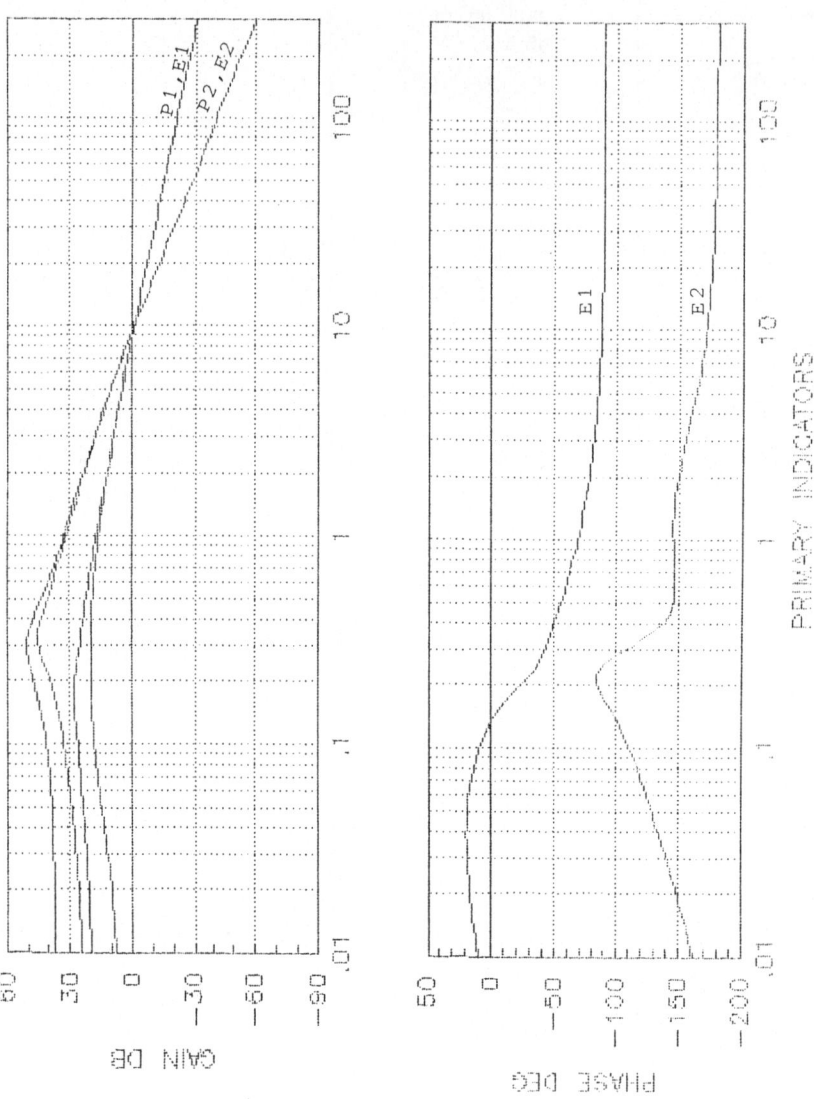

Fig. 6.9.2 The primary indicators after high frequency region design.

147

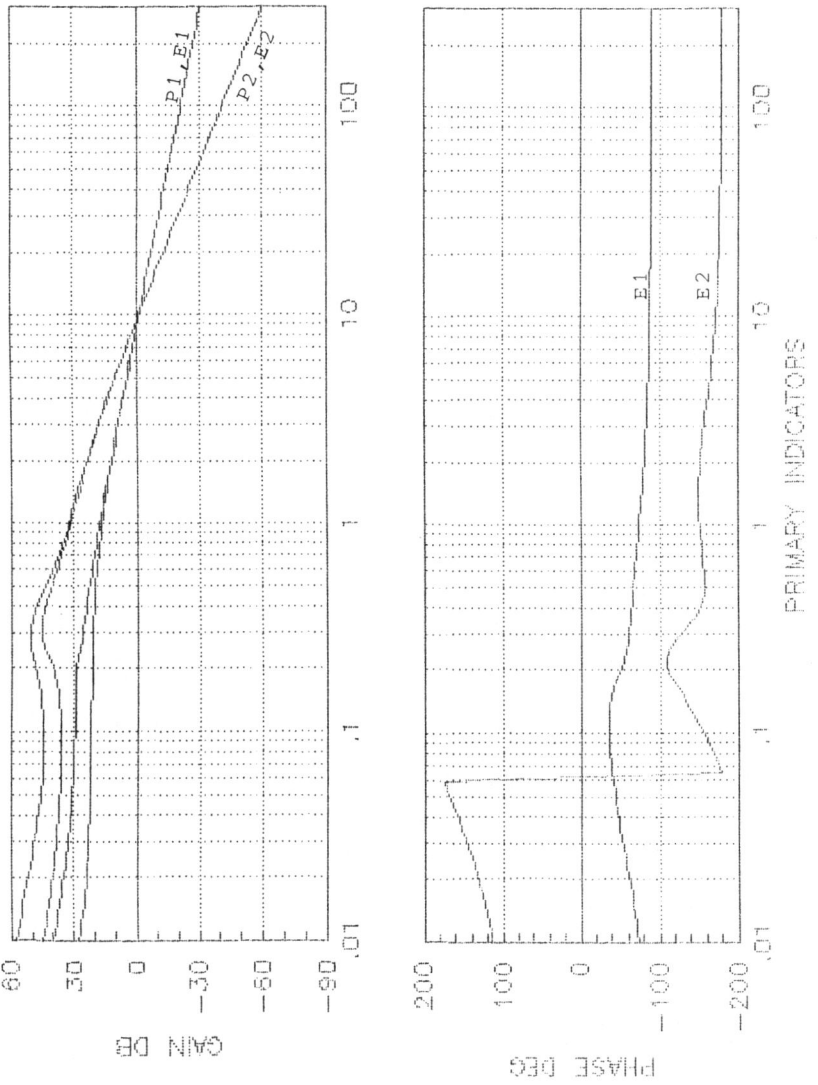

Fig. 6.9.3 The primary indicators after SDT.

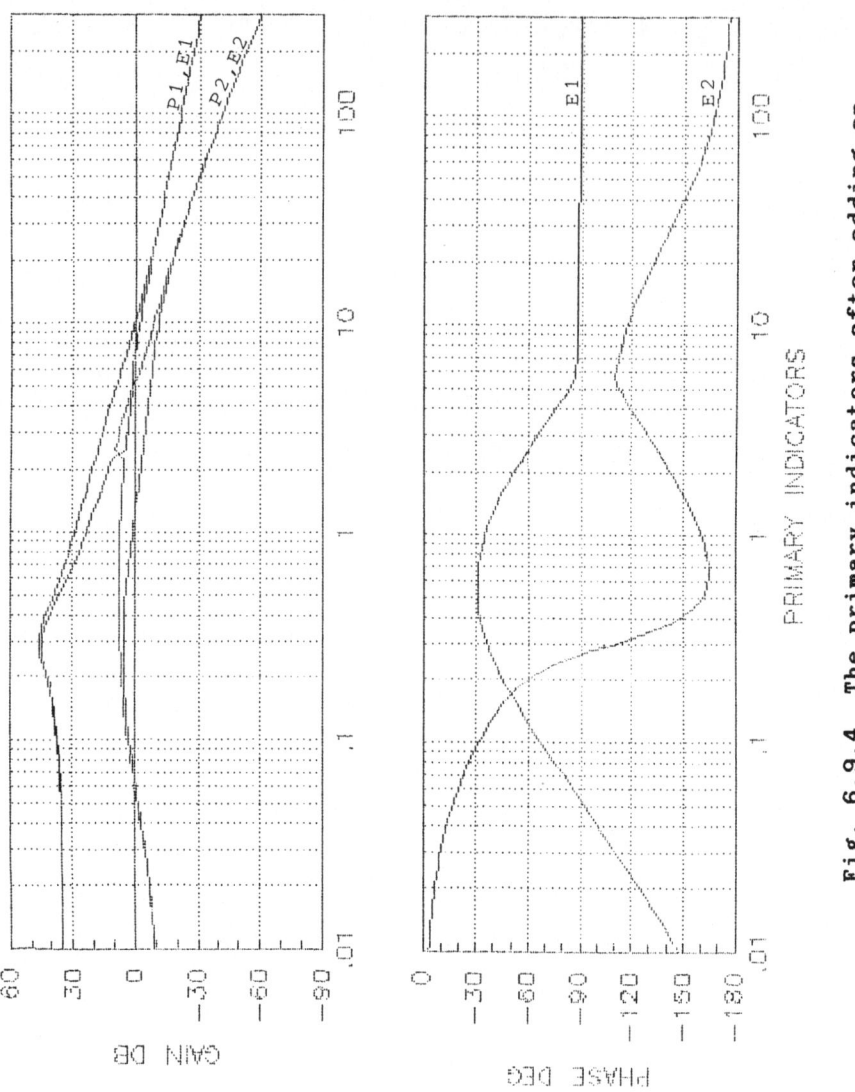

Fig. 6.9.4 The primary indicators after adding an RFA sub-controller with phase lead.

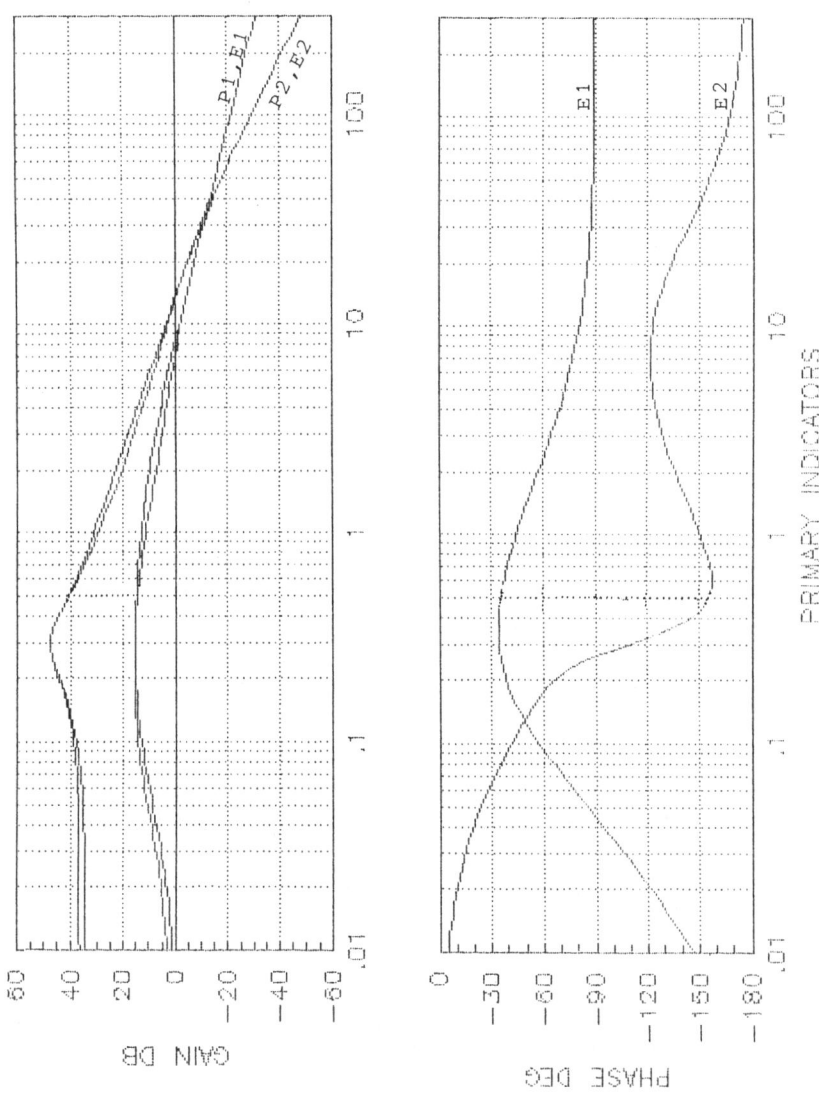

Fig. 6.9.5 The primary indicators after gain balancing at intermediate frequencies.

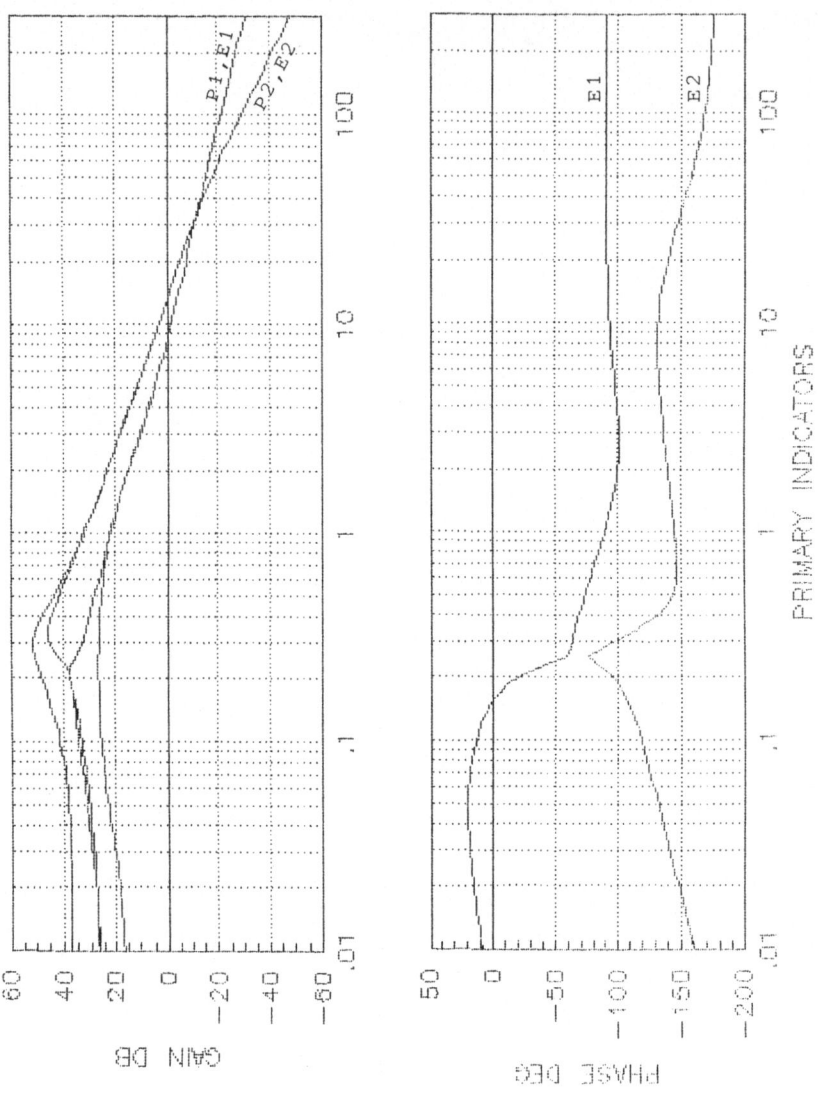

Fig. 6.9.6   The primary indicators after a second RFA sub-controller.

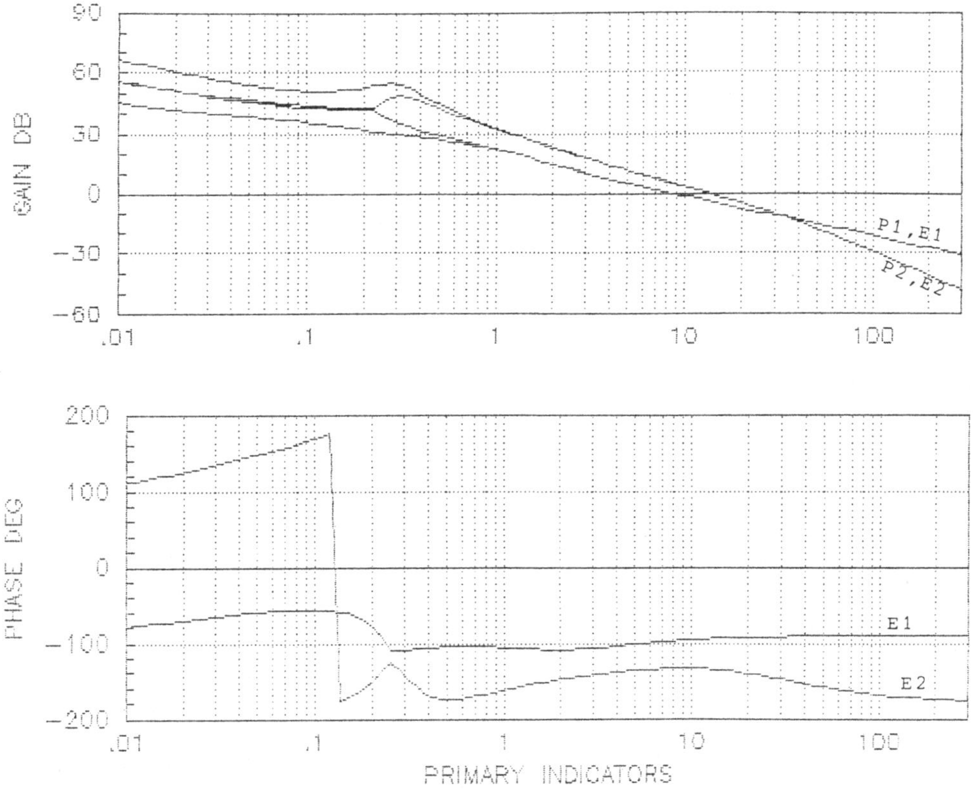

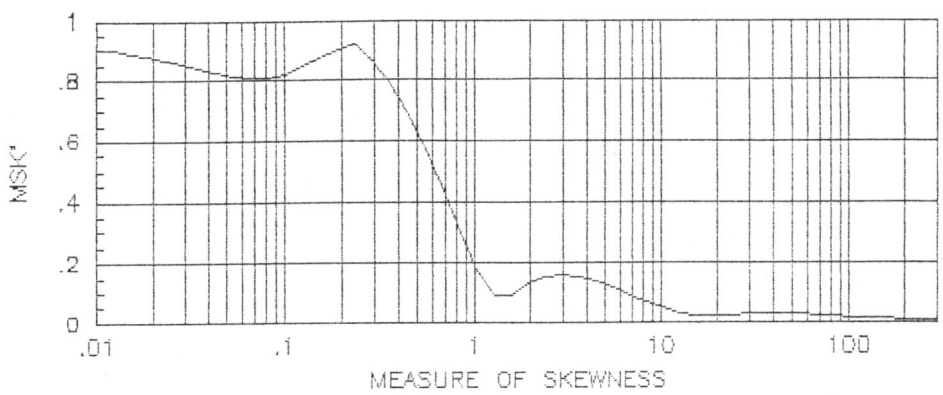

Fig. 6.9.7   The primary indicators of the final
compensated system.

Fig. 6.9.8   The normality indicator MS of the
compensated system.

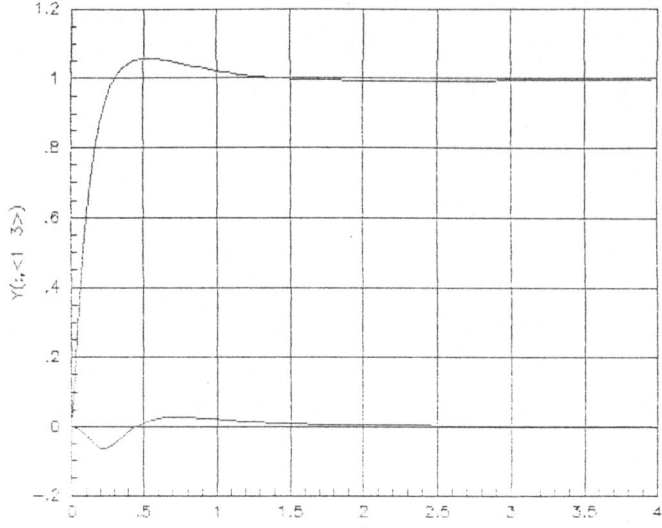

Fig. 6.9.9(a)   Closed-loop step response of the
final design (step at input 1).

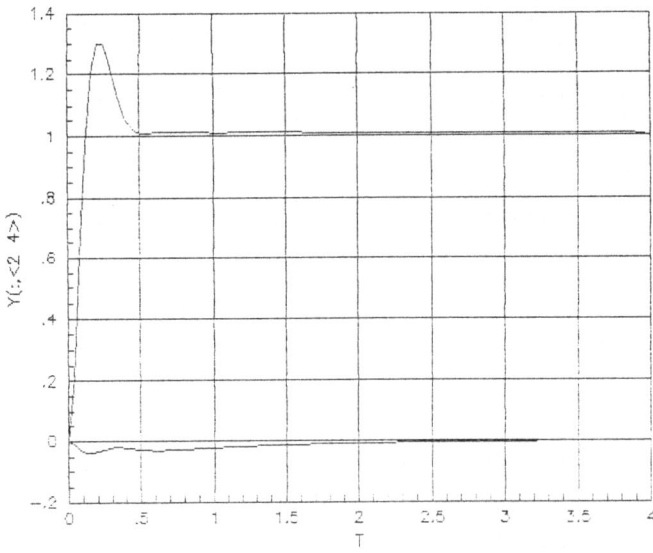

Fig. 6.9.9(b)   Closed-loop step response of the
final design (step at input 2).

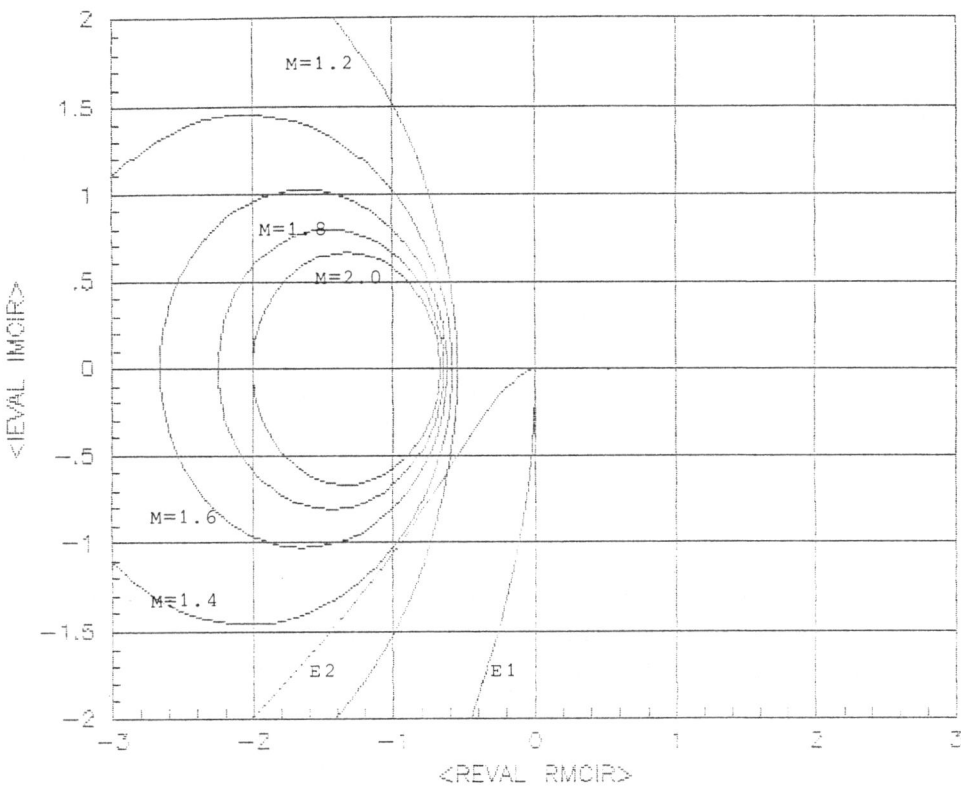

Fig. 6.9.10   Generalised Nyquist diagrams of the
compensated system.

154

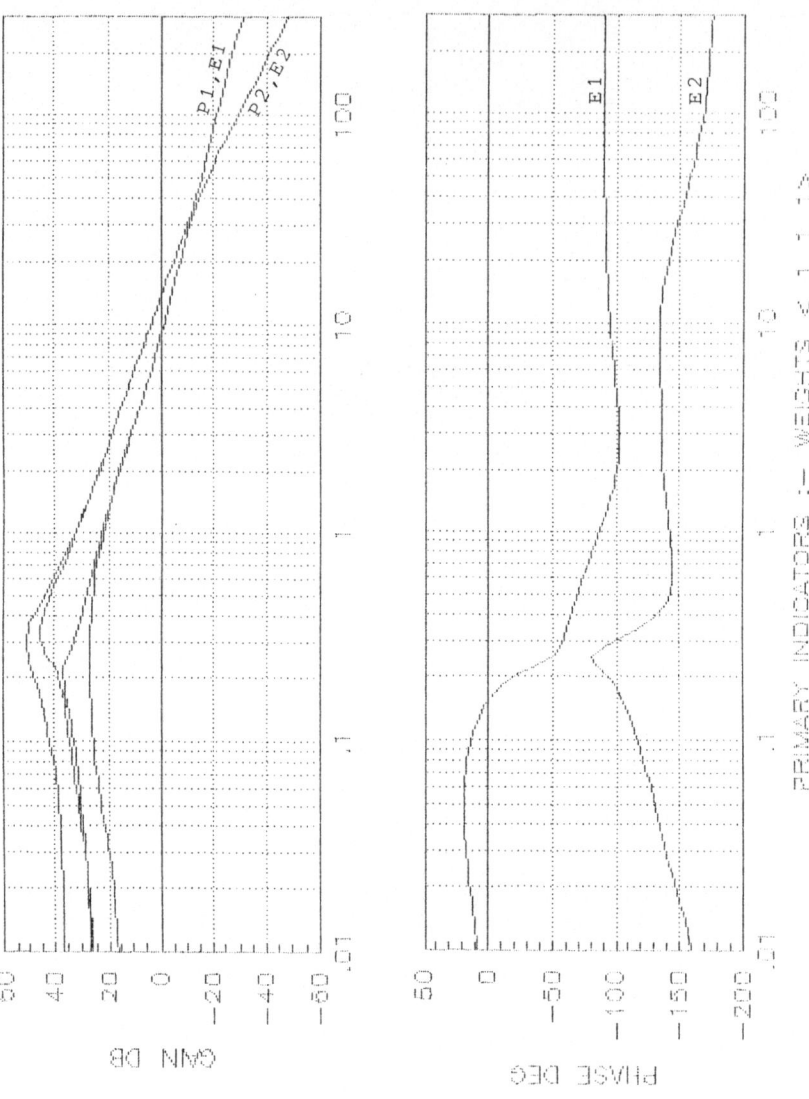

Fig. 6.9.11 The primary indicators with parameters in case 4.

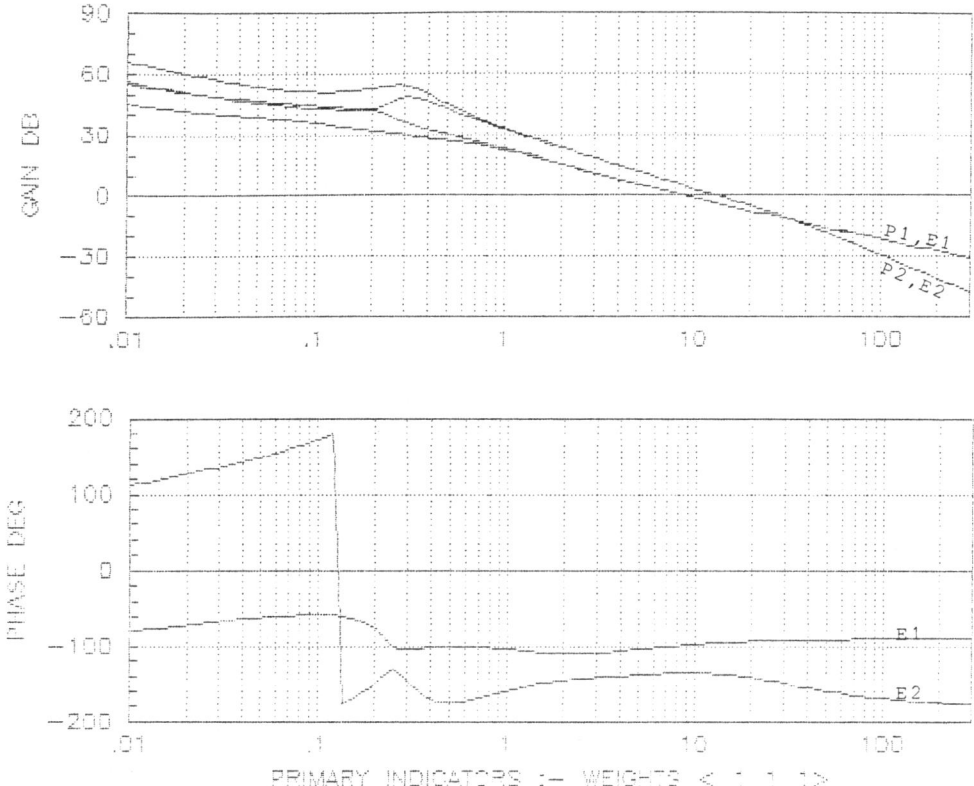

Fig. 6.9.12 The primary indicators of the design
from case 4.

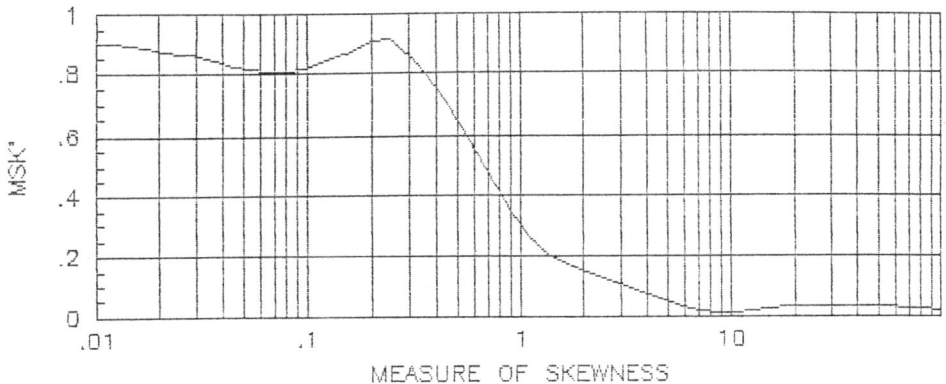

Fig. 6.9.13 The normality indicator MS of the design
from case 4.

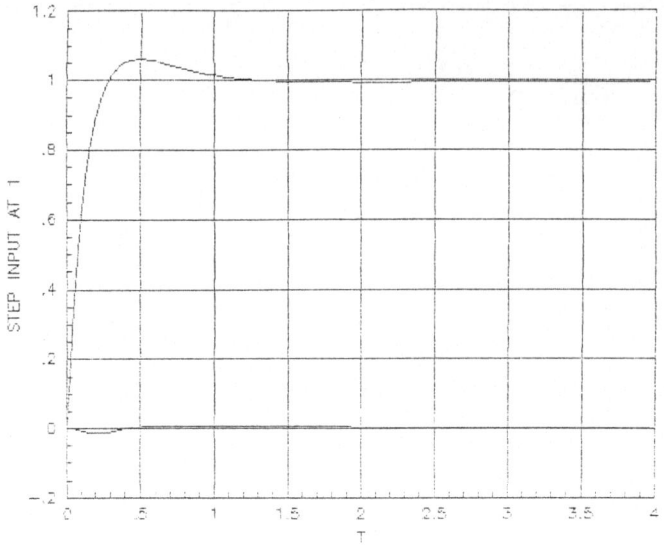

Fig. 6.9.14(a)   Closed-loop step response of the
                 design from case 4 (step at input 1).

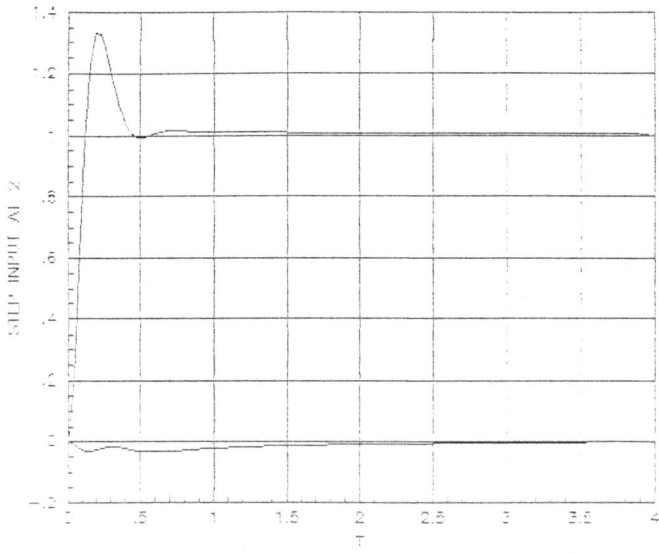

Fig. 6.9.14(b)   Closed-loop step response of the
                 design from case 4 (step at input 2).

157

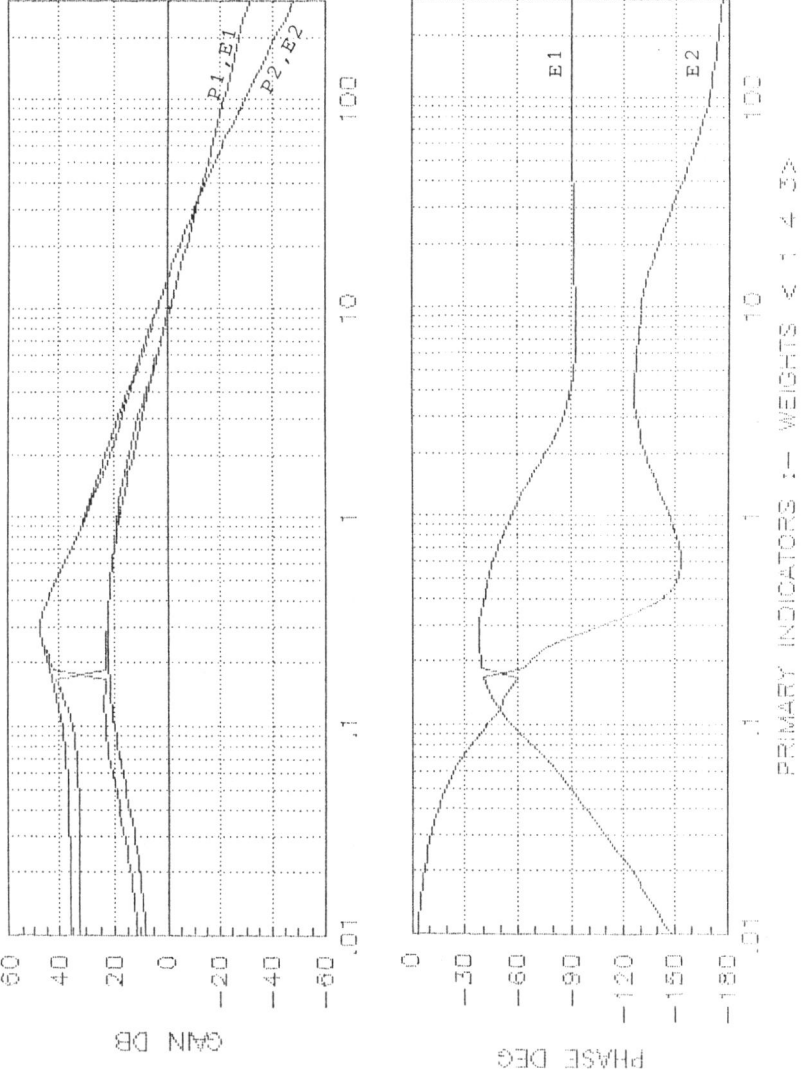

Fig. 6.9.15 The primary indicators with parameters in case 5.

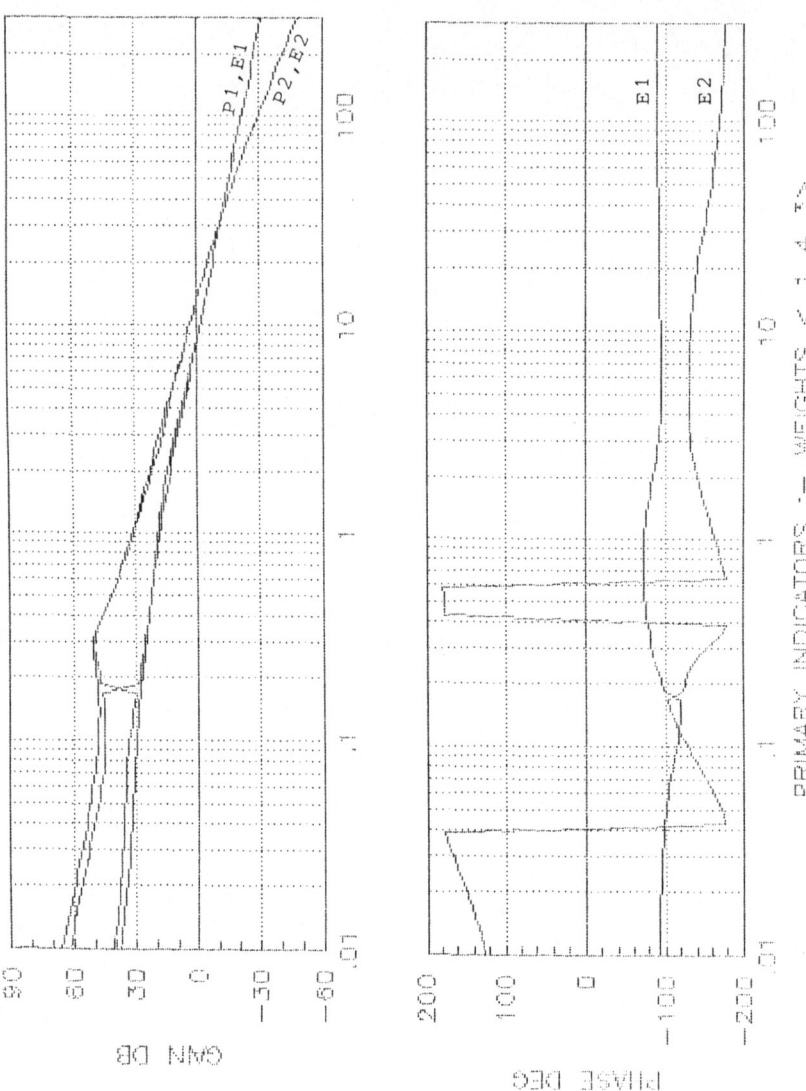

Fig. 6.9.16 The primary indicators of the design
from case 5.

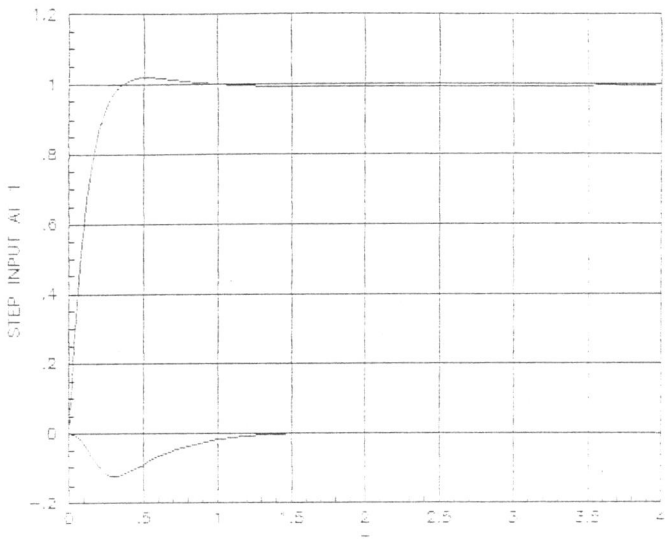

Fig. 6.9.17(a)   Closed-loop step response of the
                 design from case 5 (step at input 1).

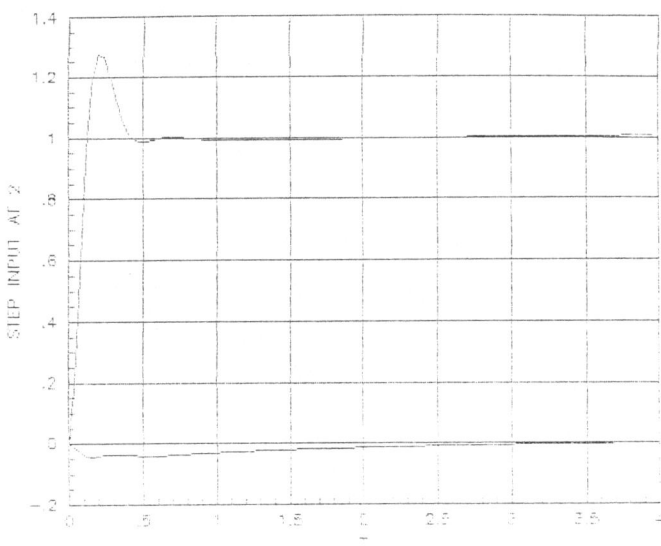

Fig. 6.9.17(b)   Closed-loop step response of the
                 design from case 5 (step at input 2).

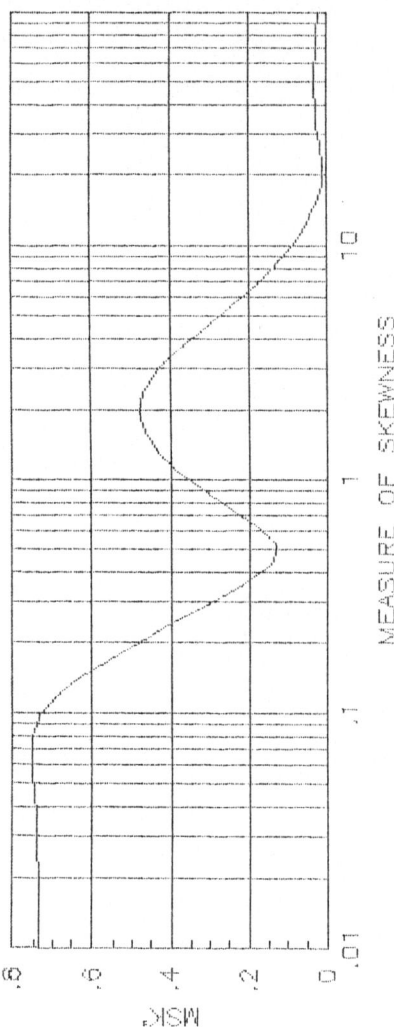

Fig. 6.9.18   The normality indicator MS of the
design from case 5.

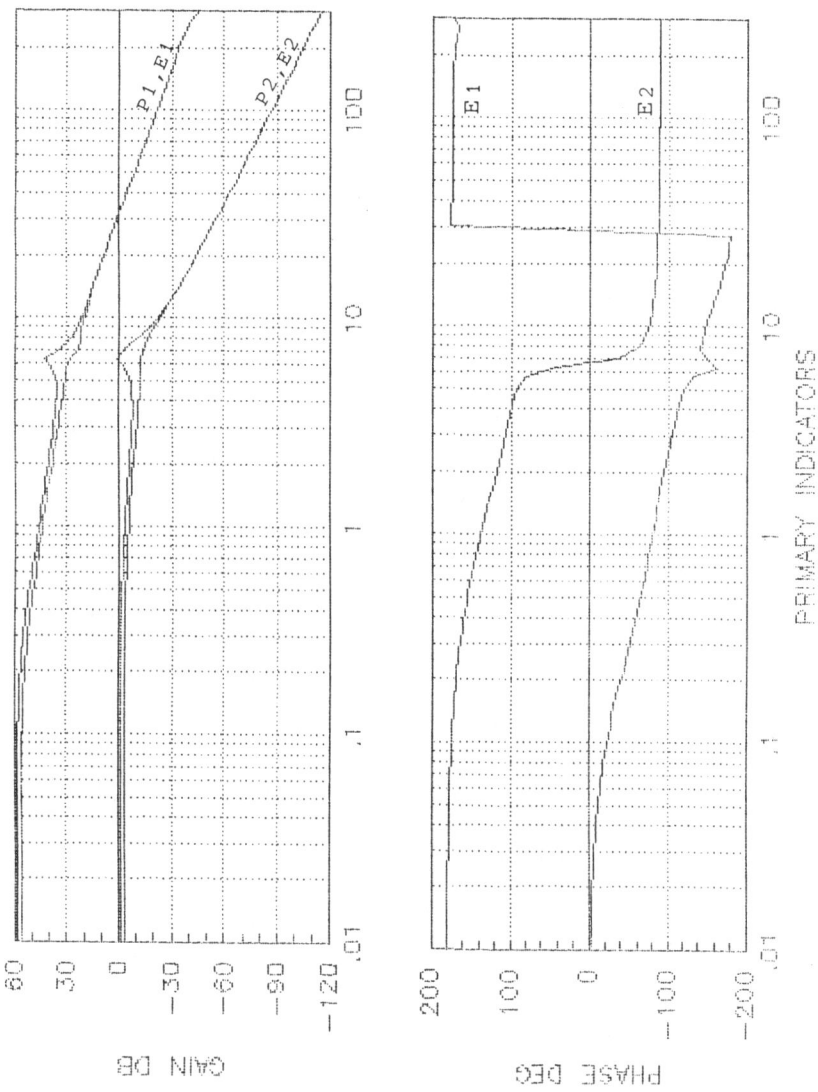

Fig. 6.10.1 The primary indicators of TGEN.

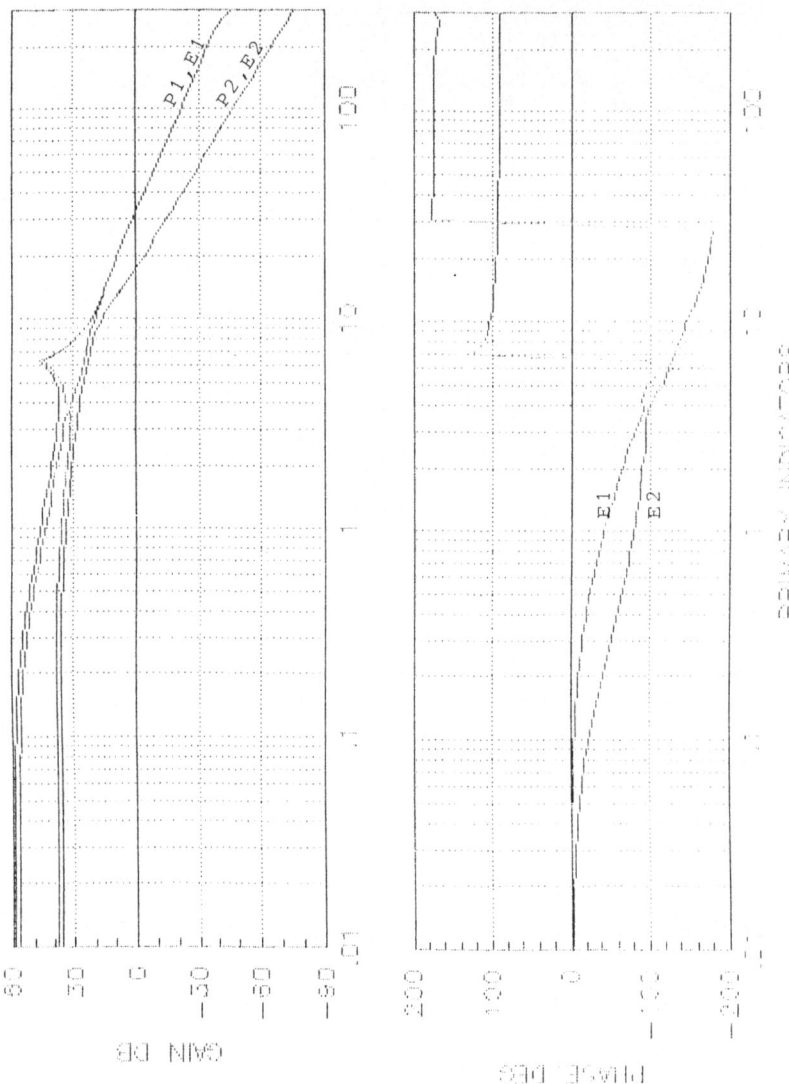

Fig. 6.10.2 The primary indicators after high frequency region design.

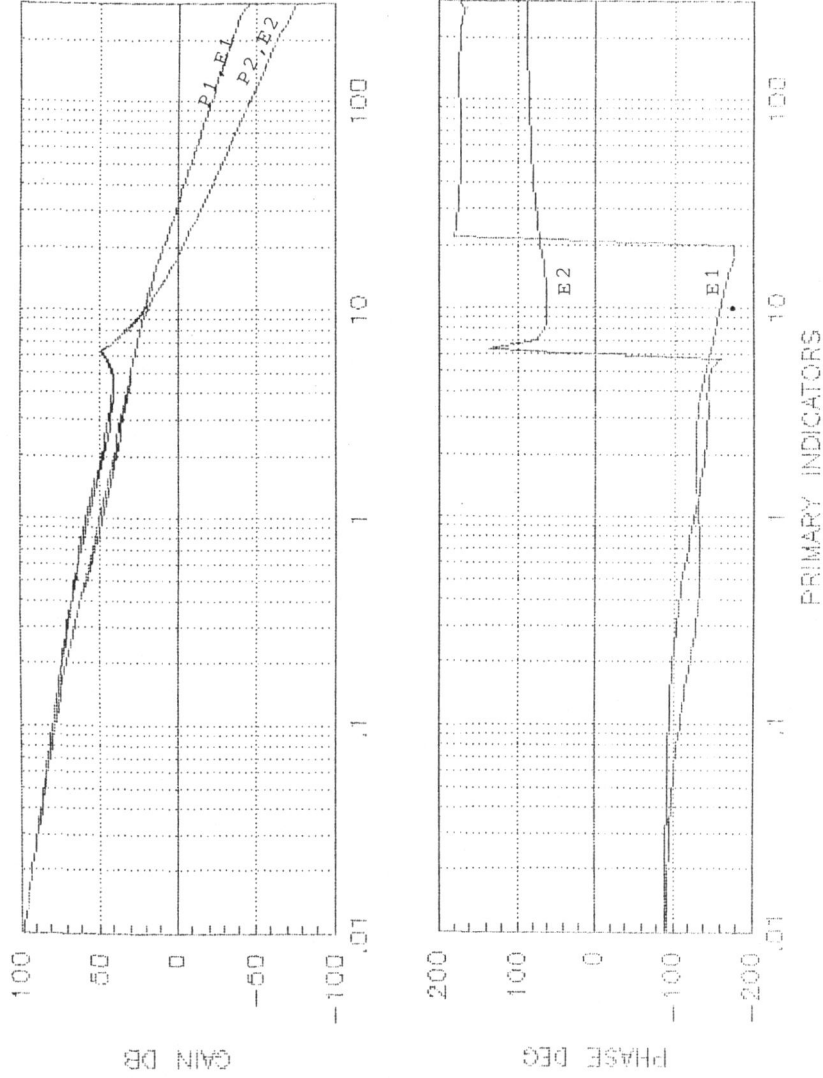

Fig. 6.10.3   The primary indicators after high and
                low frequency region design.

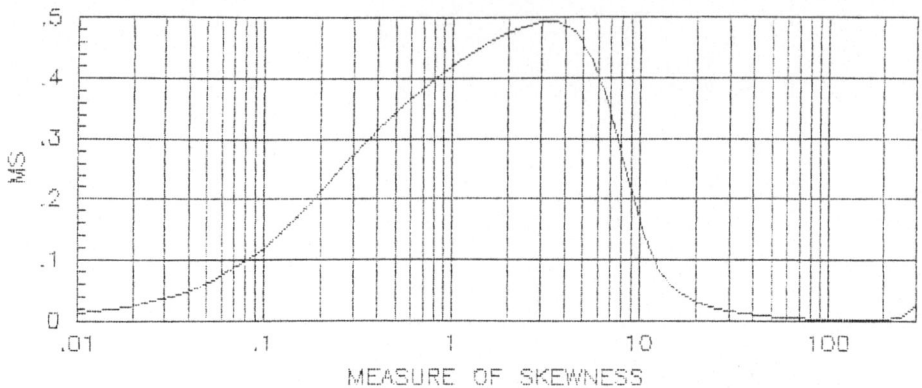

Fig. 6.10.4   The normality indicator MS after SDT.

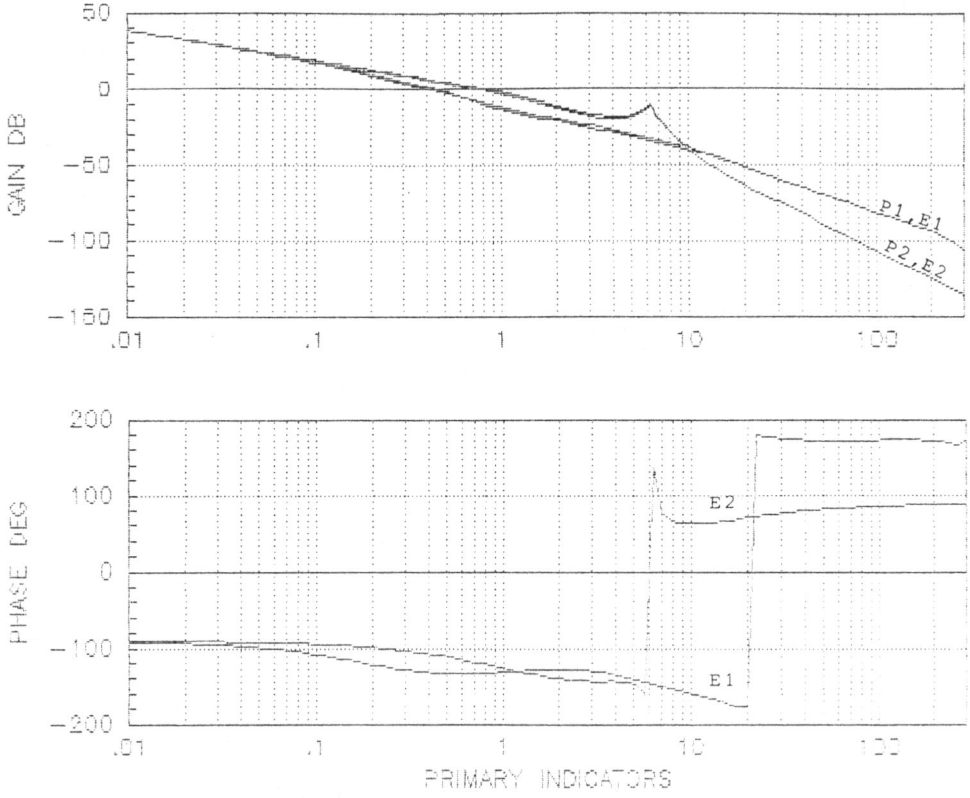

![Gain and phase plots]

Fig. 6.10.5   The primary indicators after SDT.

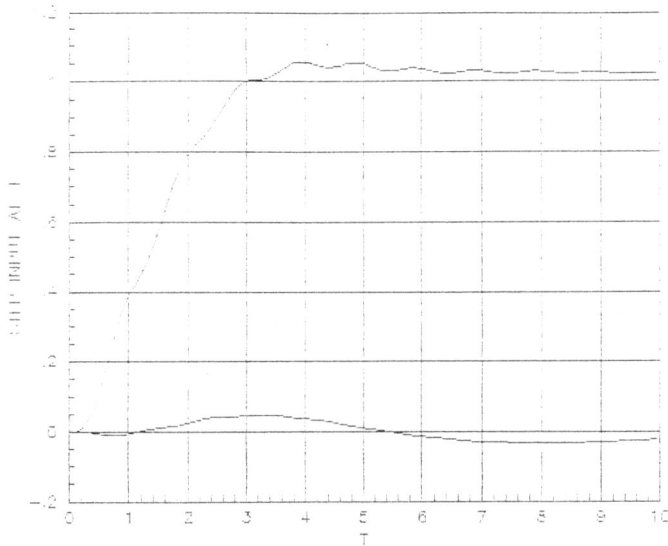

Fig. 6.10.6(a) Closed-loop step response after SDT (step at input 1).

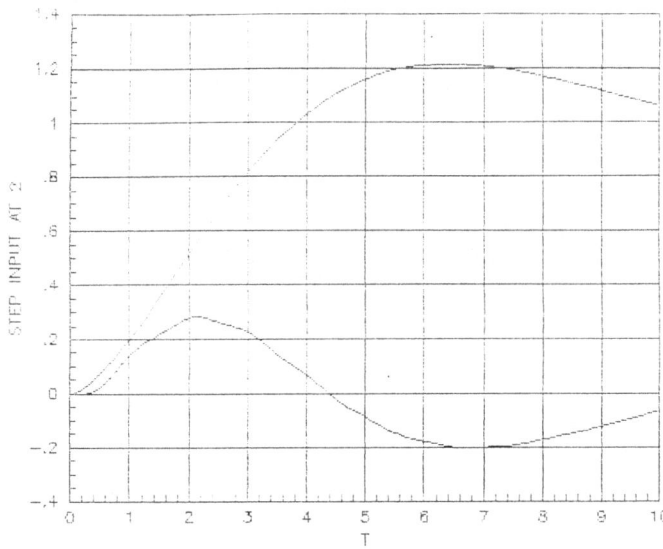

Fig. 6.10.6(b) Closed-loop step response after SDT (step at input 2).

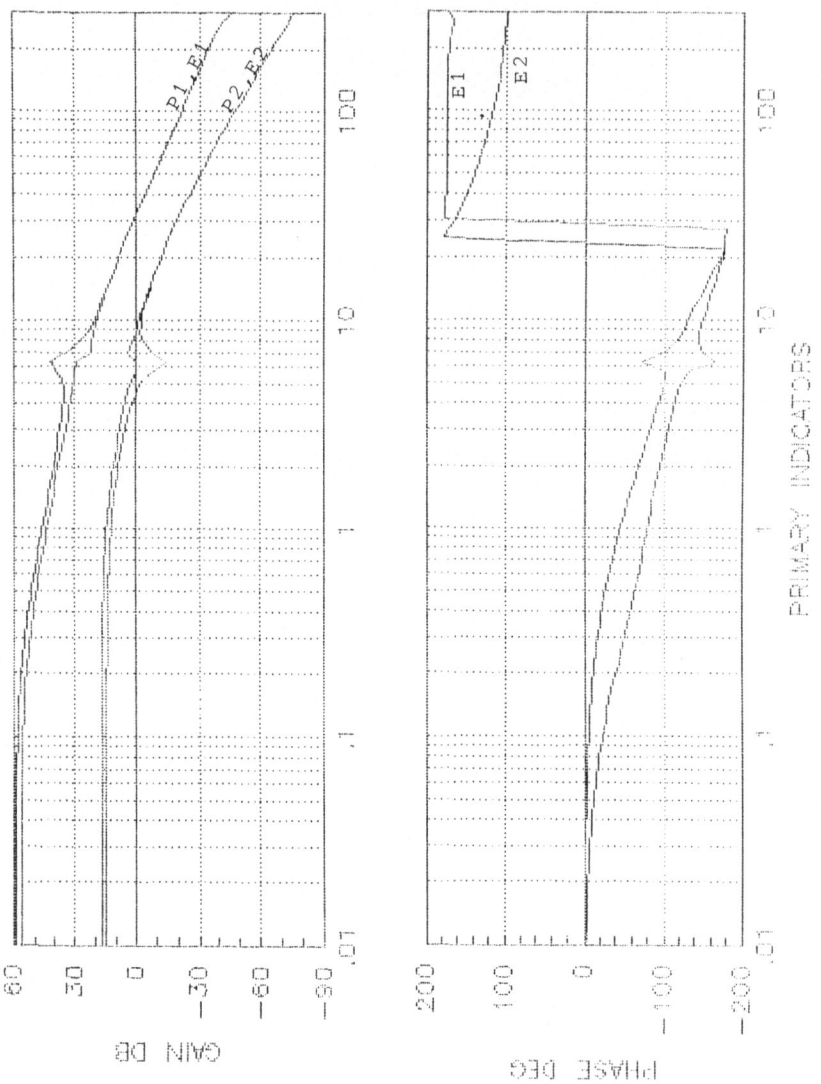

Fig. 6.10.7 The primary indicators after cascading a sub-controller with a conjugate pole-pair cancellation network.

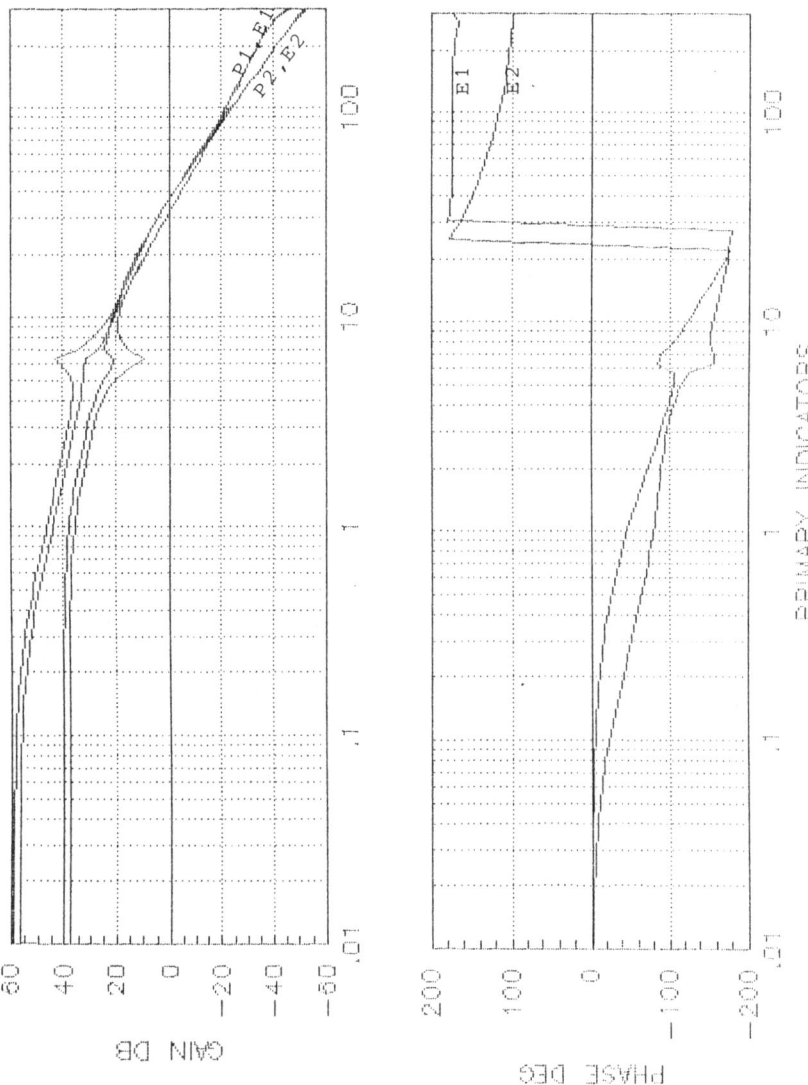

Fig. 6.10.8 The primary indicators after gain balancing at high frequencies.

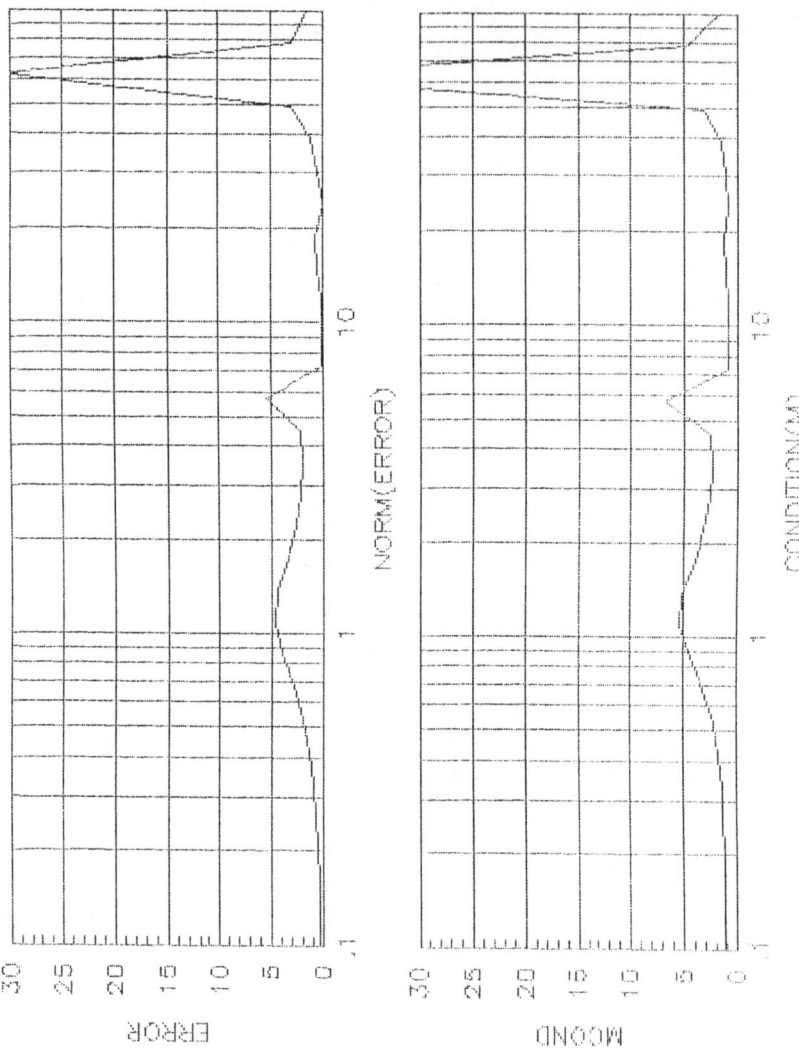

Fig. 6.10.9 The accuracy of approximation using an RFA sub-controller.

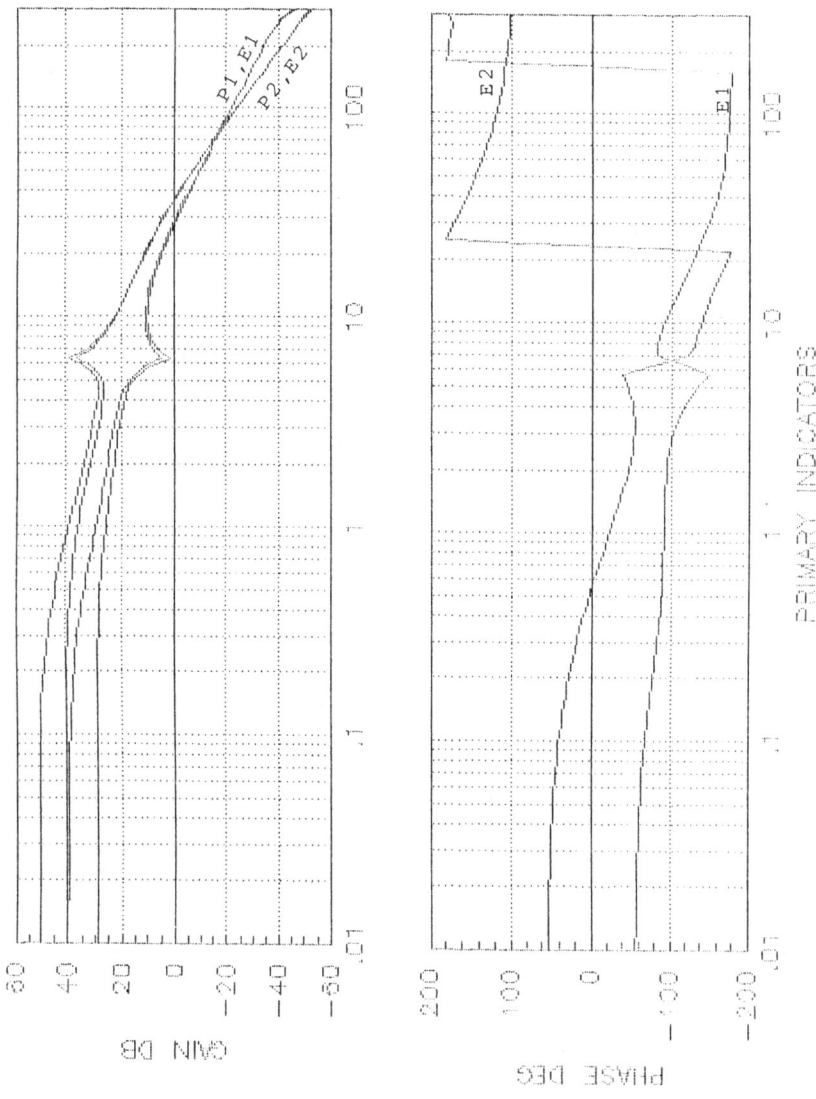

**Fig. 6.10.10 The primary indicators after an RFA
sub-controller with phase lead.**

170

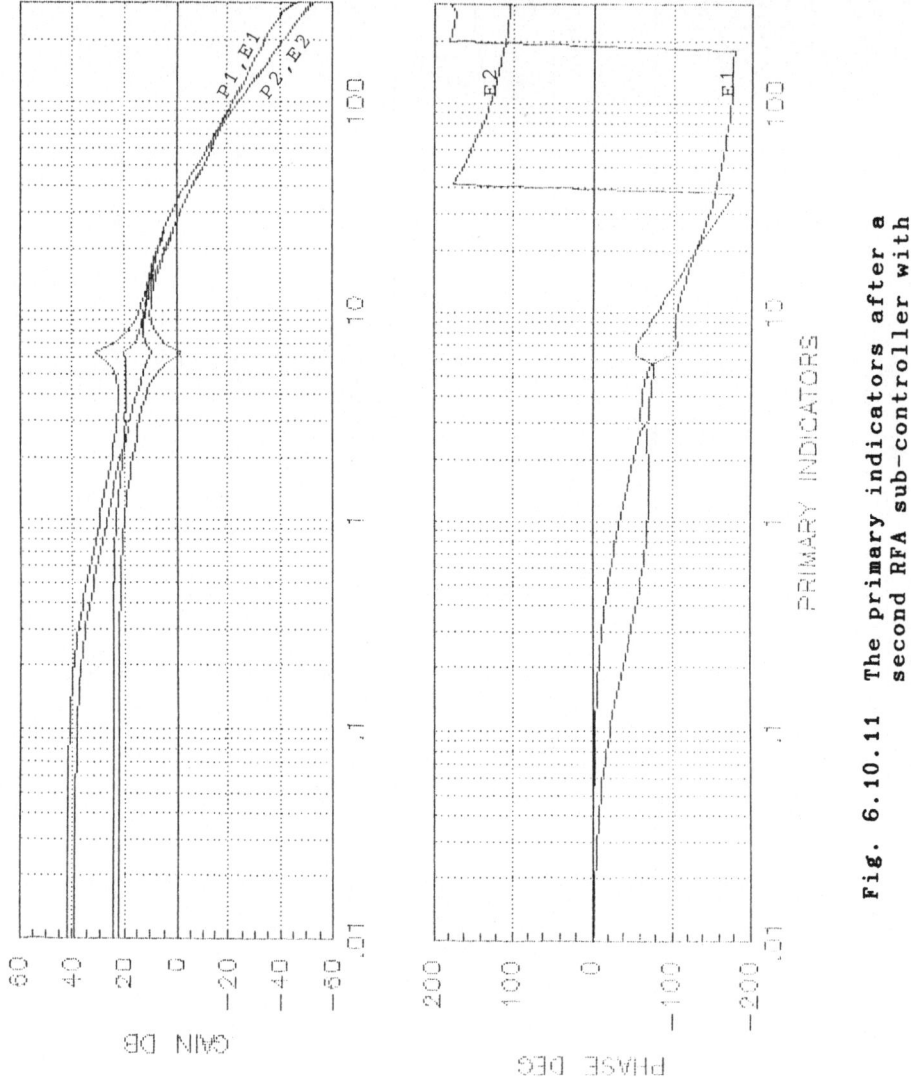

Fig. 6.10.11 The primary indicators after a second RFA sub-controller with phase lead.

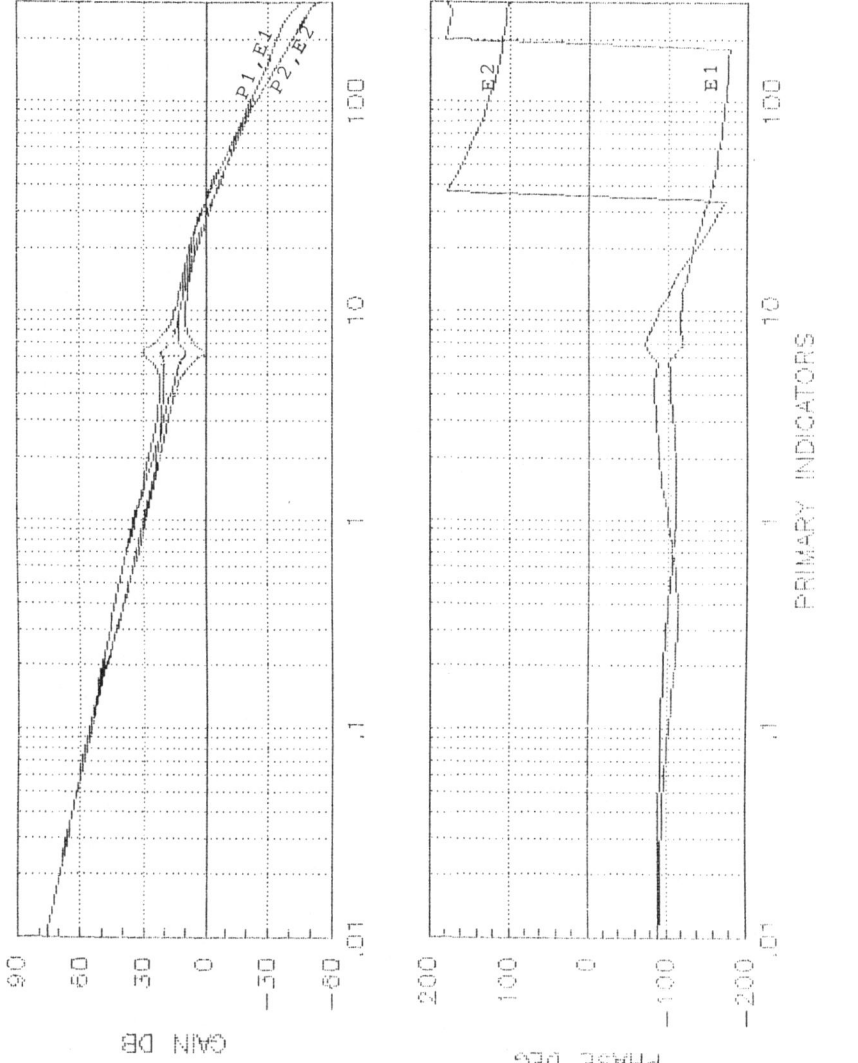

Fig. 6.10.12 The primary indicators after high, intermediate and low frequency region design.

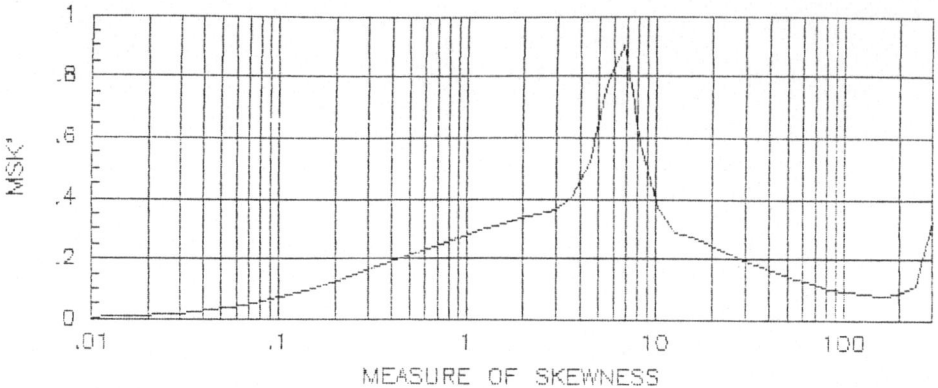

Fig. 6.10.13   The normality indicator MS of the
final design.

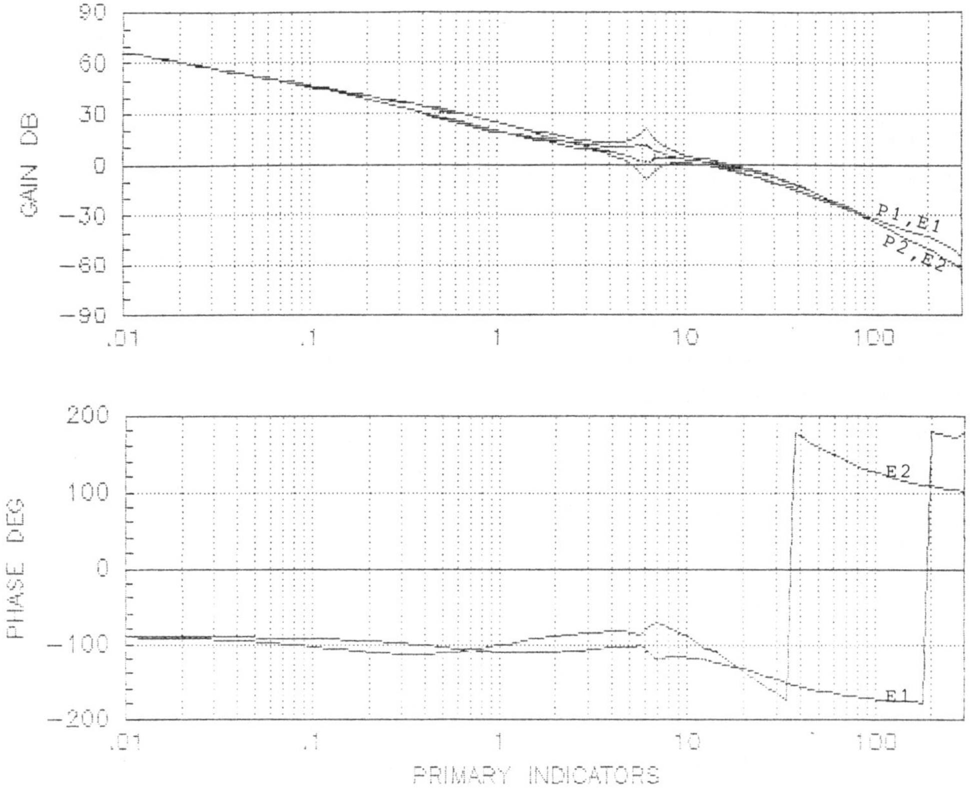

Fig. 6.10.14   The primary indicators of the final
design.

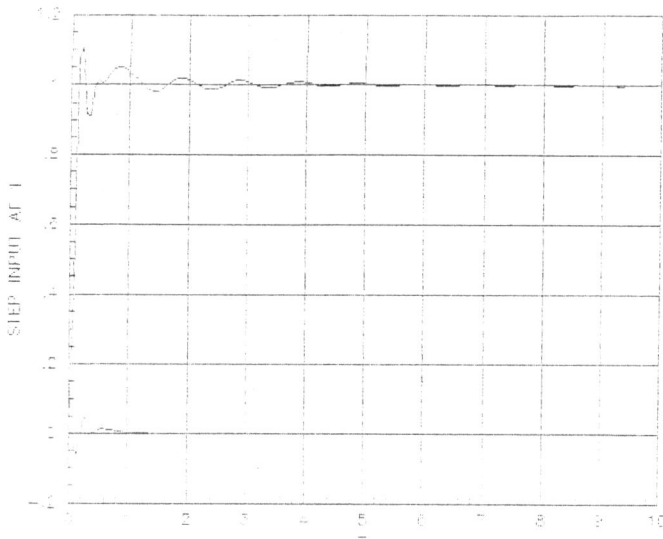

Fig. 6.10.15(a)   Closed-loop step response of the
final design (step at input 1).

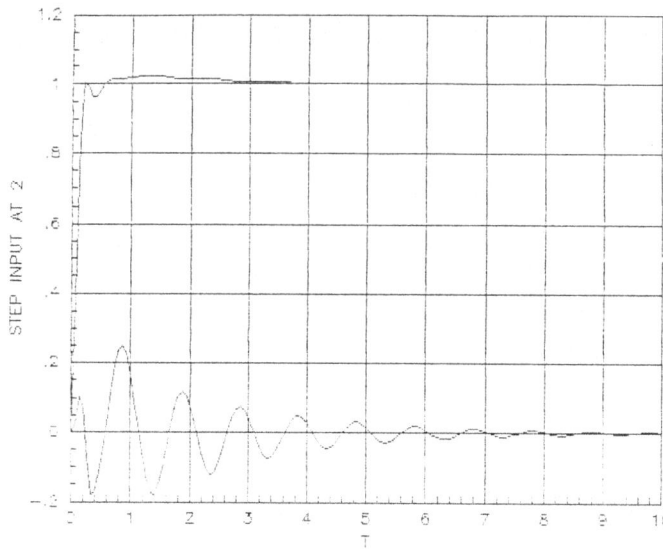

Fig. 6.10.15(b)   Closed-loop step response of the
final design (step at input 2).

174

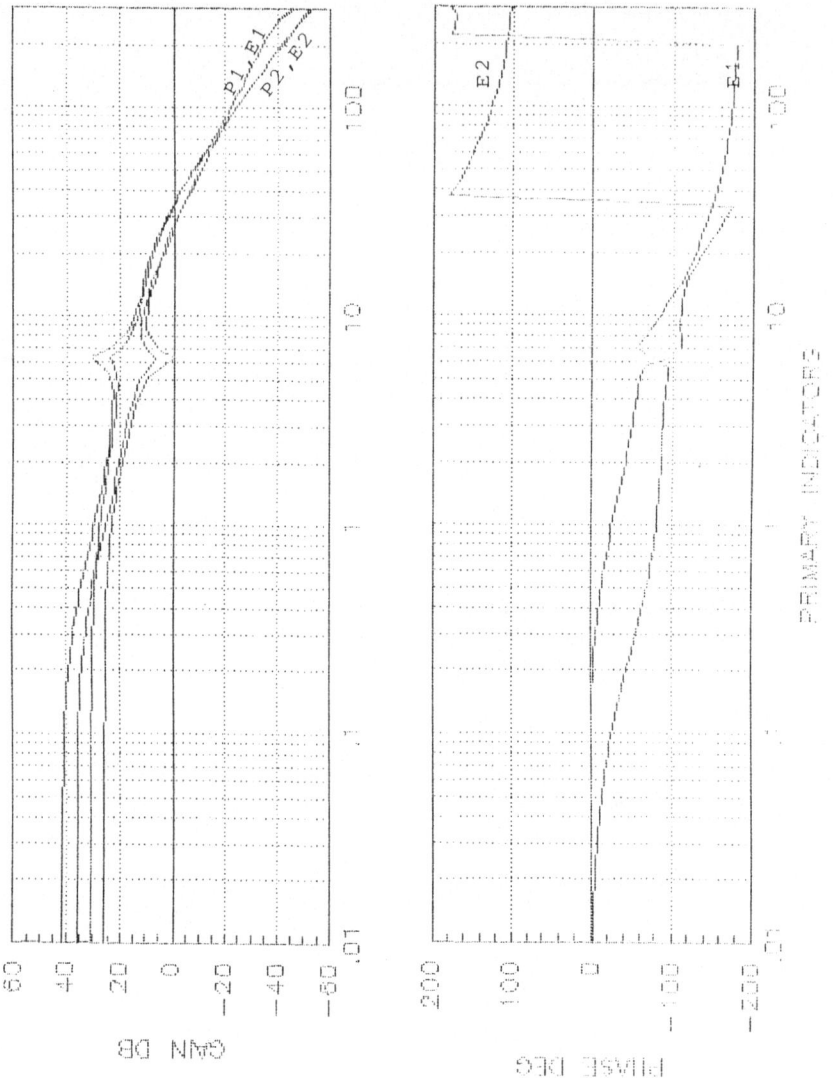

Fig. 6.10.16 The primary indicators with parameters from the optimizer.

175

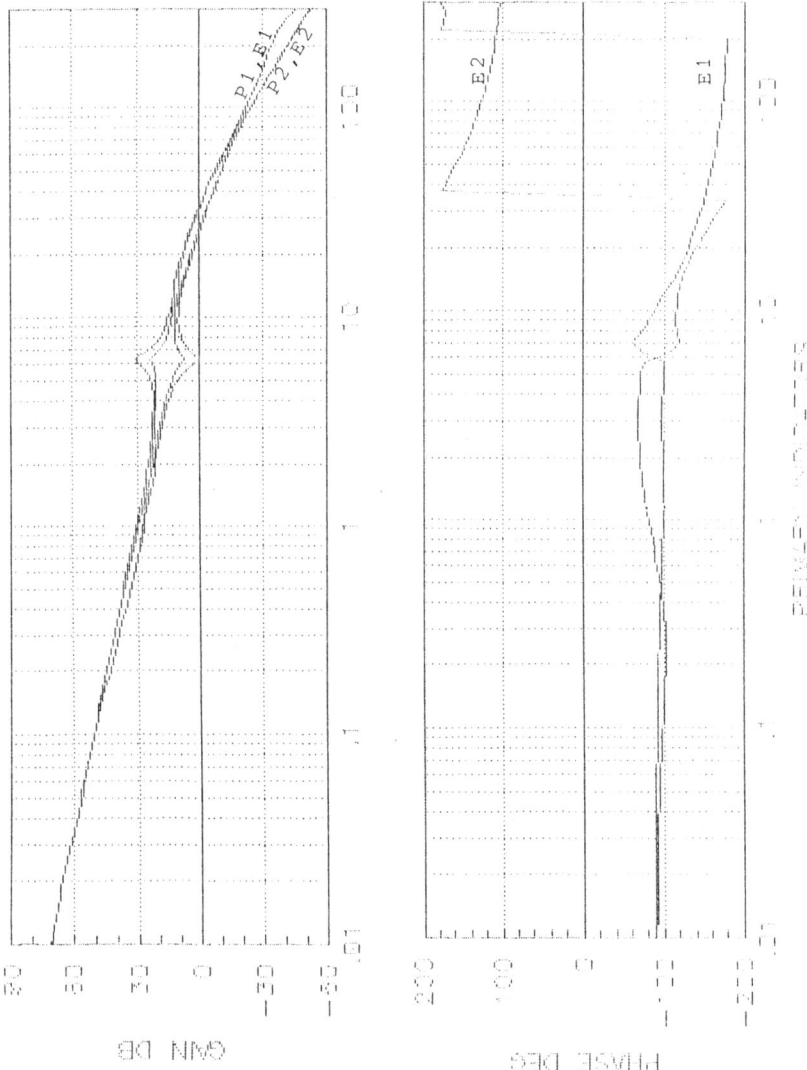

Fig. 6.10.17   The primary indicators of the design
with parameters from the optimizer.

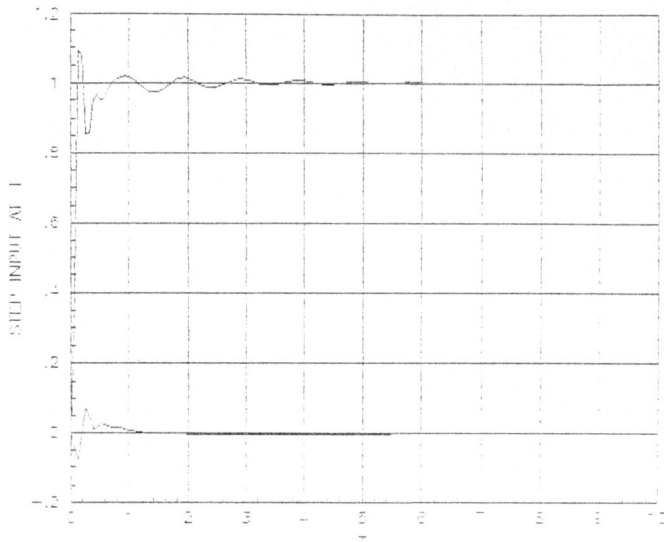

Fig. 6.10.18(a)   Closed-loop step response of the
                  final design with parameters from
                  the optimizer (step at input 1).

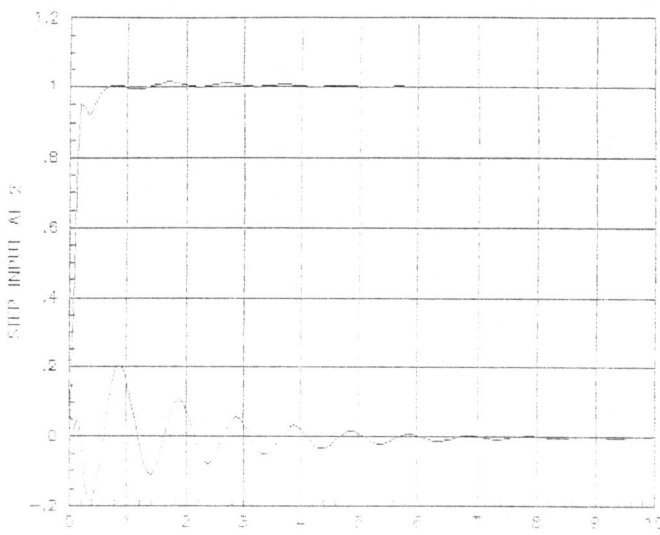

Fig. 6.10.18(b)   Closed-loop step response of the
                  final design with parameters from
                  the optimizer (step at input 2).

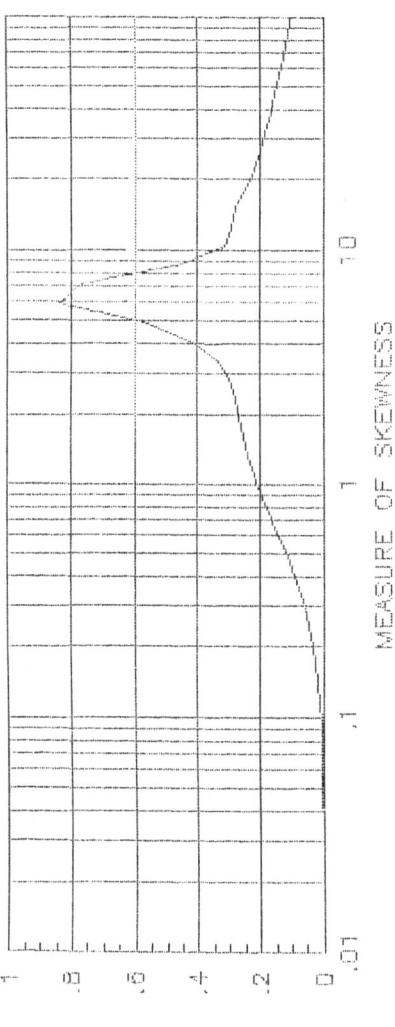

Fig. 6.10.19   The normality indicator MS of the
final design with parameters from
the optimizer.

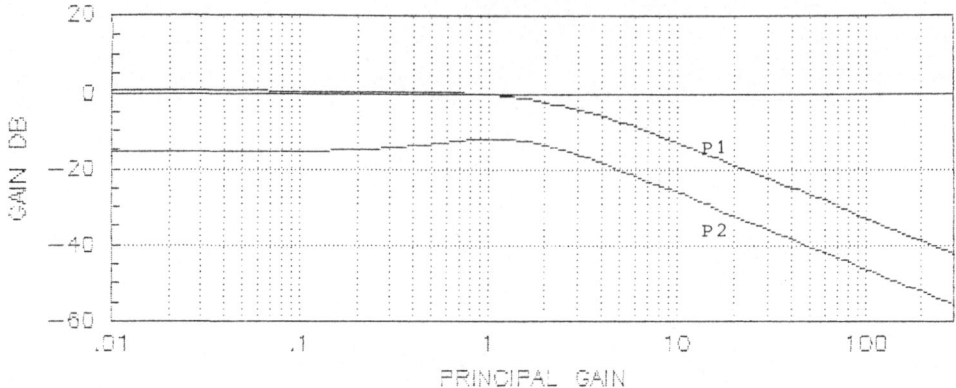

Fig. 6.11.1   The principal gains of the
uncompensated system.

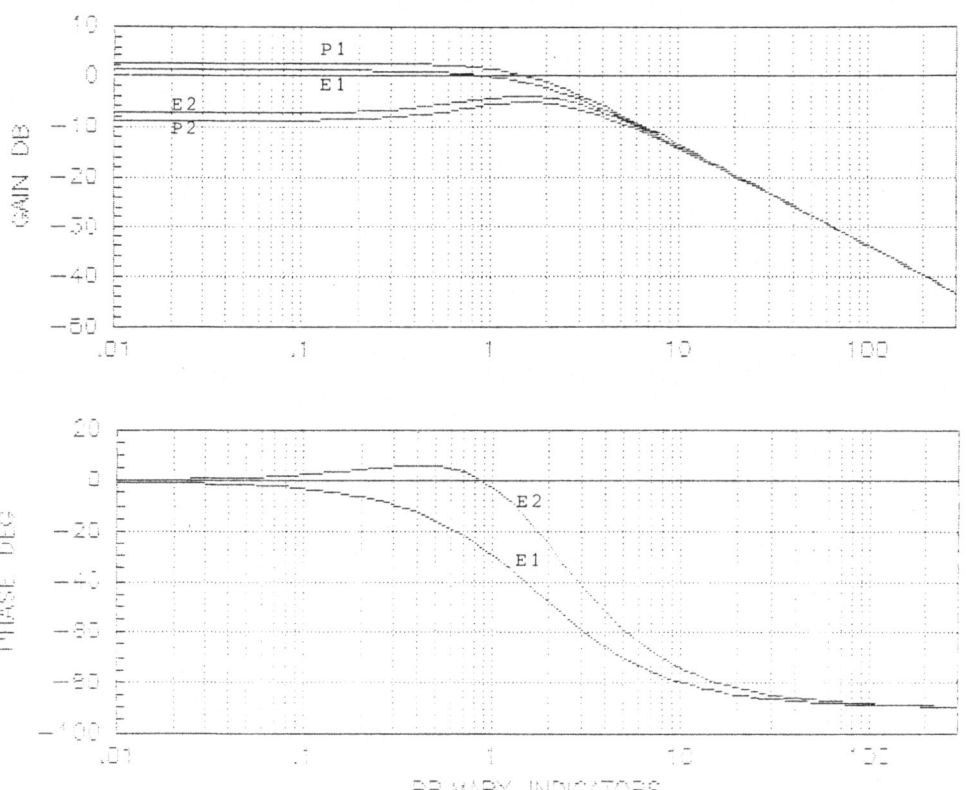

Fig. 6.11.2   The primary indicators after high
frequency region design.

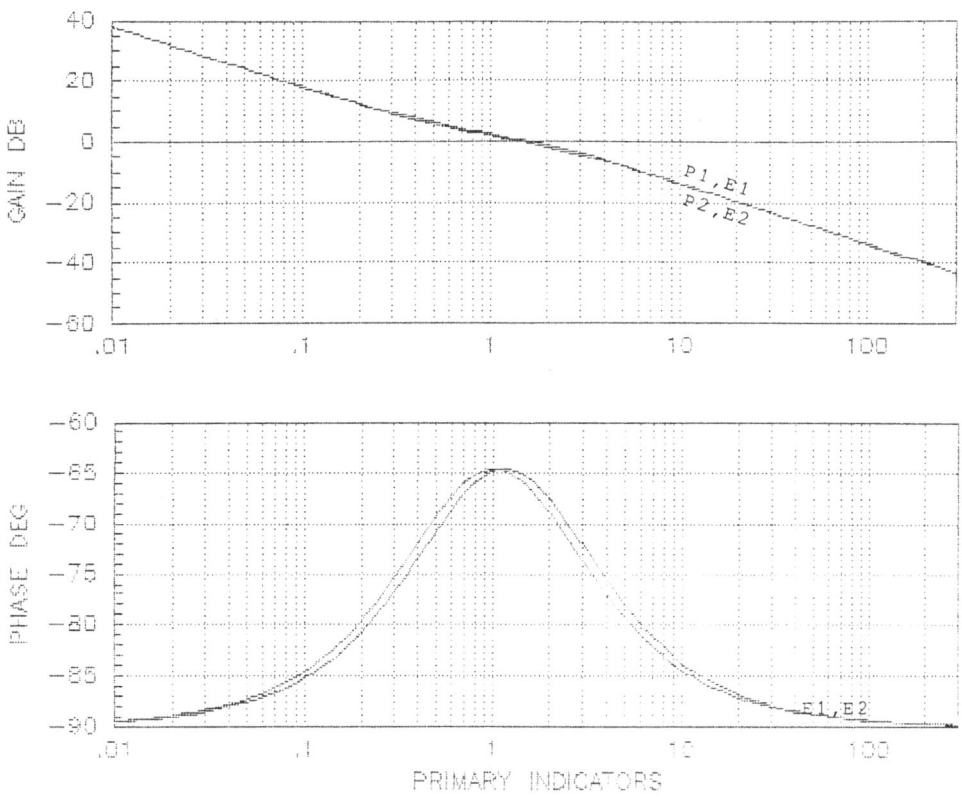

Fig. 6.11.3 The primary indicators after high
and low frequency region design.

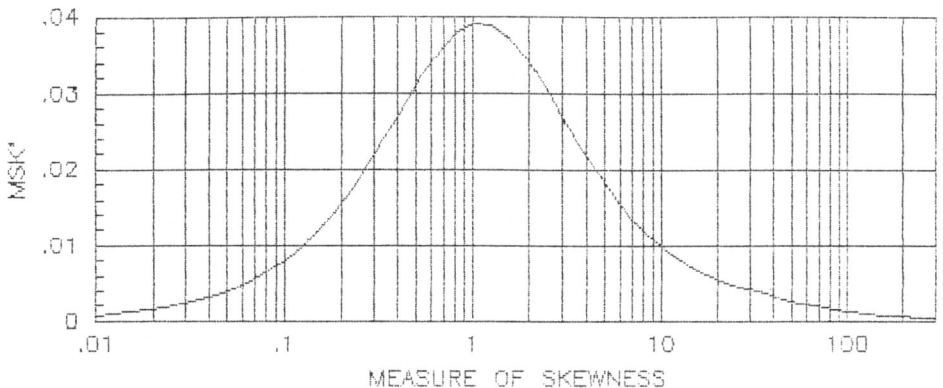

Fig. 6.11.4 The normality indicator MS of the
final design.

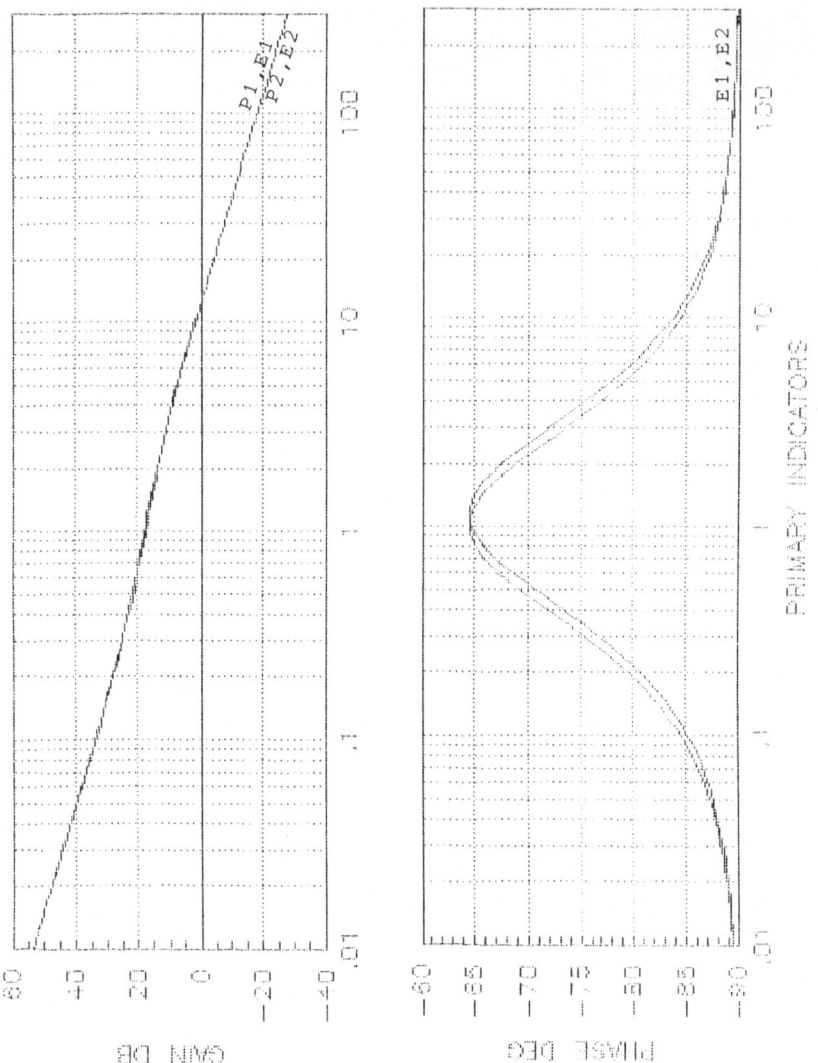

Fig. 6.11.5 The primary indicators of the final design.

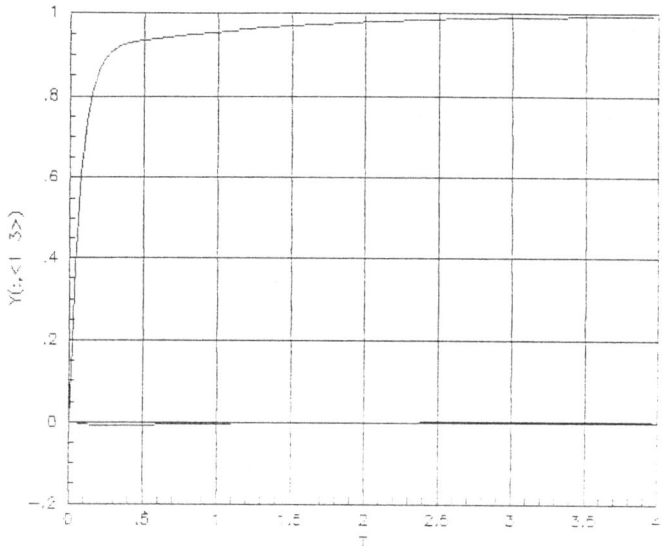

Fig. 6.11.6(a)   Closed-loop step response of the
                 final design (step at input 1).

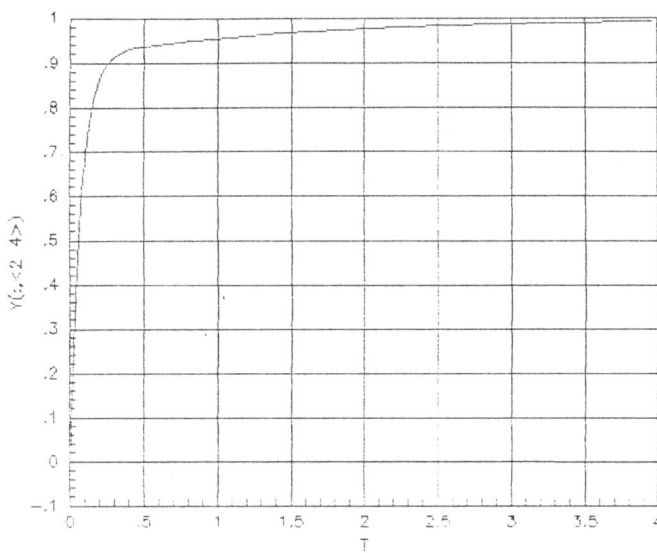

Fig. 6.11.6(b)   Closed-loop step response of the
                 final design (step at input 2).

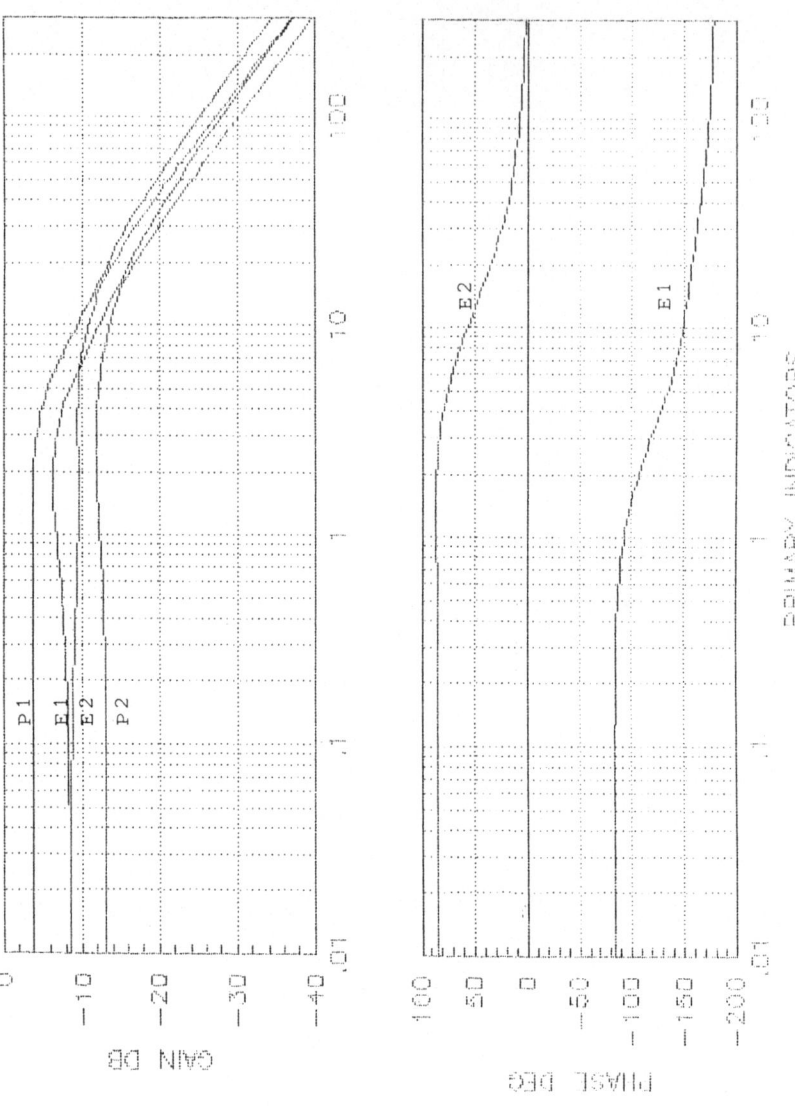

Fig. 6.13.1 The primary indicators of the system after squaring down and output feedback.

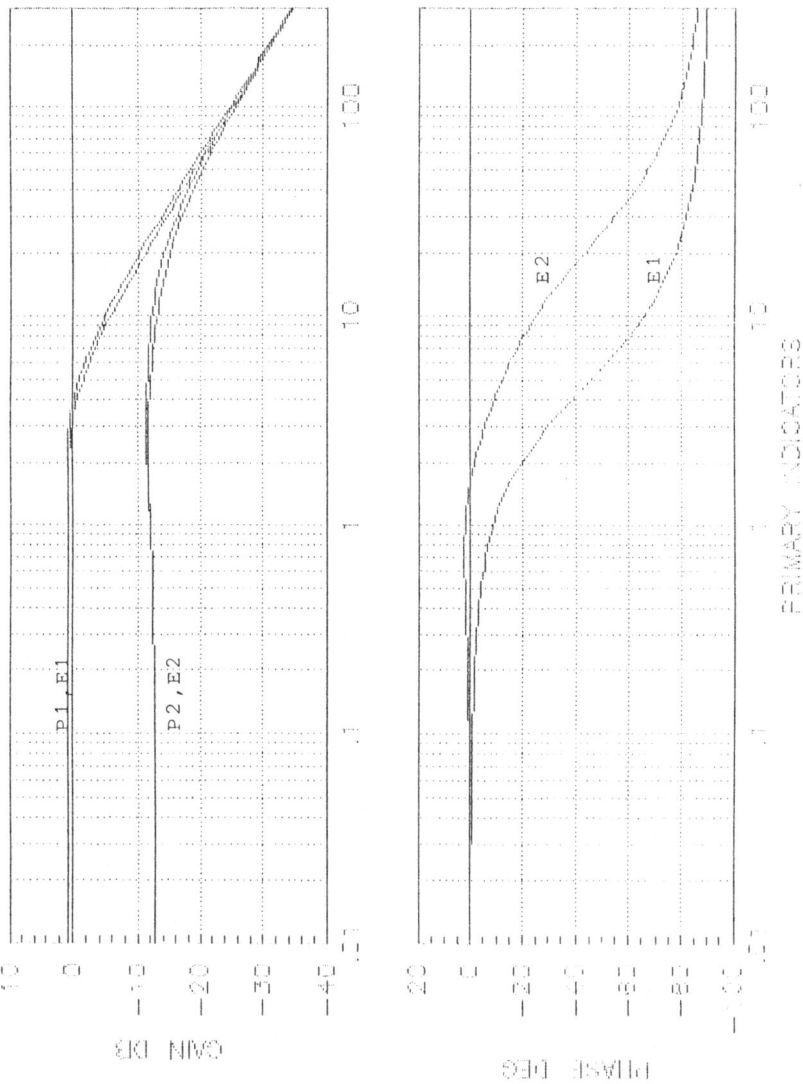

Fig. 6.13.2   The primary indicators after high
frequency region design.

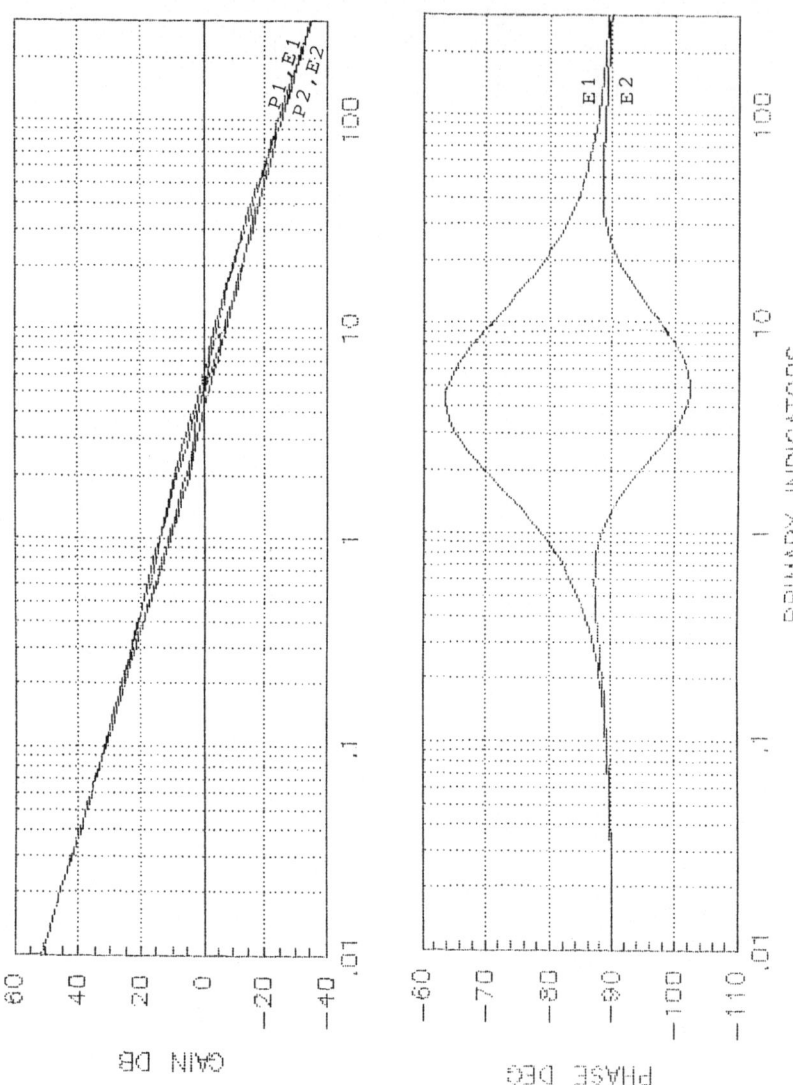

Fig. 6.13.3 The primary indicators after high and low frequency region design.

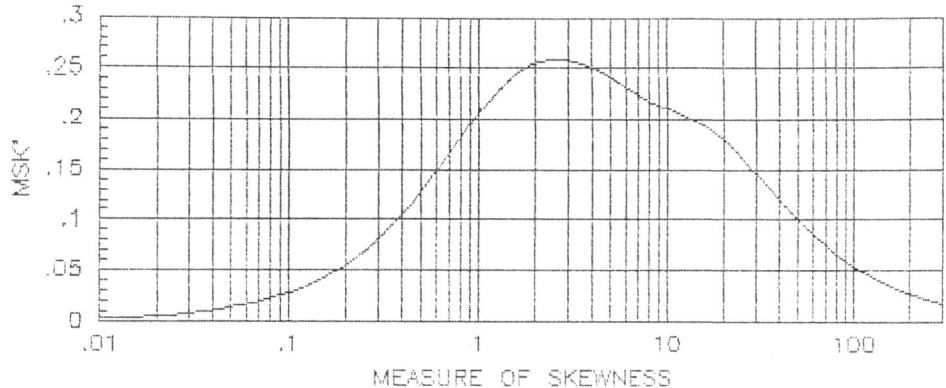

Fig. 6.13.4   The normality indicator MS of the
final design.

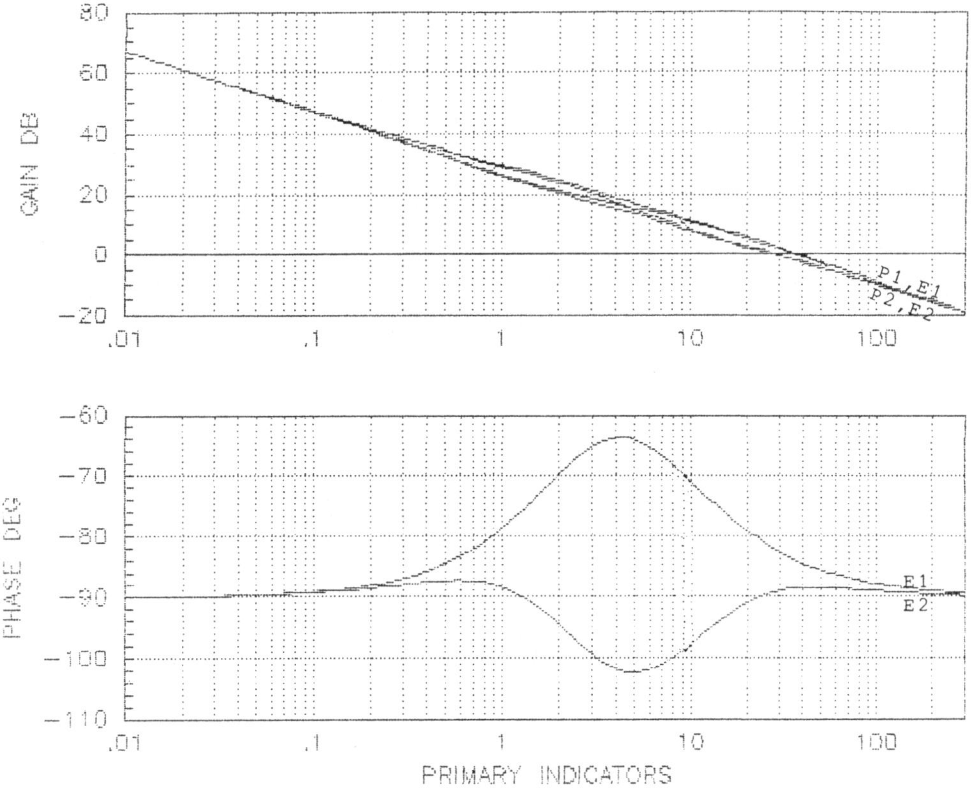

Fig. 6.13.5   The primary indicators of the
final design.

186

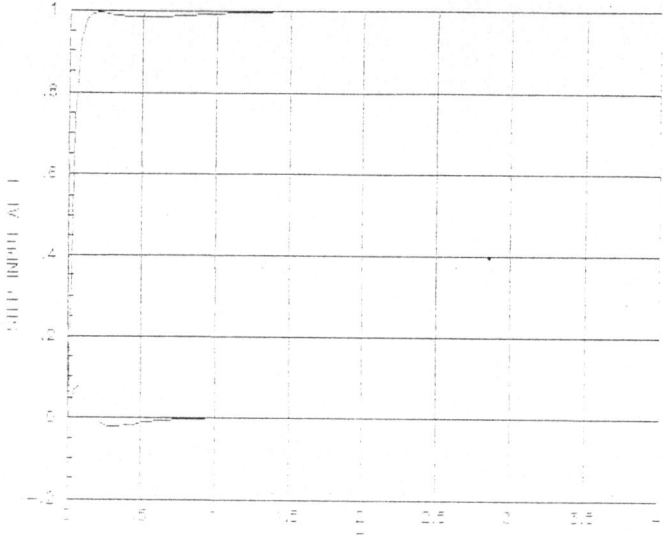

Fig. 6.13.6(a)   Closed-loop step response of the
final design (step at input 1).

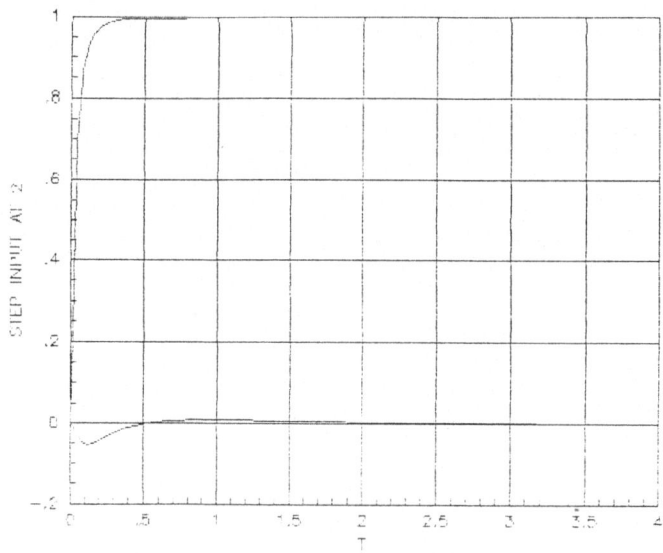

Fig. 6.13.6(b)   Closed-loop step response of the
final design (step at input 2).

Table 6.1(a) : Results on using the parameter optimizer

Case No.	WEIGHTINGS			FUNCTION VALUES		TOTAL NO.	RESULTS	
	1st term	2nd term	3rd term	Intial	Final	Function Evaluation by Simplex	Zero	Pole
1	3	5	1	2.0397	1.9399 (1.9405)	25 (54)	-6.5685 (-6.4439)	-1.9027 (-1.8441)
2	5	3	1	1.8103	1.5482 (1.5477)	15 (27)	-6.1832 (-6.2523)	-0.8412 (-0.8359)
3	1	3	5	3.7591	3.6908 (3.6952)	19 (54)	-6.3971 (-5.8692)	-1.8904 (-1.6312)
4	1	1	1	0.9282	0.8907 (0.8909)	25 (36)	-6.1998 (-6.1339)	-1.0600 (-1.0739)
5	1	5	0	1.2261	0.7469 (0.7278)	17 (18)	-2.0126 (-1.1184)	-3.0312 (-2.1885)
6	1	4	3	2.7906	2.6383 (2.6673)	21 (54)	-7.9445 (-5.7508)	-3.6498 (-2.1227)

Note : The results by alternative variable method
are in brackets.

Table 6.1(b) : Average MS, minimum phase margin and maximum
condition number of various cases from parameter optimization

	Average MS	Minimum phase margin	Maximum condition number ($\bar{\sigma} / \underline{\sigma}$)
Initial guess	0.1088	43.4845	2.4752
Case 1	0.1278	46.8601	2.4866
Case 2	0.0268	41.8213	2.4825
Case 3	0.1342	47.1933	2.4906
Case 4	0.0454	43.8585	2.4634
Case 5	0.5017	51.1910	5.5882
Case 6	0.2398	48.8010	2.7112

Table 6.2(a) : Results on using the parameter optimizer

Case No.	WEIGHTINGS 1st term	2nd term	3rd term	FUNCTION VALUES Initial	Final	TOTAL NO. Function Evaluation	Zero	RESULTS Pole	ω
1	3	5	1	2.9161	2.6761	17	-6.4942	-27.2504	7.5587
2	5	2	1	4.2101	3.7519	27	-7.3744	-27.5781	5.4517
3	1	3	5	5.5224	5.4191	17	-6.4942	-27.2504	7.5587
4	1	1	1	1.6221	1.5387	17	-6.4942	-27.2504	7.5587

Table 6.2(b) : Average MS, minimum phase margin and maximum condition number of various cases from parameter optimization

	Average MS	Minimum phase margin	Maximum condition number ($\bar{\sigma}$ / $\underline{\sigma}$)
Initial guess	0.647	71.9935	40.1376
Case 1	0.5687	67.5617	33.4199
Case 2	0.4177	16.9155	336.477
Case 3	0.5687	67.5617	33.4199
Case 4	0.5687	67.5617	33.4199

CHAPTER SEVEN

AN OBSERVER-BASED APPROACH TO DESIGN

7.1 Introduction

The design of multivariable control systems using negative-unity feedback has been described in the previous sections. The Simple Design Technique and the Reverse Frame Alignment technique were developed with an aim of manipulating the primary indicators of the open-loop system into a satisfactory form, especially over the intermediate frequency region. This technique works for a wide class of systems. If the compensated system is still not satisfactory, then we propose the use of a new compensation scheme which consists of an observer and other dynamic and constant compensators. The objective of using an observer is to obtain information about the states of the system and use this information to obtain a good gain-phase characteristic over the high and intermediate frequency regions. The reason that some systems are complicated and have a poor gain-phase characteristic is due to a loss of information when all the information in the states is transferred to the outputs. As a result, we cannot obtain a good gain-phase characteristic just by operating directly on the outputs. The use of observers will put feedback around a model of a plant to recover the inaccessible states. Therefore, the function of the observer is analogous to taking more measurements from the plant.

7.2 Structure of the Observer-based Controller (OBC)

The structure of the compensation scheme consists of two closed loops. (see Fig. 7.1) LOOP-1 consists of an observer and a constant gain state feedback controller. It is an <u>inner stabilising loop</u> (Kouvaritakis et al., 1979), its purpose being to set the positions of the closed-loop poles of LOOP-1 using the observer and state feedback controller. Notice that these closed-loop poles will become the open-loop poles of the forward path. Hence, if there are unstable poles in the original plant, they can be shifted to the left-half plane (LHP). If all the poles are already in the LHP, the dominant ones can be shifted even further to the left to obtain better gain-phase characteristics. The outer loop, LOOP-2, consists of LOOP-1 and a dynamic compensator K(s). It is an <u>outer gain-injection loop</u> and its purpose is to enhance the performance of the system by injection gains and/or integral action.

7.3 Implementation of the Observer-based Controller

7.3.1 Design of the state feedback controller

The designer should start off by designing a constant state feedback matrix F. We pose the problem of designing F as a regulator problem in the following way:

To find F such that the closed-loop feedback system of Fig. 7.2 is stable and has a satisfactory dynamic response.

The techniques for obtaining F include

(i)   pole placement method, and

(ii)   linear quadratic regulator technique.

The problem of linear state feedback pole placement for multivariable systems has received much attention from researchers in the past ten years (Sprinathkumar and Rhoten, 1975; Shah et al., 1975; Moore, 1976; Gourishankar and Ramar, 1976; Klein and Moore, 1977; Porter and D'Azzo, 1978a&b, Fahmy and

O'Reilly, 1982, Kautsky et al., 1985, Hassan and Amin, 1985). In the paper by Kautsky, Nichols and Van Dooren (1985), four algorithms for computing robust solutions to the problem are described. As normal matrices are insensitive to perturbations, their algorithms work by making the eigenvector frame of the closed-loop system as well-conditioned as possible for the specified set of poles. They have also shown that for these solutions, upper bounds on the norm of the feedback matrix and on the transient response are also minimized. In addition, a lower bound on the stability margin is maximized.

The design of the state feedback matrix using linear quadratic regulators has many advantages (Patel and Munro, 1982). Its use for pole placement has also attracted the attention of some researchers (Solheim, 1972; Juang and Lee, 1984, Shieh et al., 1986). Juang and Lee (1984) propose a method to determine the state weighting matrix which can shift one real or two complex eigenvalues to the desired locations for the optimal closed-loop system. Shieh and his coworkers (1986) aim to place the poles within a vertical strip.

Similar to the two-stage design technique by Kouvaritakis et al. (1979), the aim of the inner-loop is to obtain a set of closed-loop poles which are stable and have good damping properties. This will result in a suitable form of transmittance from &a to &b in the arrangement shown in Fig. 7.3, which will have good gain and phase margins.

So far, we have assumed that we can access all the states of the system. This may not be so in practice. Therefore, the next task is to design an observer to construct an estimate of the states.

7.3.2 Full-order observer (O'Reilly, 1983)

A full-order observer is an auxiliary system of order n, which is equal to the state dimension of the original system. Let the linear time-invariant system be described by

$$\dot{x} = A \, x + B \, u \qquad\qquad [7.3.1]$$

$$y = C \, x \qquad\qquad [7.3.2]$$

where

$x \in \mathbb{R}^n$ is the state, $u \in \mathbb{R}^r$ is the control input, and

$y \in \mathbb{R}^m$ is the system output.

We assume that the system is completely controllable and observable. A full-order observer may be described by

$$\dot{\hat{x}} = A \, \hat{x} + B \, u - T \, (\, y - C \, \hat{x} \,) \qquad\qquad [7.3.3]$$

Figure 7.4 gives a schematic of the observer-based controller in LOOP-1.

By the well-known separation principle in observer-based controller design (e.g. see O'Reilly, 1983), the design of the observer gain matrix T and the state feedback gain matrix F may be carried out separately. The solution of the F matrix is dual to the solution of the T matrix. Therefore, the problem of designing T may be posed as follows:

To find T such that the closed-loop system of Fig. 7.5 is stable and has a satisfactory dynamic response. The technique for obtaining F can be applied to obtain T.

The linear state feedback control law gives

$$u = F \, \hat{x} + r \qquad\qquad [7.3.4]$$

Hence, the overall closed-loop composite system is described by the $2n^{th}$ order system

$$\begin{bmatrix} \dot{x} \\ \dot{\hat{x}} \end{bmatrix} = \begin{bmatrix} A & \vdots & BF \\ -TC & \vdots & A+TC+BF \end{bmatrix} \cdot \begin{bmatrix} x \\ \hat{x} \end{bmatrix} + \begin{bmatrix} B \\ B \end{bmatrix} r \qquad\qquad [7.3.5]$$

$$y = \left[\begin{array}{c|c} C & 0 \end{array} \right] \cdot \left[\begin{array}{c} x \\ \hat{x} \end{array} \right] \qquad\qquad [7.3.6]$$

The design is completed by the design of $K(s)$ in the outer loop. The inner loop design should result in good gain-phase characteristics at intermediate frequencies. Hence, the objective of $K(s)$ is to inject gain into the system for good performance. The design of $K(s)$ should be straight-forward using either the Simple Design Technique or the Reverse Frame Alignment technique.

7.3.3 Reduced-order observers of order (n-m)

With the presence of m measurements from the outputs, a full-order observer has a measure of redundancy. In practice, an observer of order no greater than (n-m) is sufficient. Also, we can transform the original system such that the system outputs yield directly m of the system state variables. The idea which arises is that one can build a reduced-order observer of order (n-m) to reconstruct the remaining (n-m) states. The inputs to the reduced-order observer are both the inputs and outputs of the system. The procedures for designing a reduced-order observer is given below. The subject is well-established and is included for completeness.

7.3.4 Design procedures for a reduced-order observer of order (n-m)

(MacFarlane and Limebeer, 1981)

1. Perform a coordinate transformation on the original system such that:

$$\dot{\bar{x}} = P^{-1} \cdot A \cdot P \, \bar{x} + P^{-1} \cdot B \, u = \bar{A} \, \bar{x} + \bar{B} \, u \qquad\qquad [7.3.7]$$

$$\bar{y} = C \cdot P \, \bar{x} = \bar{C} \, \bar{x} \qquad\qquad [7.3.8]$$

where

$$C \cdot P = \left[\begin{array}{c|c} I_m & 0 \end{array} \right]$$

and $\qquad x = \begin{bmatrix} \overline{x}_1 \\ \hline \overline{x}_2 \end{bmatrix} \begin{matrix} m \\ \\ n-m \end{matrix}$

Also, let [7.3.7] be partitioned into the following form:

$$\begin{bmatrix} \dot{\overline{x}}_1 \\ \hline \dot{\overline{x}}_2 \end{bmatrix} = \begin{bmatrix} \overline{A}_{11} & \vdots & \overline{A}_{12} \\ \hline \overline{A}_{21} & \vdots & \overline{A}_{22} \end{bmatrix} \cdot \begin{bmatrix} \overline{x}_1 \\ \hline \overline{x}_2 \end{bmatrix} + \begin{bmatrix} \overline{B}_1 \\ \hline \overline{B}_2 \end{bmatrix} \cdot u \quad .$$

2. Design a constant matrix T such that the closed-loop feedback system
    as shown in Fig.7.6 is stable and has a satisfactory dynamic response.
    The design of T can be performed using any one of the techniques for
    designing a state feedback gain matrix.

3. Implement the reduced-order observer as shown in Fig. 7.7.

4. The final structure of the compensation scheme is shown in Fig. 7.8.

5. After the design of the inner loop, we then proceed to the design of K(s)
    such that the system has good open-loop frequency response characteristic
    between $\$$a and $\$$c (see Fig. 7.8).

7.3.5 Reduced-order observer-based controller using model reduction

        A full-order observer is of order equal to the original system. The
kind of reduced-order observer presented in the previous section is of order
(n-m). In practice, we would like to obtain a controller of minimal order,
but still capable of producing a satisfactory response from the compensated
system. Here we present a systematic way of obtaining a *reduced-order
observer-based controller* using a model reduction technique. This technique
provides a powerful means of reducing the order of an full-order observer-
based controller systematically.

        From [7.3.3], we obtain

$$\dot{\hat{x}} = (A + TC)\, \hat{x} + \left[\begin{array}{c|c} B & -T \end{array}\right] \left[\begin{array}{c} u \\ y \end{array}\right] \qquad [7.3.9]$$

$$\hat{y} = F\, \hat{x} \qquad [7.3.10]$$

The equations describe a full-order observer-based controller with n states, (m + r) inputs and r outputs. By using a model reduction method (e.g. see Glover (1984) or Silverman (1980)), we can reduce its order to η where $\eta < n$. The reduced-order observer-based controller is described by

$$\dot{\hat{x}}_r = A_r \cdot \hat{x}_r + \left[\begin{array}{c|c} B_r & -T_r \end{array}\right] \cdot \left[\begin{array}{c} u \\ y \end{array}\right] \qquad [7.3.11]$$

$$\hat{y}_r = F_r \cdot \hat{x}_r \qquad [7.3.12]$$

Together with the linear state feedback law

$$u = \hat{y}_r + r \qquad [7.3.13]$$

the new composite system is of order n + η. It is described by

$$\left[\begin{array}{c} \dot{x} \\ \dot{\hat{x}}_r \end{array}\right] = \left[\begin{array}{c|c} A & BF_r \\ -T_r C & A_r + B_r \cdot F_r \end{array}\right] \cdot \left[\begin{array}{c} x \\ \hat{x}_r \end{array}\right] + \left[\begin{array}{c} B \\ B_r \end{array}\right] \cdot r \qquad [7.3.14]$$

$$y = \left[\begin{array}{c|c} C & 0 \end{array}\right] \cdot \left[\begin{array}{c} x \\ \hat{x}_r \end{array}\right] \qquad [7.3.15]$$

The choice of η has a significant effect on the properties of the new composite system after model reduction. If the order of the full-order observer-based controller is reduced too much, one may naturally expect the composite system to have very poor frequency response characteristics. The configuration of the compensation scheme is shown in Fig. 7.10.

7.3.6 Design procedures for a reduced-order observer-based controller

1. Assuming all the states of the system are known, design a state feedback gain matrix F such that the closed-loop system ,LOOP-1, is stable and has good gain-phase characteristics. Go to step 2.

2. Design K(s) for the outer loop using SDT. If the design is not satisfactory, Reverse Frame Alignment should be tried. If a satisfactory
   design is obtained, then go to step 3. Otherwise, go back to step 1 to
   obtain another F.

3. Design an observer gain matrix T and implement the above design using K(s)
   and a full-order observer. A full-order observer-based controller is
   obtained.

4. If the design is again satisfactory, then go to step 5. Otherwise, go
   back to step 3.

5. Transform the full-order observer-based controller to an internally
   balanced form and reduce its order by m, which is the number of outputs.
   A reduced-order observer-based controller is obtained.

6. Implement the design using K(s) and examine the results.

7. If the design is satisfactory, then go to step 8. Otherwise, reduce the
   order of the full-order observer-based controller by one and go back to
   step 6.

8. Reduce the order of the reduced-order observer-based controller by one and
   implement the design using K(s).

9. If the design is satisfactory, save the reduced-order observer and go to
   step 8. Otherwise, go to step 10.

10. Discard the recent reduced-order observer and use the one obtained just
    before. The order of the final controller is the sum of the order of K(s)
    plus the order of the reduced-order observer-based controller.

Note that the relative magnitude of the Hankel singular values of the
full-order observer-based controller after a balanced realisation may give
some clues as to how far its order should be reduced (Glover, 1984).

7.4 Examples

7.4.1   Example 1

For the sake of comparison, we repeat the design of the unstable chemical reactor system REAC carried out in Section 5.5.5 and design an observer-based controller for it.

Design of a state feedback sub-controller

The original system has two unstable poles at 2.011 and 0.062. The two stable poles of the system are at -8.6646 and -5.0564. The robust pole placement algorithm of Kautsky et al. (1985) is used to shift the two unstable poles to the left-half plane at -3 and -4. The stable poles are kept at the same positions. The state feedback matrix is

$$
F = \begin{bmatrix} -0.1024 & -0.7340 & -0.4393 & -0.2577 \\ 2.2125 & 0.6321 & 1.3073 & -0.4012 \end{bmatrix}
$$

The primary indicators of the system assuming full-state feedback are given in Fig.7.9.1.

High frequency region design

It can be seen that the high frequency gain-phase characteristics are not correct. A high frequency sub-controller is cascaded with the system.

$$
K_\infty = \begin{bmatrix} 0 & 1 \\ -1.85 & 0 \end{bmatrix}
$$

The primary indicators (Fig.7.9.2) show that all the gain loci at high frequencies have been balanced and the system exhibits simple frequency response characteristics.

Low frequency region design

A low frequency sub-controller is cascaded.

$$
K_L(s) = (K_0/s + I_2)
$$

where $K_0 = \begin{bmatrix} 1.1463 & 2.7349 \\ 0.1171 & 6.2580 \end{bmatrix}$

The primary indicators and indicator MS are given in Fig.7.9.3 and Fig.7.9.4 respectively. The maximum value of MS is 0.15 which is small. A final scalar gain of 6 is introduced to give a bandwidth of about 12 rad/s (Fig.7.9.5). Figures 7.9.6(a) & (b) show that the closed-loop step responses are fast and non-interactive. Hence, under the assumption of full-state feedback, compensation using state feedback and the SDT is satisfactory. Next, we implement the design using observers.

Implementation using full-order observer

A full-order observer is built to implement the state feedback design. The observer gain matrix is obtained using the robust pole placement algorithm again. The observer poles are placed at −1, −2, −3 and −4. The observer gain matrix is

$$T = \begin{bmatrix} -6.8853 & 0.0812 & 4.8729 & -1.3319 \\ 2.9981 & 2.3285 & -8.7757 & -3.4970 \end{bmatrix}^t$$

The primary indicators of the system with state feedback implemented via a full-order observer (see Fig.7.4) are shown in Fig.7.9.7, and are almost identical to Fig.7.9.1. Also, it is found that the rest of the design using SDT is the same as the case assuming all the states are accessible. The order of the final controller would be the order of the full-order observer plus the order of the integrators in the SDT, which has a total of six.

Implementation using reduced-order observer-based controller

The model reduction algorithm in MATRIX$_x$ is used to reduce the order of the full-order observer-based controller. Figure 7.9.8 gives the primary indicators of the system with a reduced-order observer-based controller of order 3, as seen between point $a and $b (see Fig.7.10). The primary indicators are similar to the case assuming all the states are available. Again, the design can be concluded using high and low sub-controllers as before. The order of the final controller is five.

The design is now fifth order, but it is intended that this shall reduce to 4. Hence, the order of the observer-based controller is further reduced to 2. The primary indicators as seen between $a and $b are shown in Fig.7.9.9. Comparing with the full-state feedback case, the high frequency region is similar but the principal gains differ more at low frequencies. A high frequency sub-controller K_∞ is cascaded with

$$K_\infty = \begin{bmatrix} 0 & 1 \\ -1.805 & 0 \end{bmatrix} \ .$$

The primary indicators are given in Fig.7.9.10. A low frequency sub-controller is then added.

$$K_L(s) = (K_0/s + I_2)$$

where
$$K_0 = \begin{bmatrix} 1.7433 & 4.0916 \\ 1.4646 & 9.4701 \end{bmatrix}$$

An inspection of the primary indicators as shown in Fig.7.9.11 shows that the two principal loci do not coincide at intermediate frequencies. The primary indicators with a scalar gain of 6 injected and closed-loop step responses are shown in Fig.7.9.12 and Figs.7.9.13 (a) & (b) respectively. The design is satisfactory and the final order of the controller is 4.

For completeness, we illustrate what happens when the order of the observer is reduced further to 1. The primary indicators of the composite system (see Fig.7.9.14) are quite different from the full-state feedback case. In addition, the composite system has a right-half plane pole at 1.9186. However, the design can be completed using high and low frequency sub-controllers given below:

$$K_\infty = \begin{bmatrix} 0.000 & 1 \\ -1.805 & 0 \end{bmatrix}$$

$$K_L(s) = (5 \cdot I_2/s + I_2)$$

The primary indicators after high and low frequency region design are given in Fig.7.9.15 and Fig.7.9.16. A final scalar gain of 6 is used to bring the gain loci up to give a bandwidth of 12 rad/s (Fig.7.9.17). Indicator MS and closed-loop step responses are given in Fig.7.9.18 and Figs.7.9.19 (a) & (b) respectively. The above design is similar to the one in Section 5.5.5, except that the compensation scheme is more complicated here. Therefore, one has to be careful as to how far the order of the full-order observer-based controller is reduced.

Example 7.4.2

The example we consider now is the turbo-generator system TGEN in Section 6.9.3. Here, an observer-based approach is used to design a controller for the system.

Design of a state feedback sub-controller

The robust pole placement algorithm of Kautsky et al.(1985) is used to find a state feedback matrix which places the closed-loop poles of the system at -3, -4, -5, -10, -10.74, -15, -15, -17.67 and $-29 \pm 313.96j$. Note that out of the ten poles of the original system, the five which are furthest away from the origin are retained. The state feedback matrix is

$$F = \begin{bmatrix} 2.7527 & -0.0023 \\ 9.1466 & 0.0008 \\ -36.0868 & -0.0233 \\ -35.4100 & -0.0268 \\ -34.5429 & -0.0231 \\ 74.3056 & -0.0054 \\ 88.9149 & -0.0067 \\ 76.0119 & -0.0053 \\ -22.1904 & -0.0021 \\ -5.4586 & -0.1379 \end{bmatrix}^t$$

The primary indicators of the system with state feedback are displayed in Fig. 7.11.1. The system exhibits very simple frequency response characteristics and we cannot observe any resonance peak, which has been a difficulty in Section 6.9.3. Assuming that all the states are available, we design the system using the SDT. The high frequency gains are balanced first (Fig.7.11.2) and integral action is added (Fig.7.11.3).

$$K_\infty = \begin{bmatrix} -13 & 0.00 \\ 0 & 0.01 \end{bmatrix}$$

$$K_L(s) = (2 \cdot K_0/s + I_2)$$

where

$$K_0 = \begin{bmatrix} 2.1456 & 0.2277 \\ 0.1454 & 1.0298 \end{bmatrix}$$

An additional gain of 5 is introduced in each loop (see Fig.7.11.4) and the resulting closed-loop time responses are examined to assess the performance of the controller configuration. The responses as shown in Figs.7.11.5 (a) & (b) are fast, non-interactive, accurate in the steady-state, with no significant overshoots. The design is robust as indicated by the small value of MS in Fig.7.11.6.

Implementation using full-order observer

The design using state feedback is satisfactory. A full-order observer is then built to test the design. We have attempted to obtain an observer gain matrix using the robust pole placement algorithm. However, very high gains (from several hundreds to several thousands) are required to place the observer poles at positions similar to the system poles. Finally, a linear regulator design was carried out with $R = I_{10}$ and $Q = I_2$. The gain matrix obtained is

$$T = \begin{bmatrix} -4.1876 & 1.6490 \\ -9.6275 & 26.0862 \\ 1.4991 & 14.6805 \\ -1.0001 & -36.2692 \\ 0.7016 & 15.2137 \\ -2.6056 & 0.7088 \\ 0.3567 & -0.0186 \\ 1.2972 & -0.8173 \\ 0.1203 & 0.0050 \\ -0.0182 & -0.2845 \end{bmatrix}$$

The observer poles are at -1.43, -1.79, $-2.06 \pm 7.05j$, $-9.48 \pm 7.82j$, -10.72, -19.31, $-29.54 \pm 313.97j$. It was found that the primary indicators of the system with F and the full-order observer are almost identical (see Fig.7.11.7). The rest of the design follows in the same way. The order of the final controller is twelve.

Implementation using reduced-order observer-based controller

The full-order observer-based controller has 10 states and its balanced realization is formed. The Hankel singular values are calculated to be

$\sigma_1 = 1.2028 \times 10^5$ $\sigma_2 = 2.2808 \times 10^4$ $\sigma_3 = 1.5850 \times 10^3$

$\sigma_4 = 2.7843 \times 10^2$ $\sigma_5 = 1.9401 \times 10^2$ $\sigma_6 = 3.9226$

$\sigma_7 = 0.3139$ $\sigma_8 = 0.6108 \times 10^{-4}$ $\sigma_9 = 0.4321 \times 10^{-4}$

$\sigma_{10} = 0.6390 \times 10^{-7}$

The model reduction algorithm in MATRIX$_x$ is then used to reduce the order of the full-order observer-based controller. Figures 7.11.8 to 7.11.16 give the primary indicators of the system with the reduced-order observer-based controller when its order is reduced from 9 to 1. It can be seen that complications arise when the order is below 6. Also, right-half plane poles start to appear in the composite system when the order is below six. Therefore, a design was performed with a reduced-order observer-based controller of order 6. The poles of the composite system are at -1.41, -1.89,

-2.03 $\pm$ 7.04j, -2.97, -3.28, -5.17, -7.72$\pm$ 8.48j, -7.99 $\pm$ 4.84j, -11.35,

-18.29, -26.99, -29.46 $\pm$ 313.96j.   The same high frequency sub-controller

diag{ -13, 0.01 } was used and the primary indicators are given in Fig.7.11.7.

A low frequency sub-controller is then added.

$$K_L(s) = (2 \cdot K_0/s + I_2)$$

where

$$K_0 = \begin{bmatrix} 2.0657 & 0.2233 \\ 0.1435 & 1.0310 \end{bmatrix}$$

The primary indicators with a scalar gain injection of 5 and the normality

indicator are given in Fig.7.11.19 and Fig.7.11.20 respectively.   As shown in

Figs.7.11.21 (a) & (b), the closed-loop step responses are similar to the

responses under the assumption of full-state feedback.   Note that the design

is now of order 8, but the design using Reverse Frame Alignment technique in

Section 6.8.3 requires a controller of order 6.   Hence, the observer-based

approach requires a slightly more complicated controller but the design is

much better than the one using Reverse Frame Alignment technique.

7.4.3 Example 3

The system considered here is a linearized state-space model of an

aircraft denoted AIRC (see Appendix B for details).   The system has 3 inputs,

5 states and 3 outputs.   All the states are available for measurement.

Therefore, we have no need to use an observer to implement any state feedback

for the system.   The design objective is to obtain closed-loop step responses

of the output that are fast, non-oscillatory, non-overshooting non-interactive

and have zero steady-state error.   A bandwidth of no more than 10 rad/s is

assumed.

The primary indicators of the plant are given in Fig.7.12.1.   The

system has been studied by Kouvaritakis et al. (1979) using the two-stage design method. A reverse-frame-normalizing controller was designed and given by Hung (1982). Both design methods have resulted in a satisfactory design. Here we present a design using the design technique described in this chapter.

Design of a state feedback sub-controller

First, we design a state feedback gain matrix for the plant. The system has poles at 0, -0.7801 ± 1.0296j and -0.0176 ± 0.1826j. The robust pole placement algorithm is used to place the poles at -3, -4, -5, -12 and -15. The state feedback controller obtained is

$$
F = \begin{bmatrix}
31.7610 & 0.2123 & 15.6630 & -0.1876 & -11.4000 \\
3.8113 & -4.9207 & 2.0507 & -0.0225 & -1.4385 \\
85.7830 & 0.5924 & 70.7300 & 9.5492 & -30.3990
\end{bmatrix}
$$

An inspection of the primary indicators (Fig.7.12.2) shows that the inner-loop state feedback sub-controller has improved the system dynamics and resulted in a simple gain-phase behaviour in terms of frequency response criteria. The system resonance observed at about 0.18 rad/s in the original system no longer appears in Fig.7.12.2. Next, we use the Simple Design Technique to complete the design.

High frequency region design

A high frequency sub-controller is used to balance up the gains at ω = 10. The input and output frames are aligned as well at high frequencies.

$$
K_\infty = \begin{bmatrix}
-11.1350 & -0.1192 & -0.1866 \\
-1.3362 & 0.9929 & -0.0224 \\
-29.2660 & 0.0000 & -10.5470
\end{bmatrix}
$$

The primary indicators (Fig.7.12.3) show that the objectives are achieved and the system has correct gain-phase characteristics at high frequencies.

Low frequency region design

The design is concluded by designing a low frequency sub-controller

$$K_L(s) = (3 \cdot K_0/s + I_2)$$

where
$$K_0 = \begin{bmatrix} 1.0534 & -0.0206 & 0.0558 \\ 0.0000 & 1.8359 & 0.0000 \\ 0.0849 & 0.0572 & 1.0939 \end{bmatrix}$$

The primary indicators (Fig.7.12.4) show that all the gain loci are well-balanced from low frequencies up to about 10 rad/s. The divergences between characteristic and principal gains are small over the entire frequency spectrum and thus MS is small (see Fig.7.12.5). A scalar gain of 10 is introduced to give a bandwidth of 10 rad/s (Fig.7.12.6). The closed-loop step responses of the overall feedback configuration to unit steps applied to the three inputs are shown in Figs.7.12.7 (a), (b) & (c). The system response can be seen to be fast, accurate in the steady state and non-oscillatory, with no significant overshoots (around 12%). Furthermore, the system response exhibits no appreciable interaction in all the loops. Therefore, the strategy of designing a state-feedback sub-controller followed by the SDT procedures works very well in this design example.

Before we finish, we examine the effect of placing the poles further away from the origin in the first stage of design. The robust pole placement algorithm is used again to shift the system poles to −4.77, −8.34, −11.4, −18 and −19, which are very close to those pole positions in Kouvaritakis et al. (1979). The state feedback matrix is

$$F = \begin{bmatrix} 92.1615 & 0.1864 & 48.5556 & 2.2494 & -17.7273 \\ 12.9417 & -11.3238 & 7.6993 & 0.3968 & -2.3659 \\ 265.9262 & 0.5087 & 201.0410 & 20.5242 & -48.9671 \end{bmatrix}$$

We observe that higher gains are required to place the poles more to the left. Design procedures of the SDT similar to above are then carried out. We can now draw the comparisons between the primary indicators of the compensated system using this design shown in Fig.7.12.8 on the one hand, and the corresponding plots for the previous design given in Fig.7.12.6 on the other. There is no appreciable difference in the gain plot but there are more phases in the design where poles are shifted more to the left. A phase margin of more than 60 degrees is observed at the crossover frequency. The generalised Nyguist diagram with the M circles (Fig.7.12.9) indicates that the closed-loop step responses should be non-overshooting. As shown in Figs.7.12.10 (a), (b) & (c), the closed-loop step responses are similar to the responses before except that no overshoot can be seen.

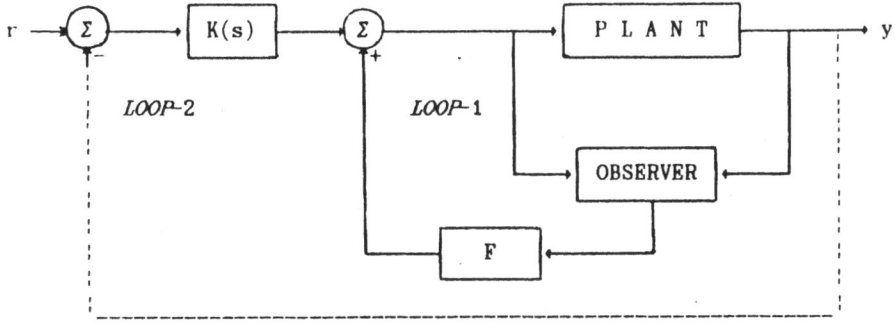

Fig. 7.1 Structure of an Observer-based Controller

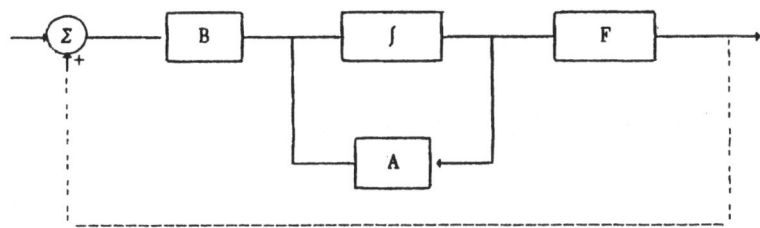

Fig. 7.2 Design of the state feedback matrix

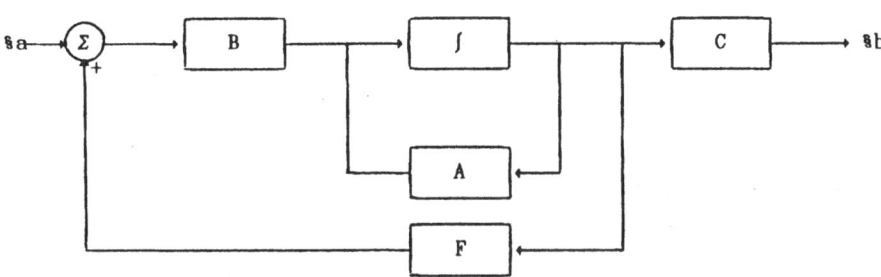

Fig. 7.3 The system with state feedback

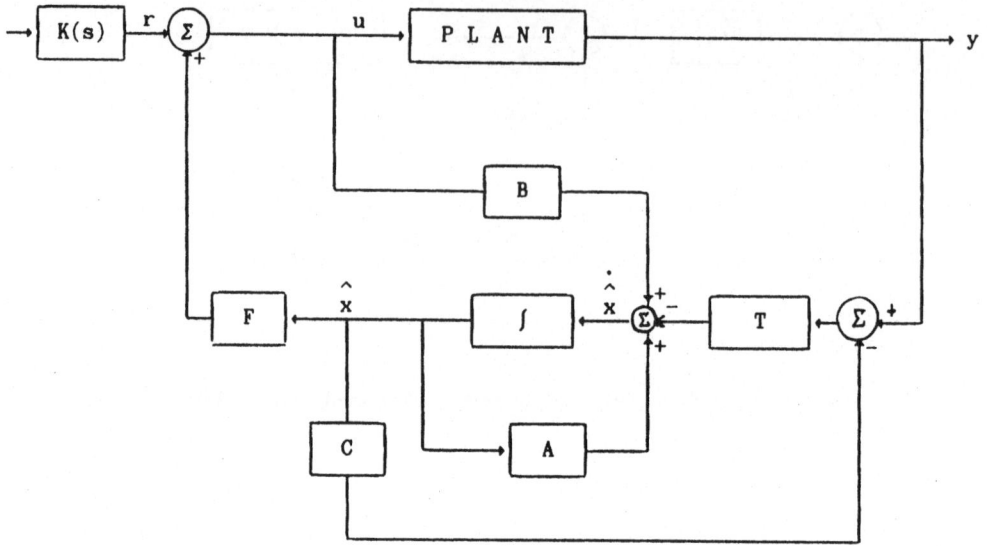

Fig. 7.4 Block diagram of an full-order observer-based controller

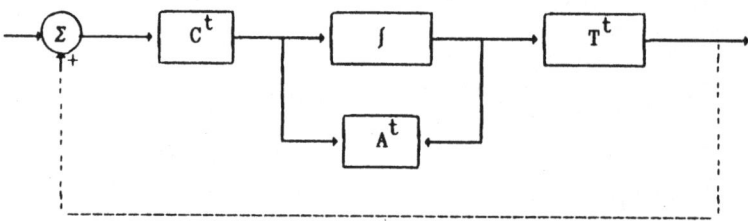

Fig. 7.5 Design of the observer gain matrix

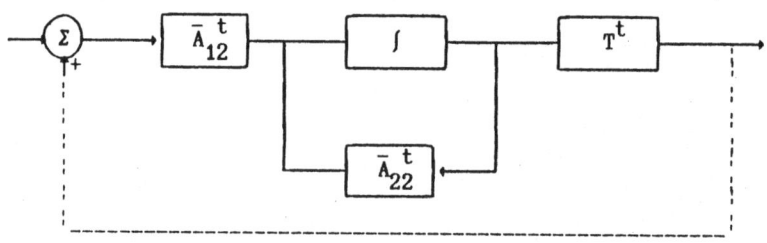

Fig. 7.6 Design for a reduced-order observer

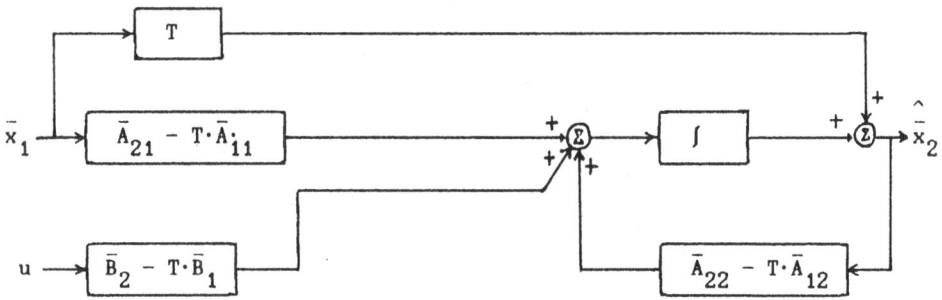

Fig. 7.7   Implementation of the reduced-order observer

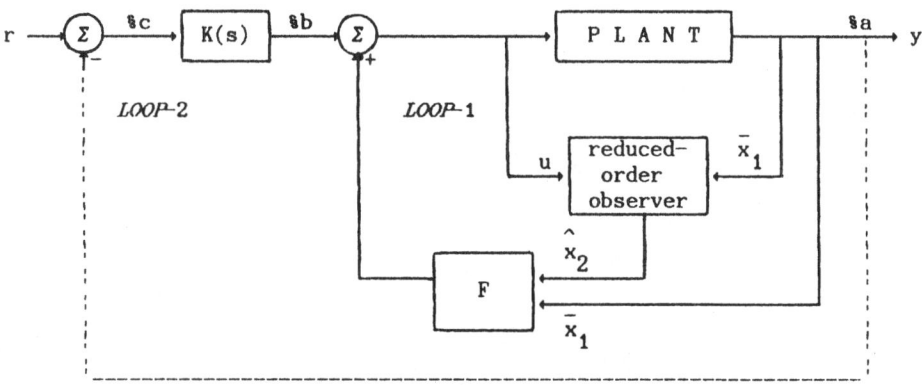

Fig. 7.8   Compensation scheme with a reduced-order observer of order (n-m)

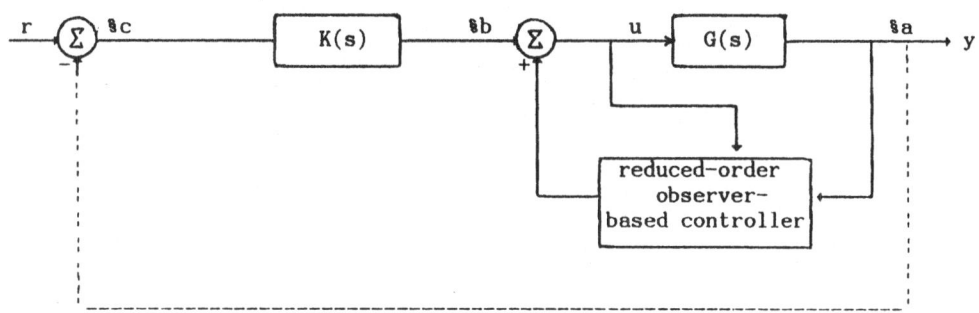

Fig. 7.10   Arrangement of the compensators

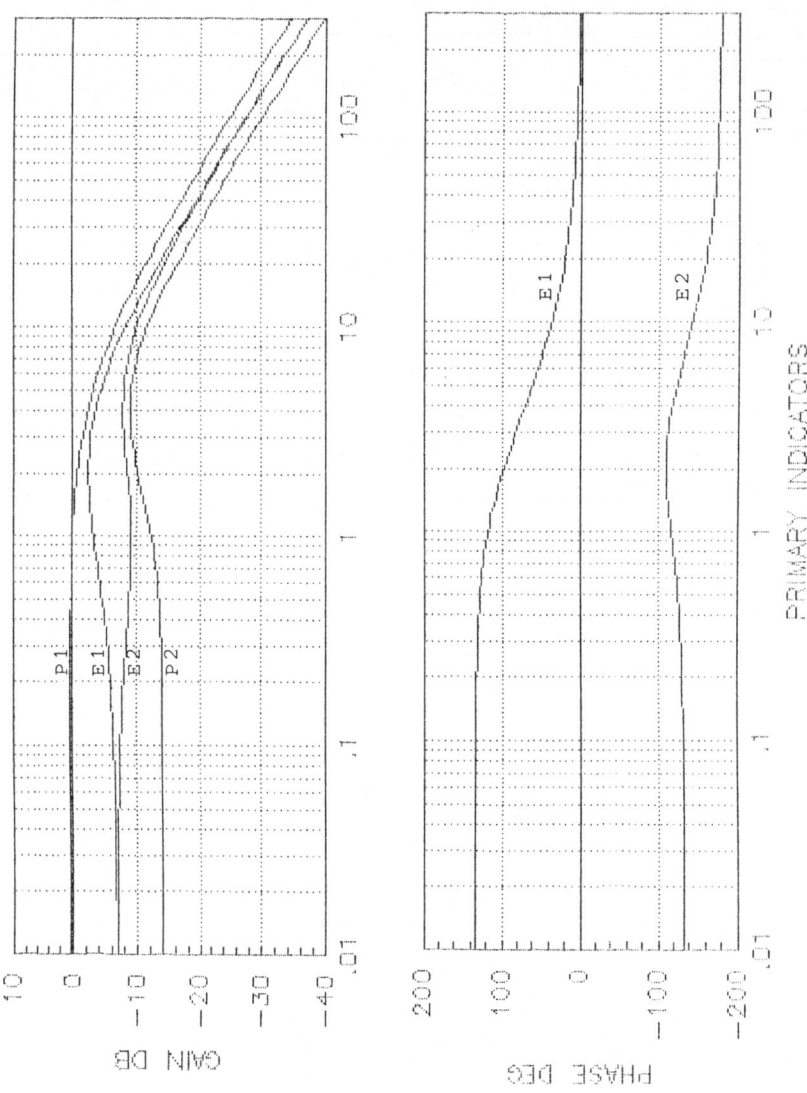

Fig. 7.9.1 The primary indicators of REAC with full-state feedback.

211

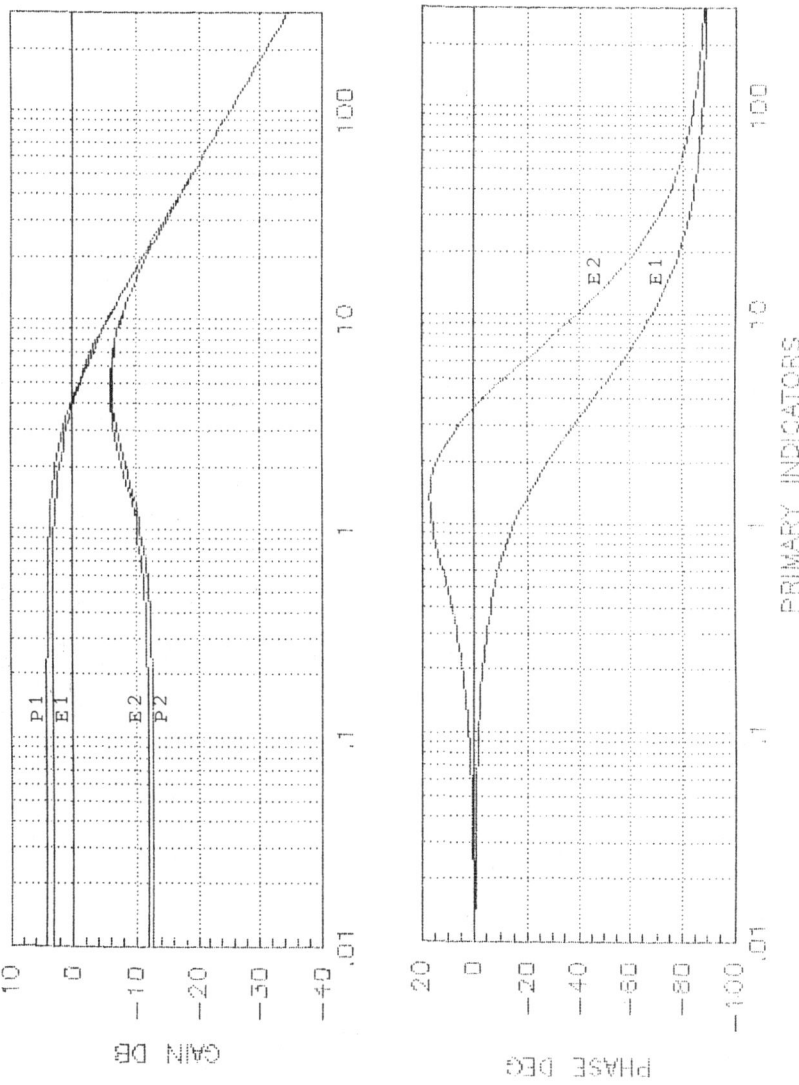

Fig. 7.9.2 The primary indicators after high frequency region design.

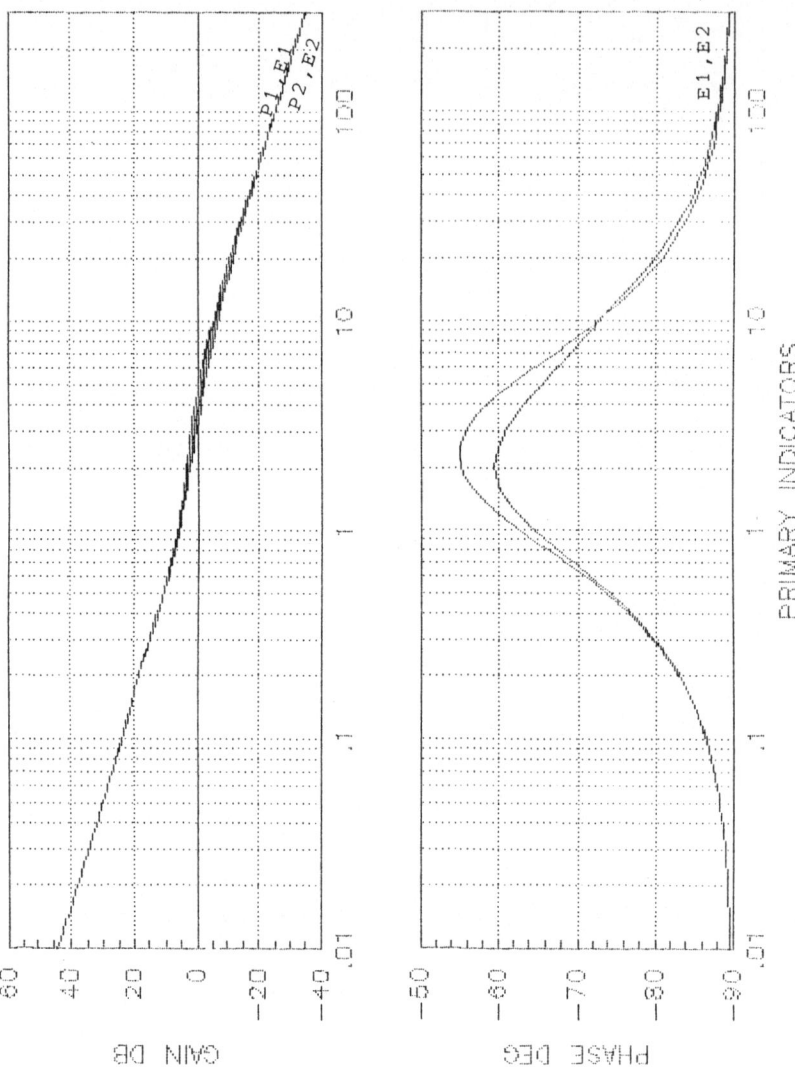

Fig. 7.9.3 The primary indicators after high and low frequency region design.

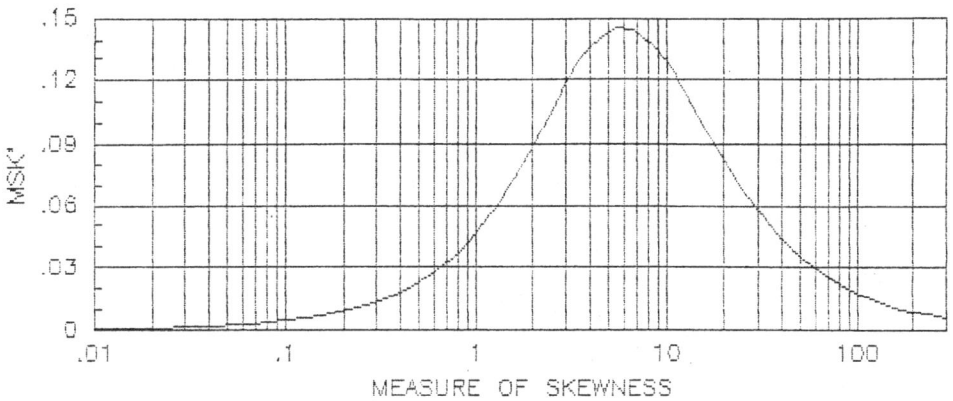

Fig. 7.9.4   The normality indicator MS of the
final design.

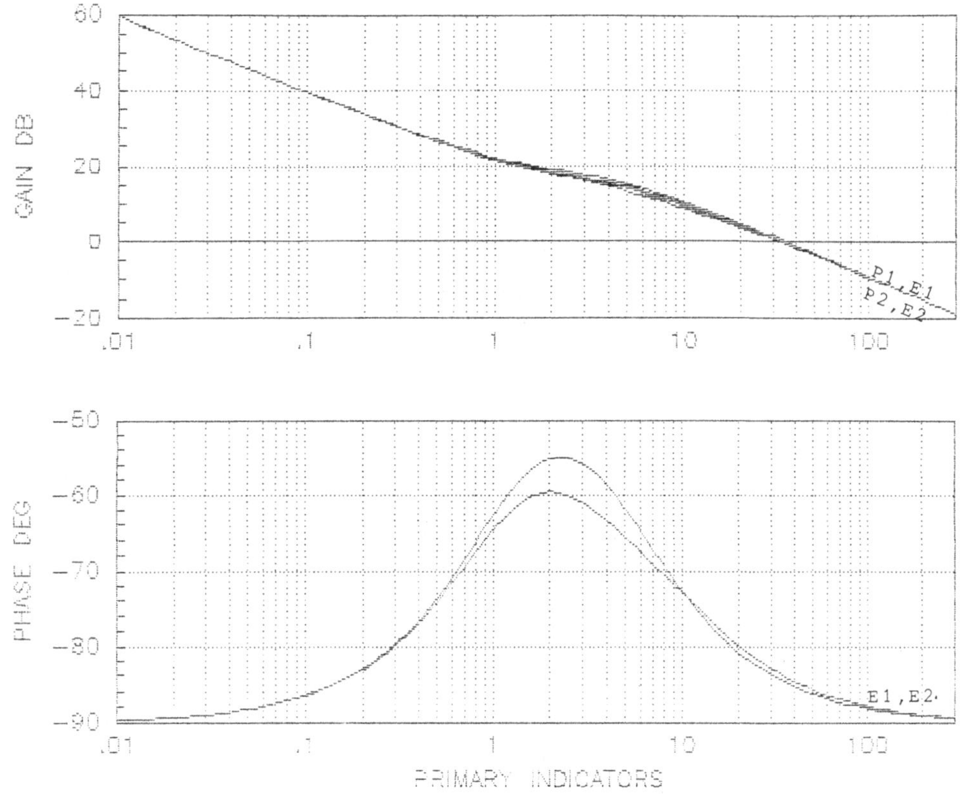

Fig. 7.9.5   The primary indicators of the
final design.

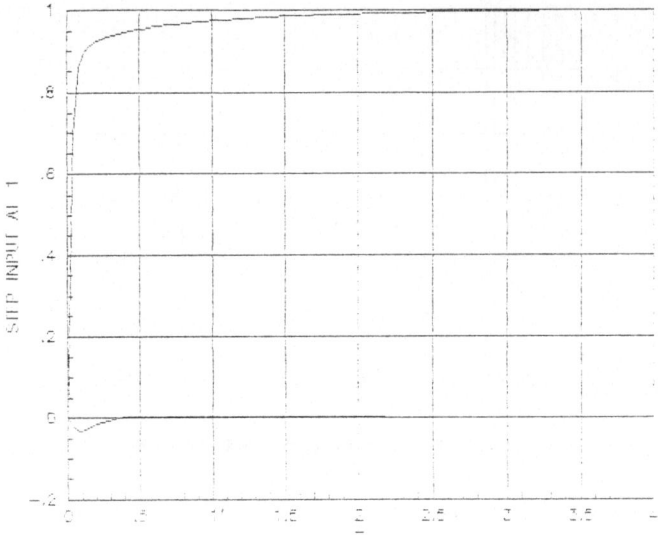

Fig. 7.9.6(a) Closed-loop step response of the
final design (step at input 1).

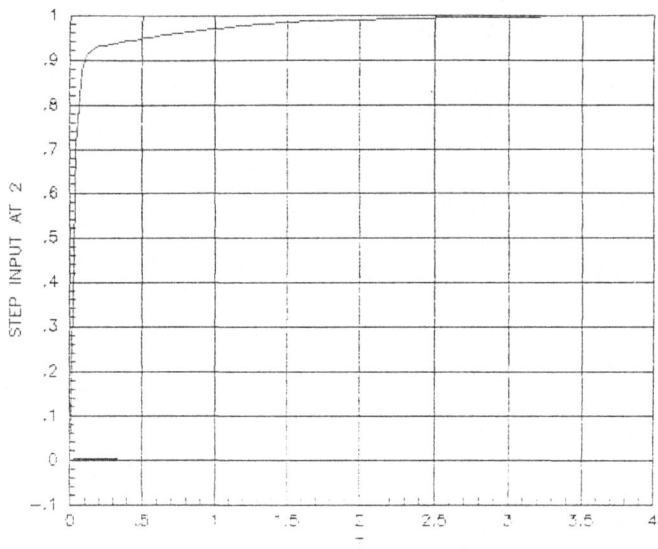

Fig. 7.9.6(b) Closed-loop step response of the
final design (step at input 2).

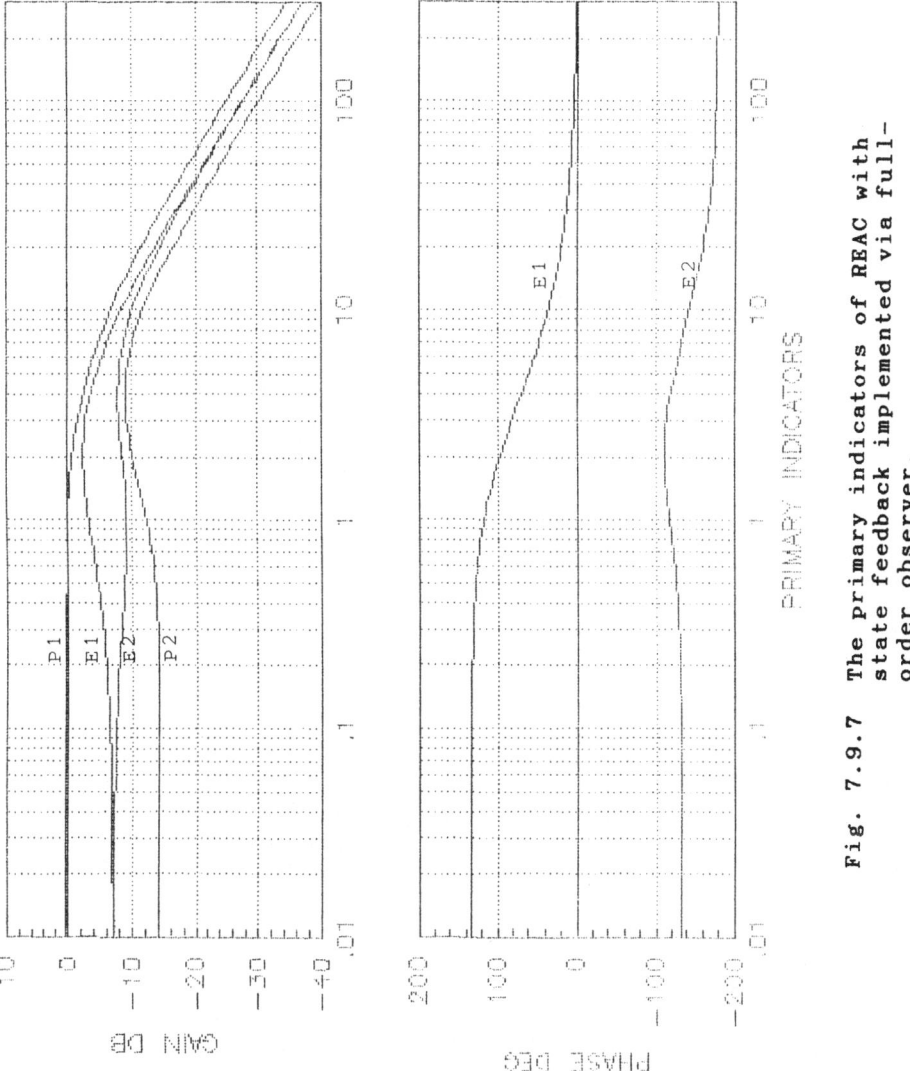

Fig. 7.9.7 The primary indicators of REAC with state feedback implemented via full-order observer.

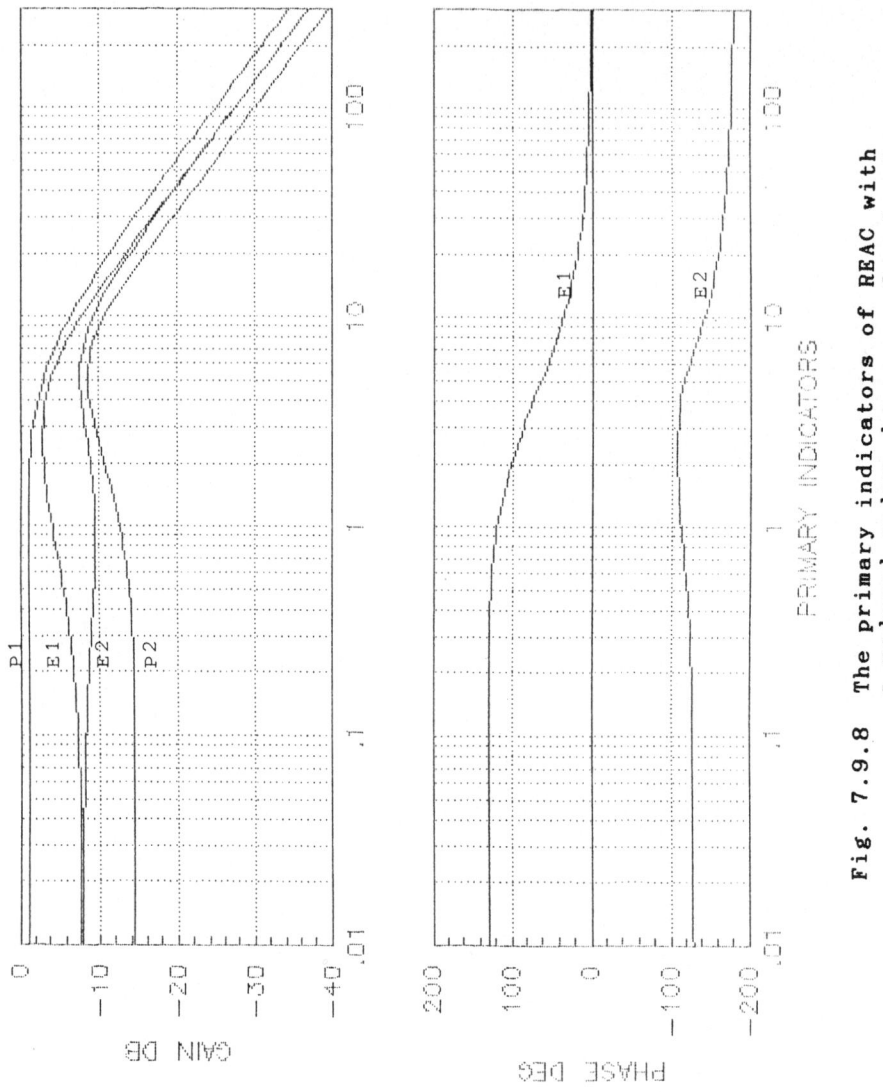

Fig. 7.9.8 The primary indicators of REAC with
a reduced-order observer of order 3.

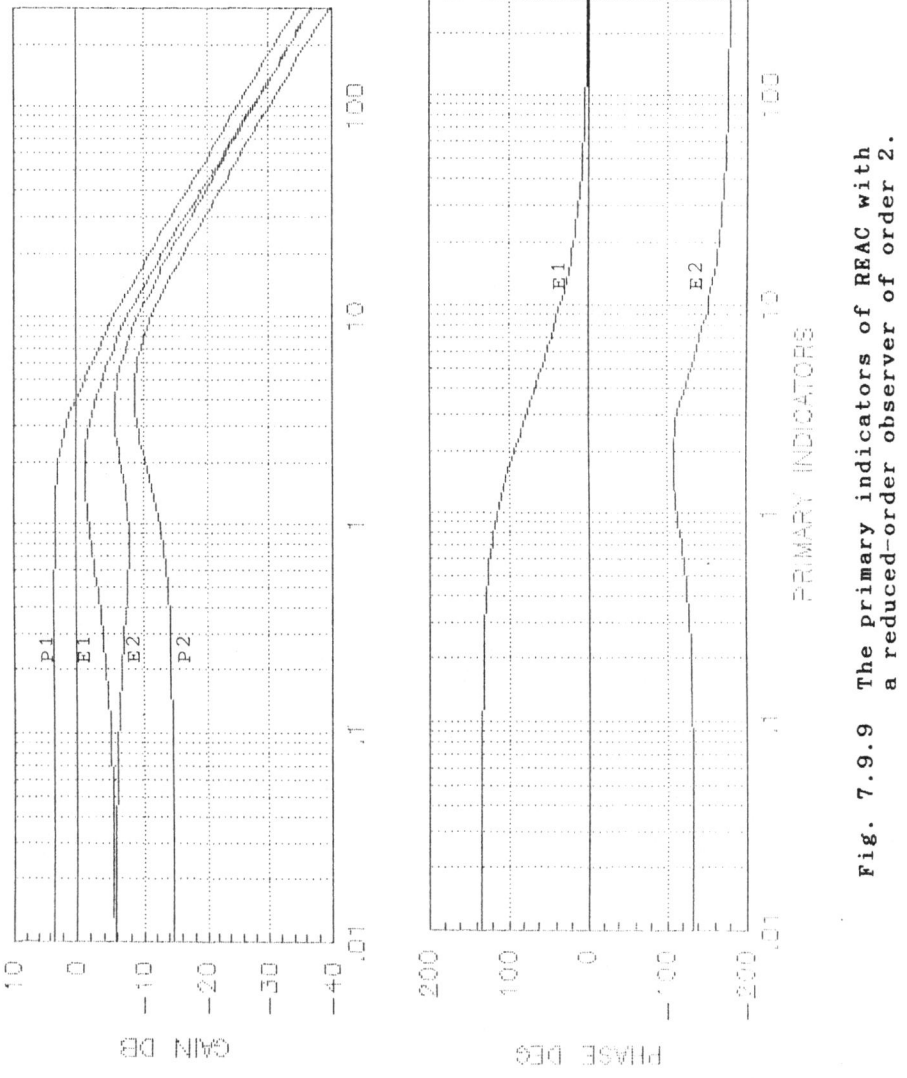

Fig. 7.9.9   The primary indicators of REAC with
a reduced-order observer of order 2.

218

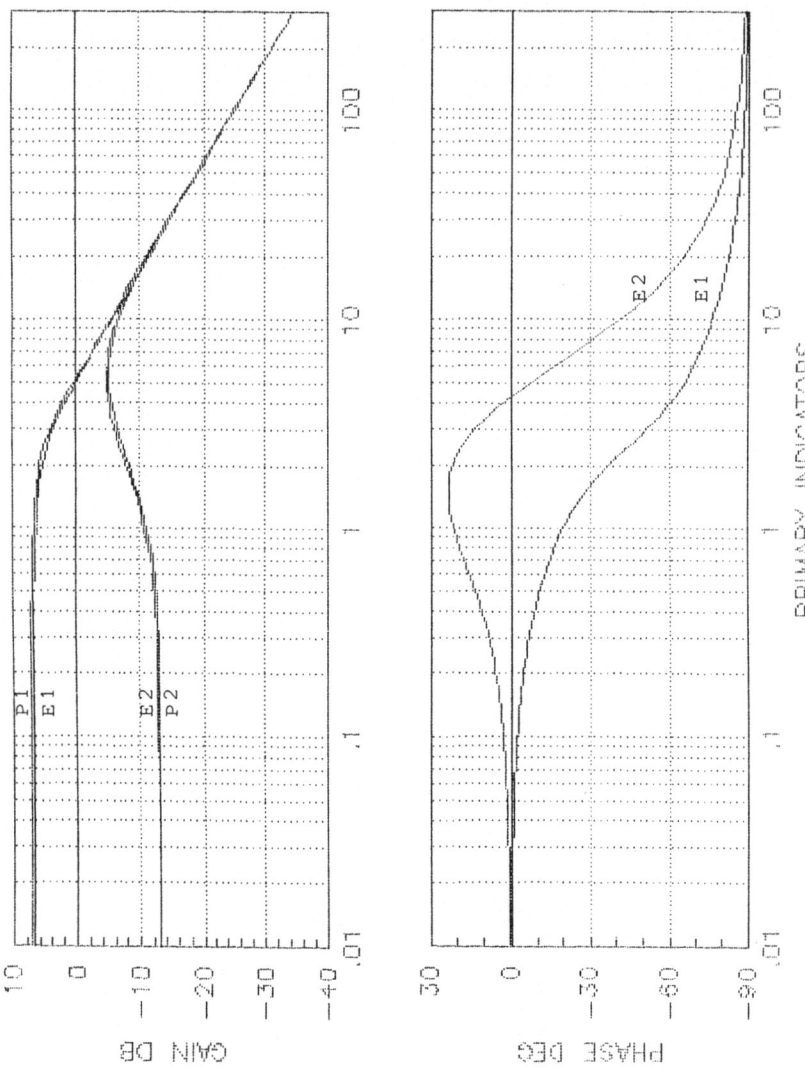

Fig. 7.9.10   The primary indicators after high
frequency region design.

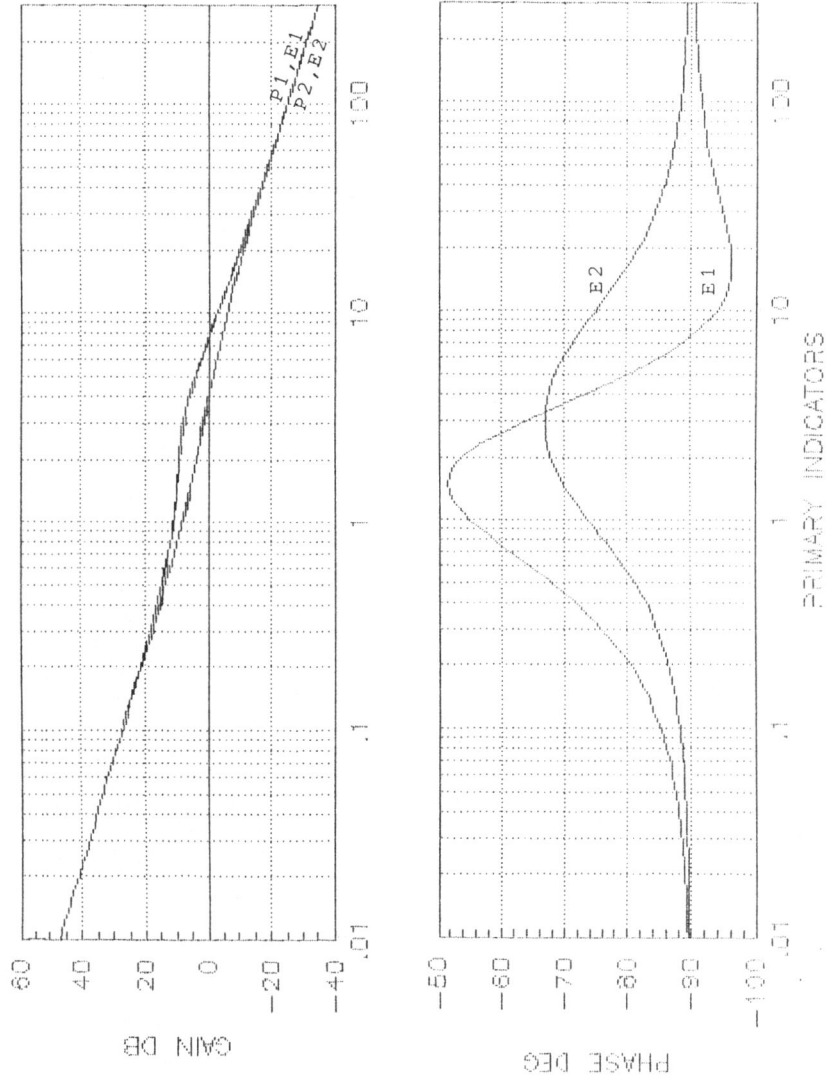

Fig. 7.9.11 The primary indicators after high
and low frequency region design.

220

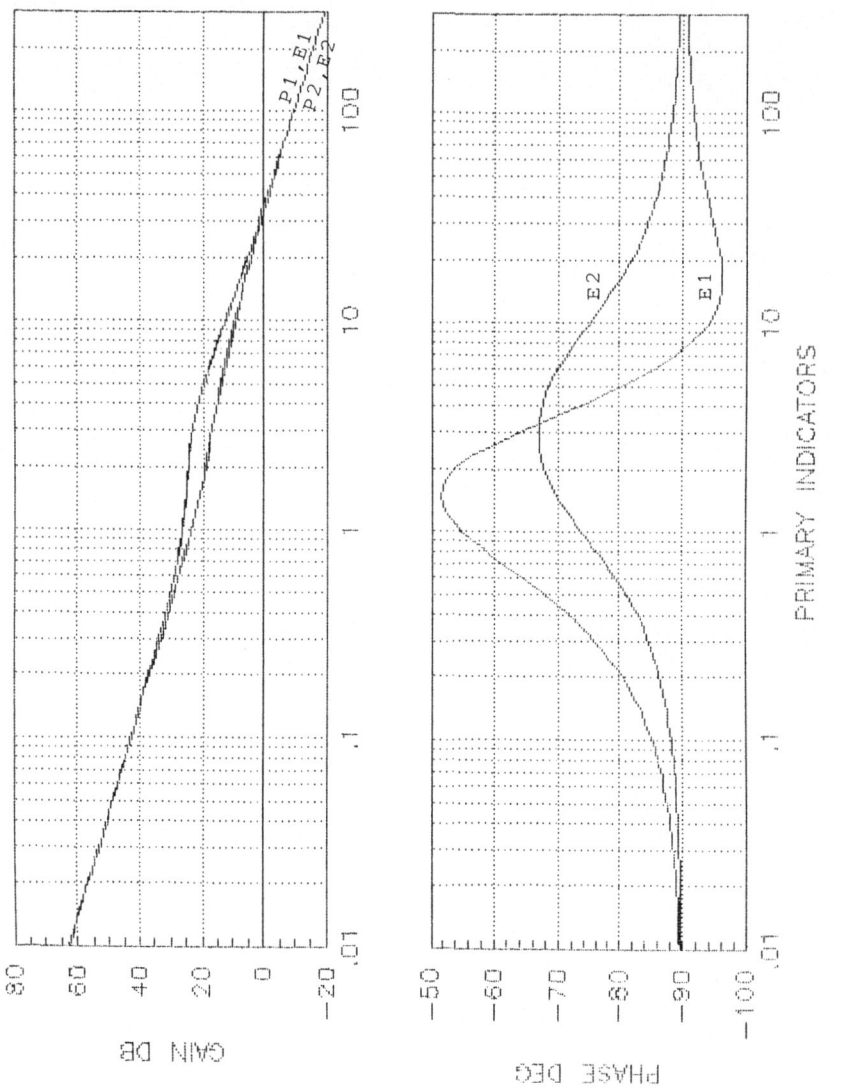

Fig. 7.9.12   The primary indicators of the final design.

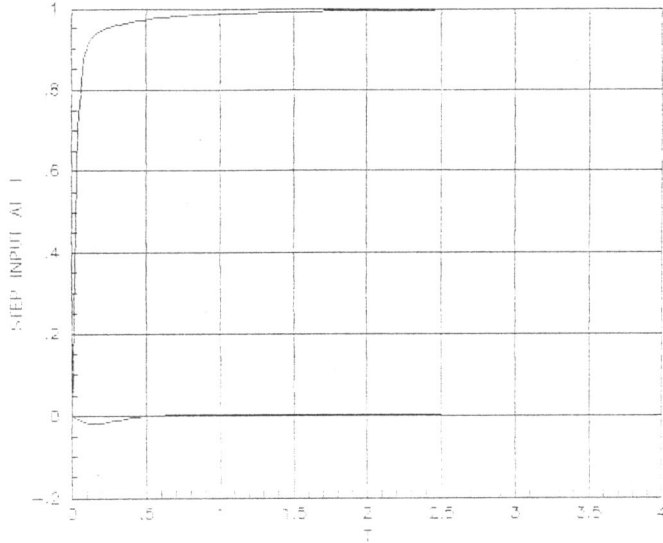

Fig. 7.9.13(a)   Closed-loop step response of the
final design (step at input 1).

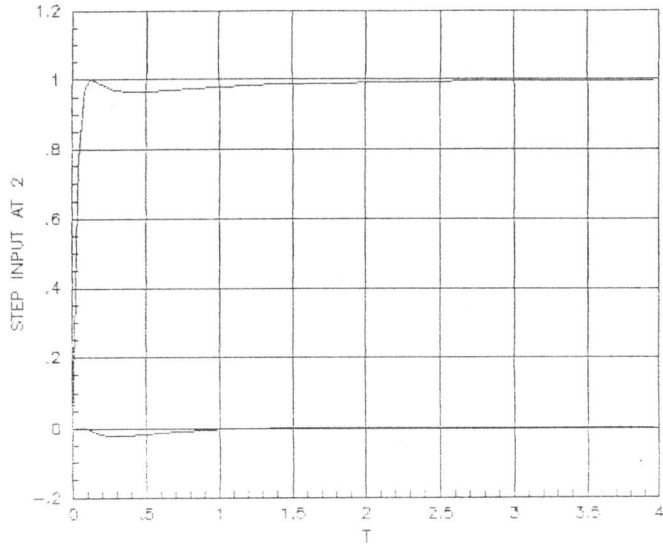

Fig. 7.9.13(b)   Closed-loop step response of the
final design (step at input 2).

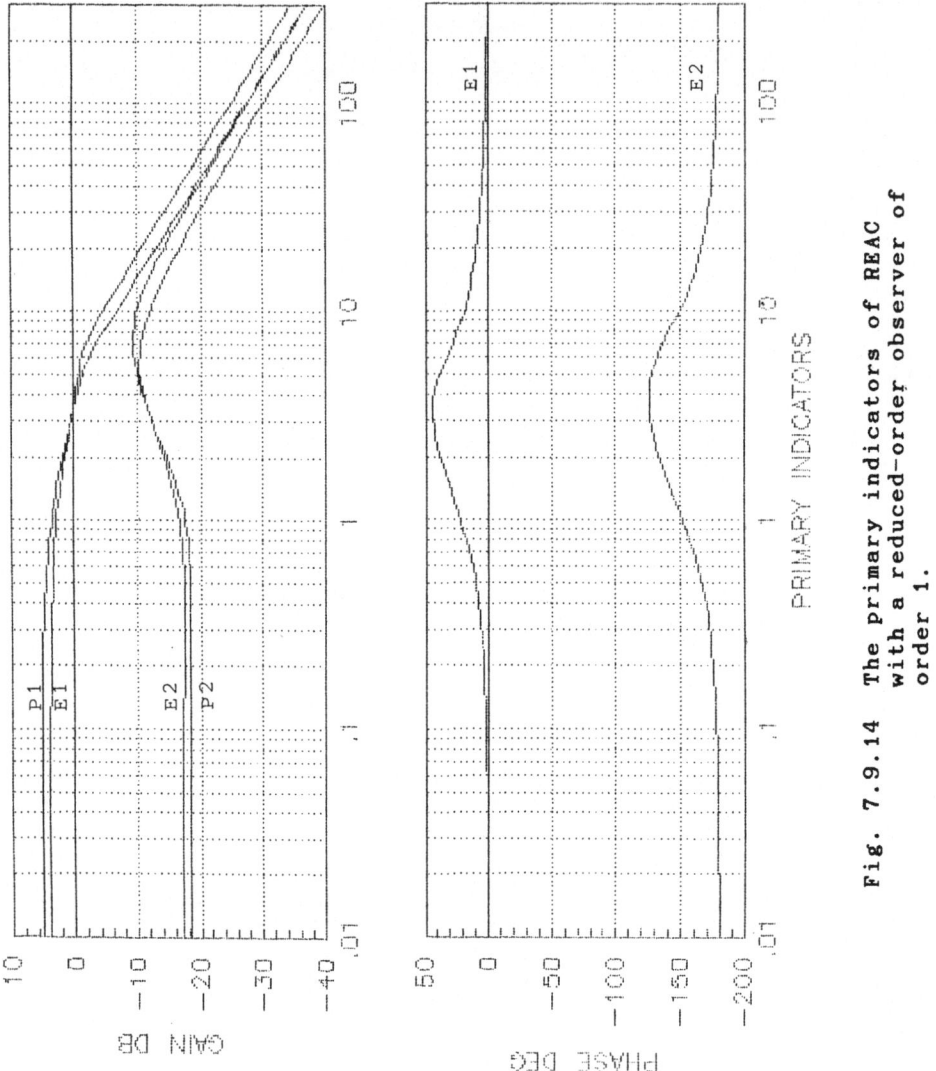

Fig. 7.9.14 The primary indicators of REAC with a reduced-order observer of order 1.

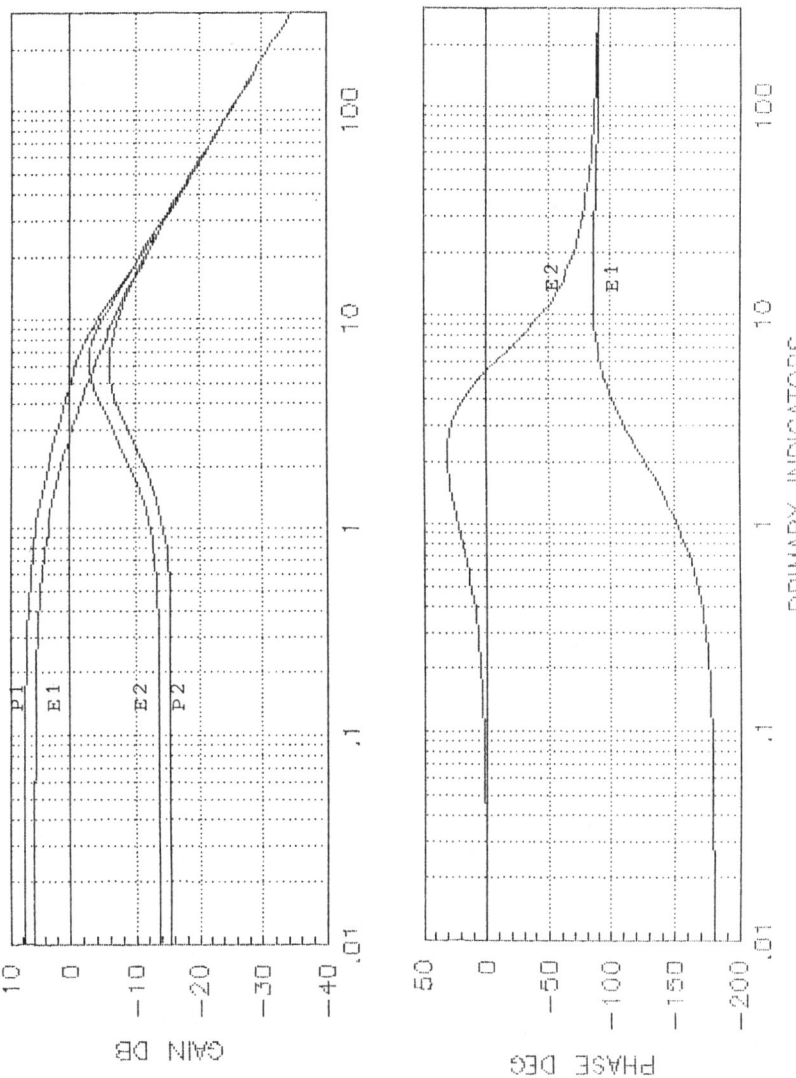

Fig. 7.9.15 The primary indicators after high frequency region design.

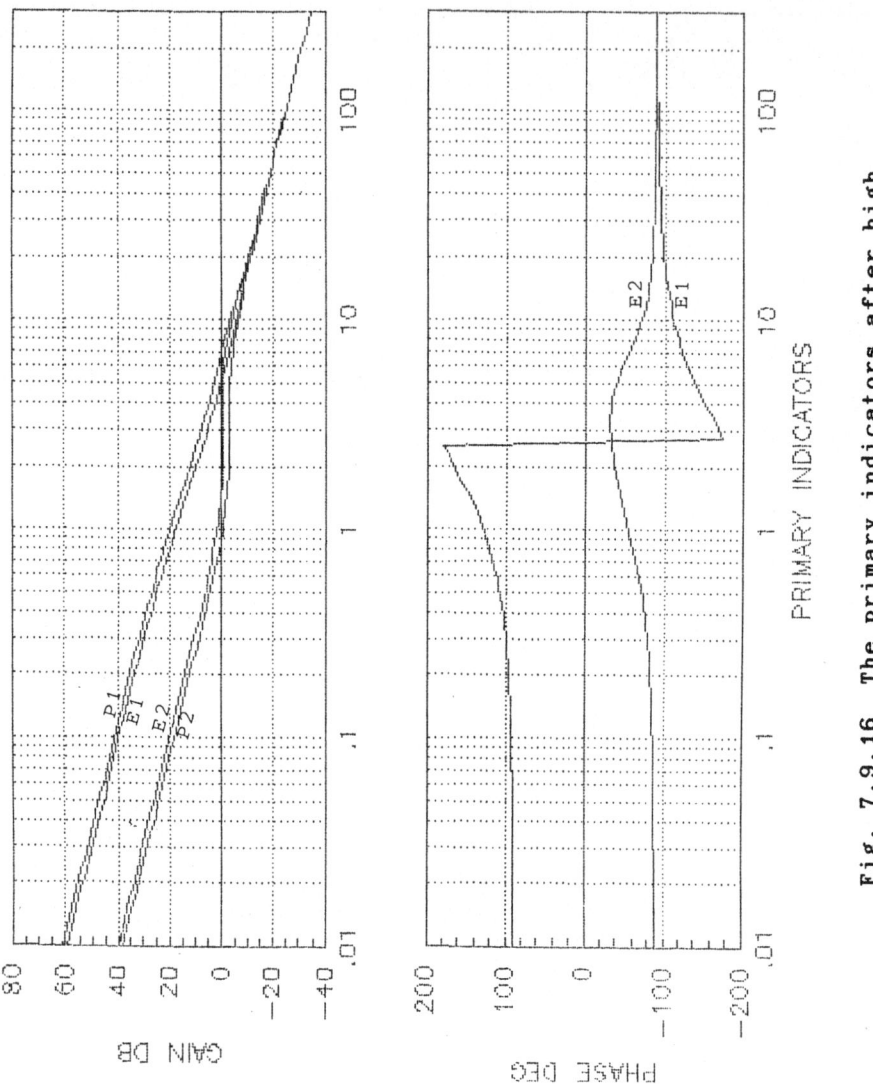

Fig. 7.9.16   The primary indicators after high
and low frequency region design.

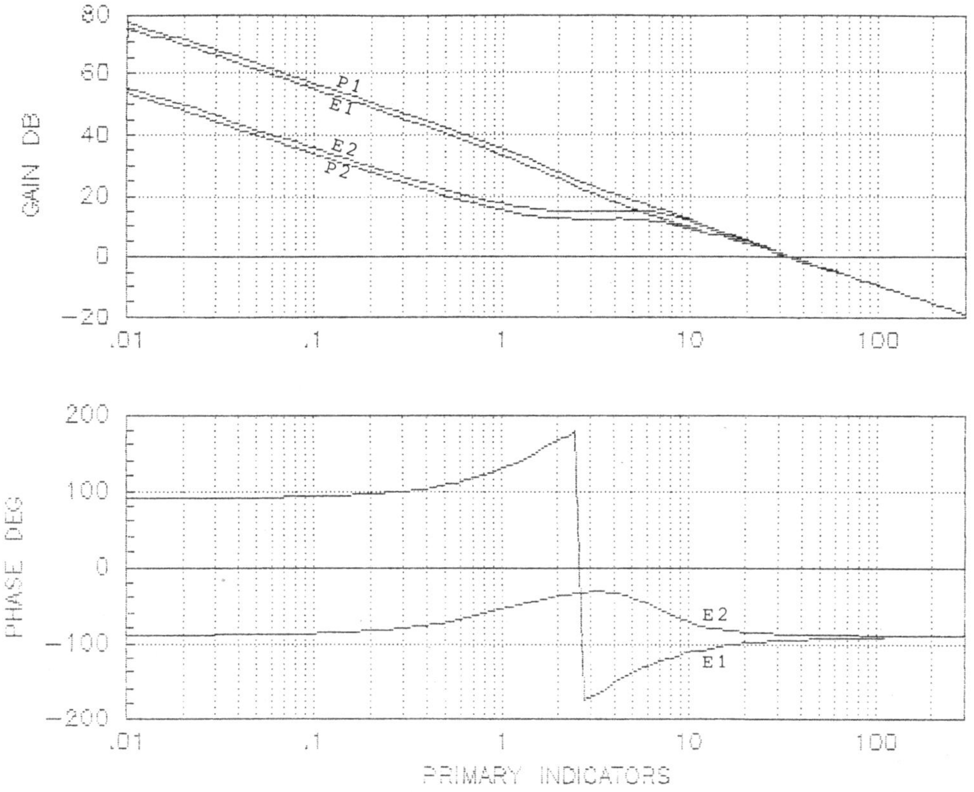

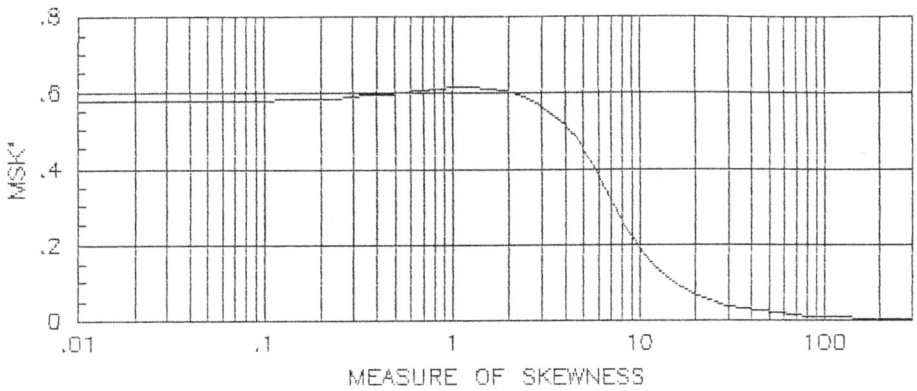

Fig. 7.9.17   The primary indicators of the
final design.

Fig. 7.9.18   The normality indicator MS of the
final design.

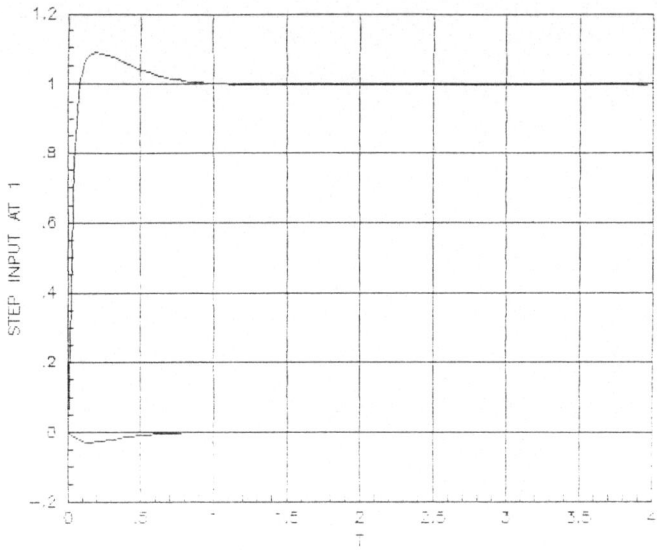

Fig. 7.9.19(a)   Closed-loop step response of the
final design (step at input 1).

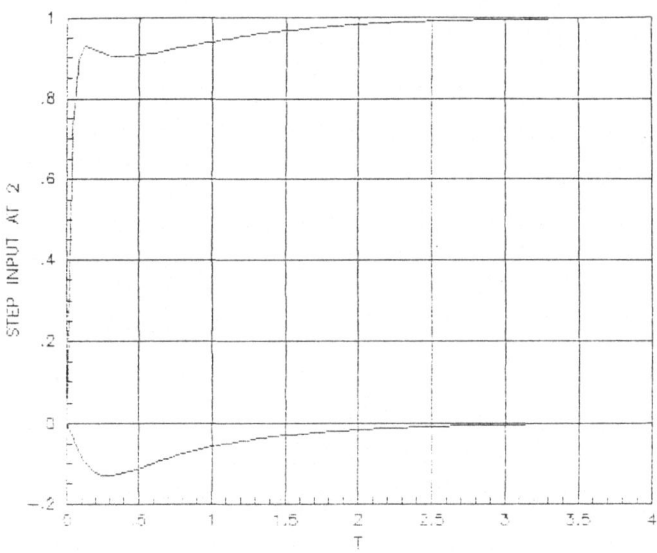

Fig. 7.9.19(b)   Closed-loop step response of the
final design (step at input 2).

227

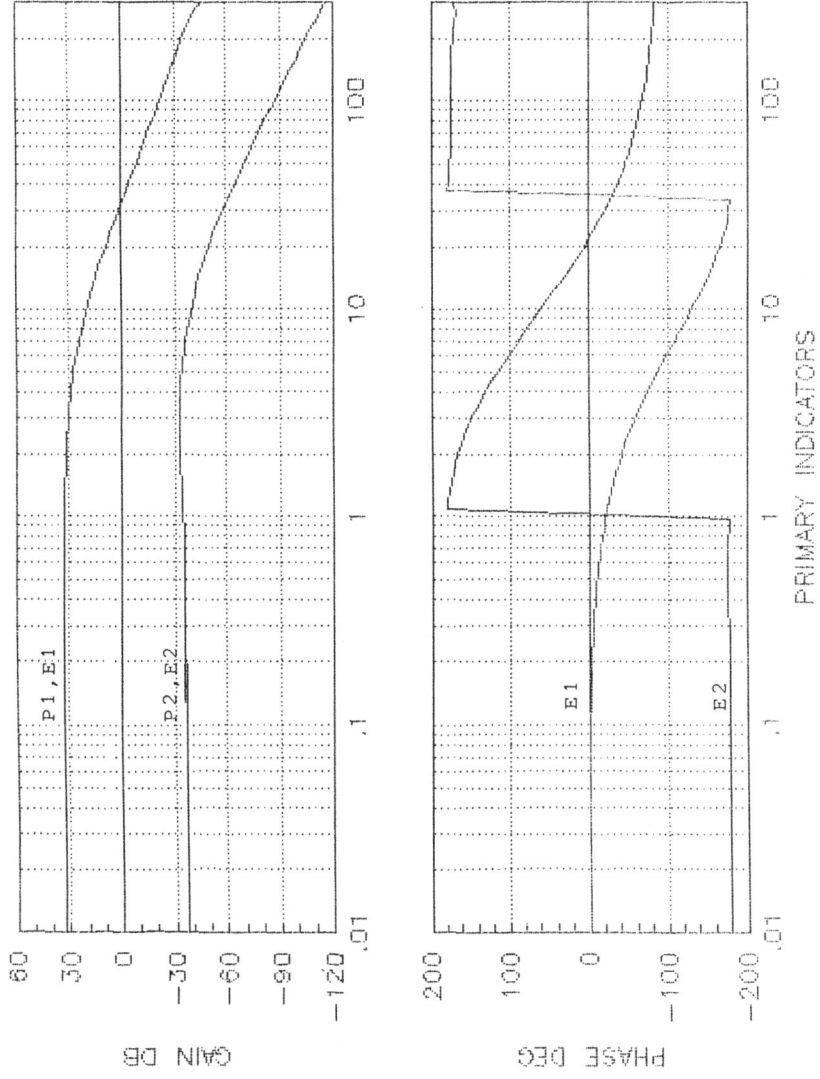

Fig. 7.11.1 The primary indicators of TGEN with full state feedback.

228

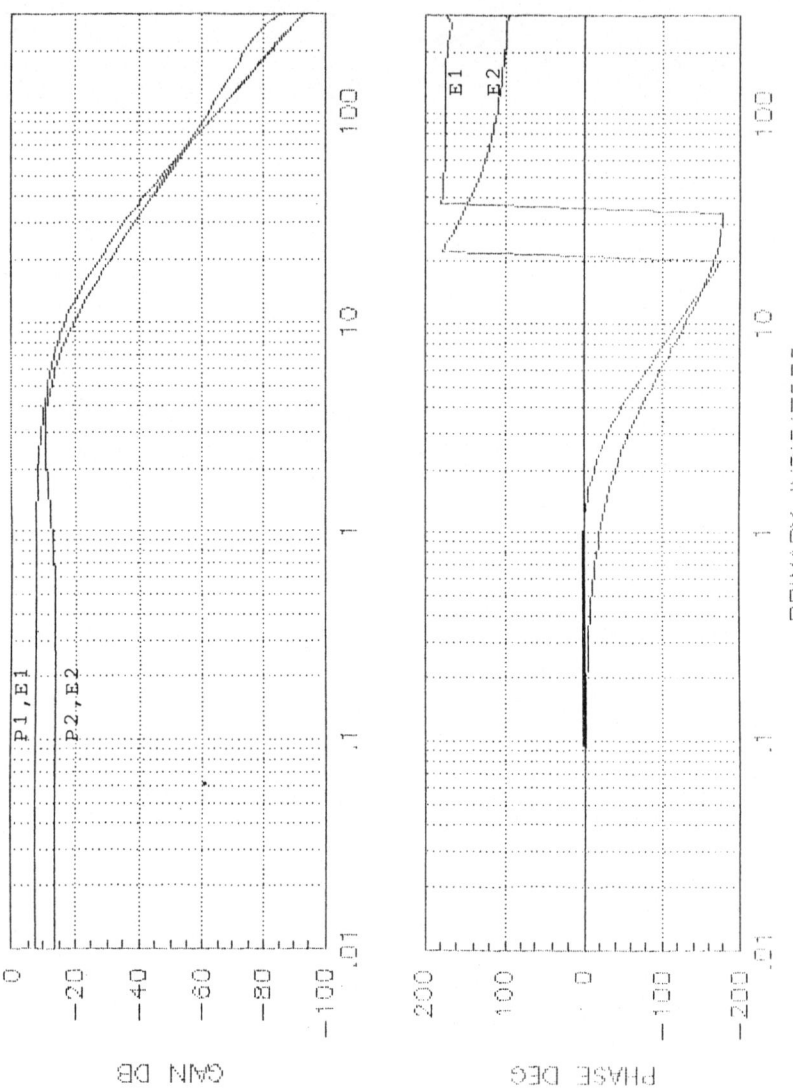

Fig. 7.11.2 The primary indicators after high
frequency region design.

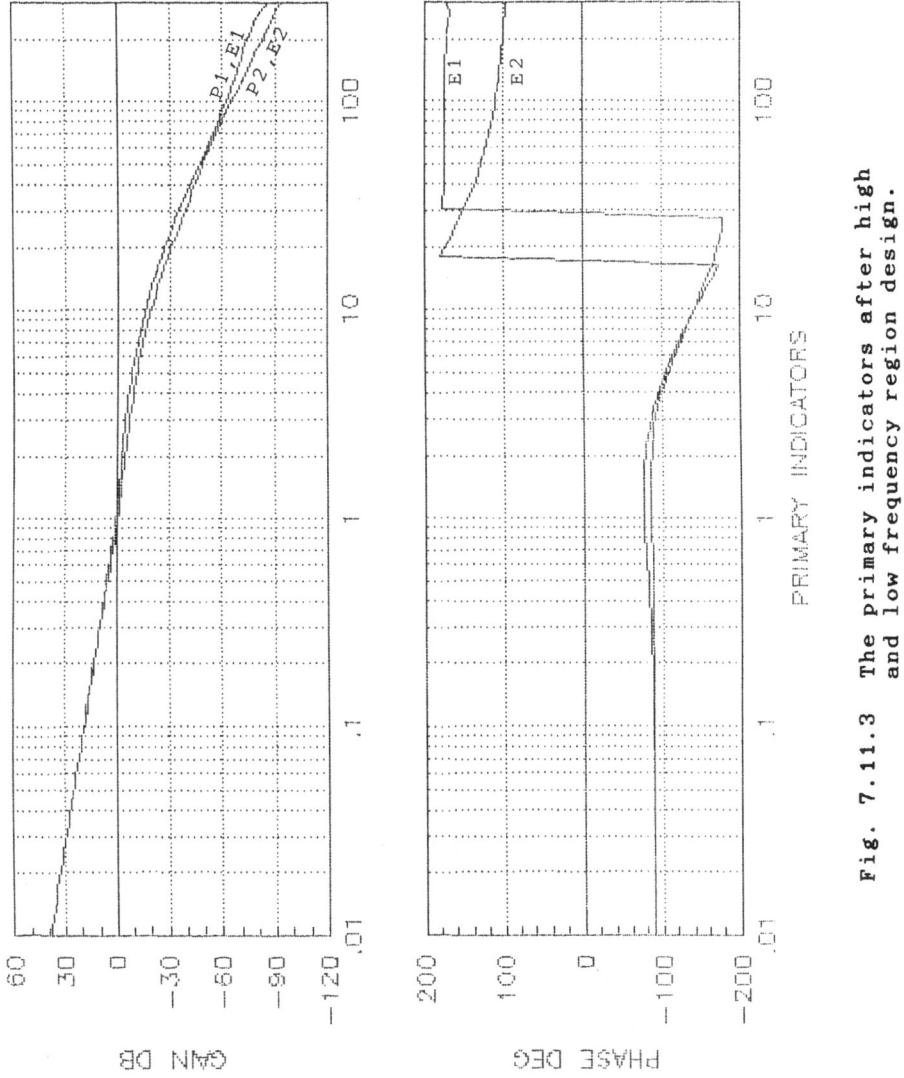

Fig. 7.11.3   The primary indicators after high
and low frequency region design.

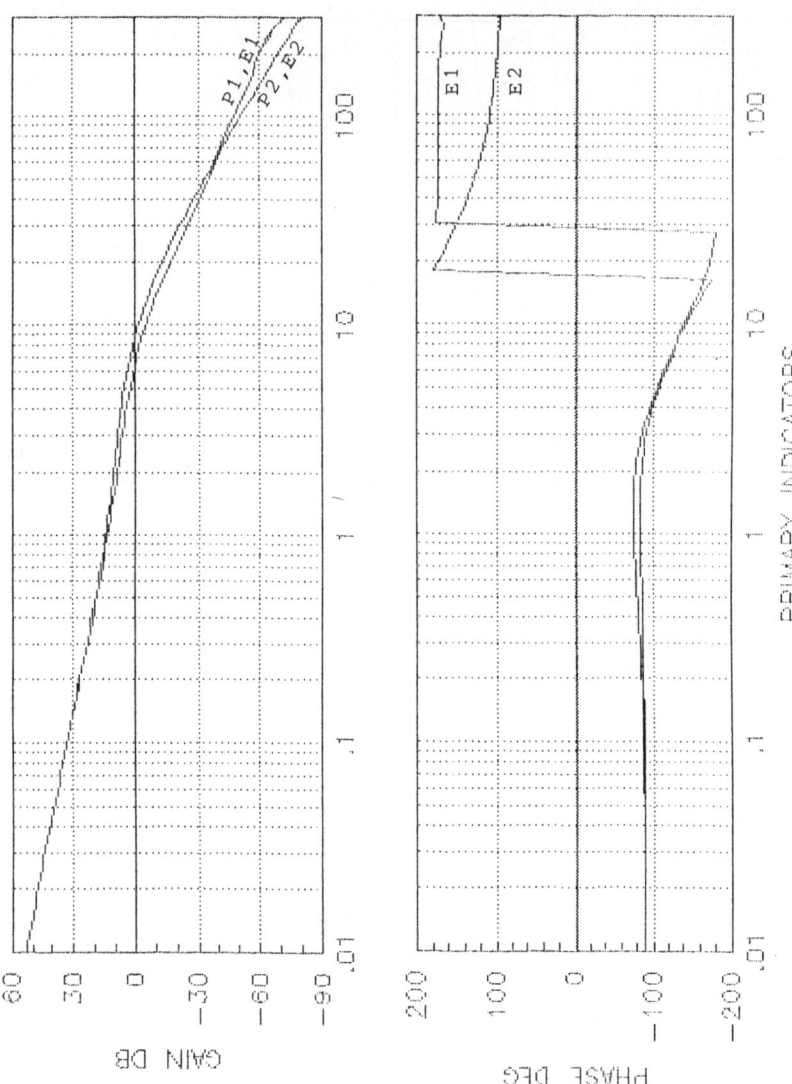

Fig. 7.11.4 The primary indicators of the final design.

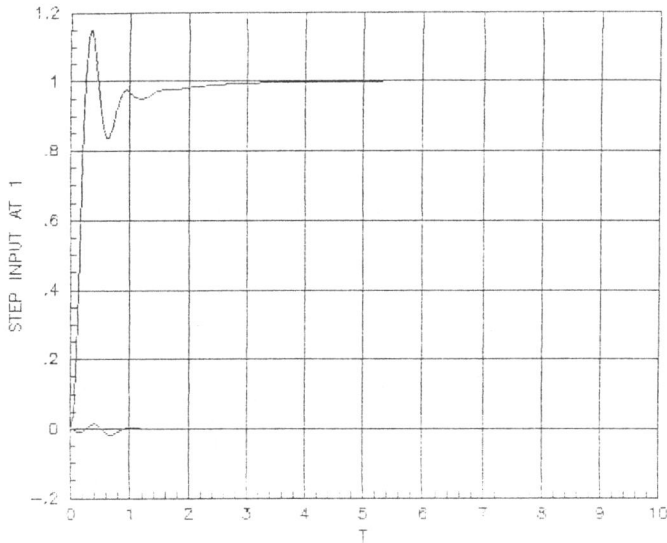

Fig. 7.11.5(a)   Closed-loop step response of
                 final design (step at input 1).

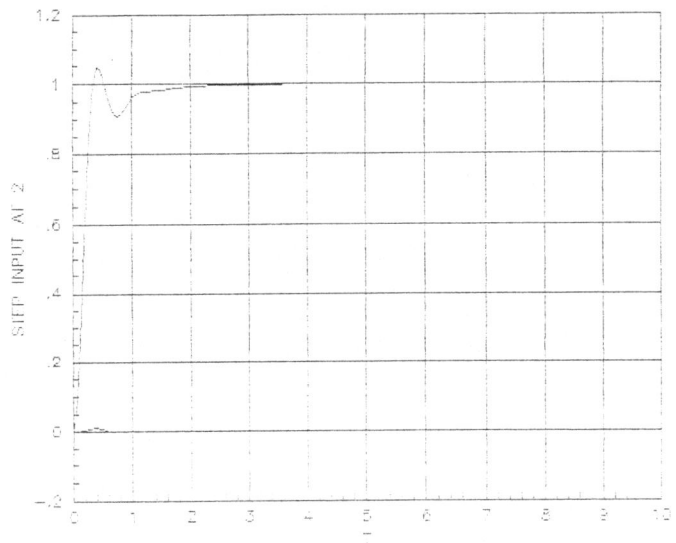

Fig. 7.11.5(b)   Closed-loop step response of
                 final design (step at input 2).

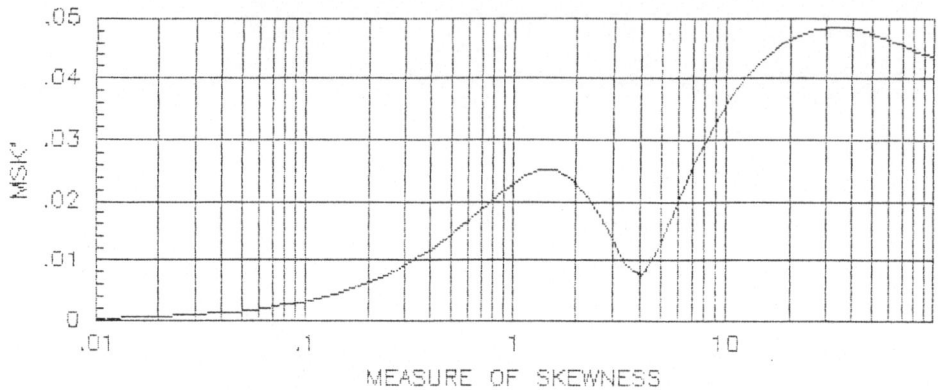

Fig. 7.11.6   The normality indicator MS of
the final design.

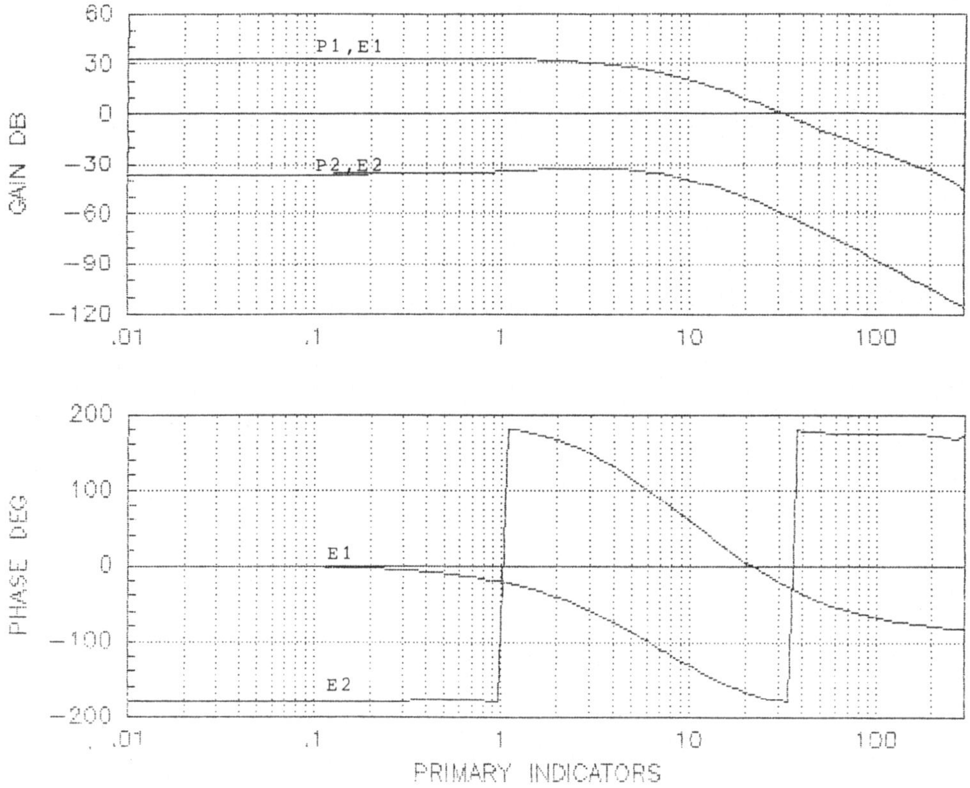

Fig. 7.11.7   The primary indicators of TGEN
with state feedback implemented
via full-order observer.

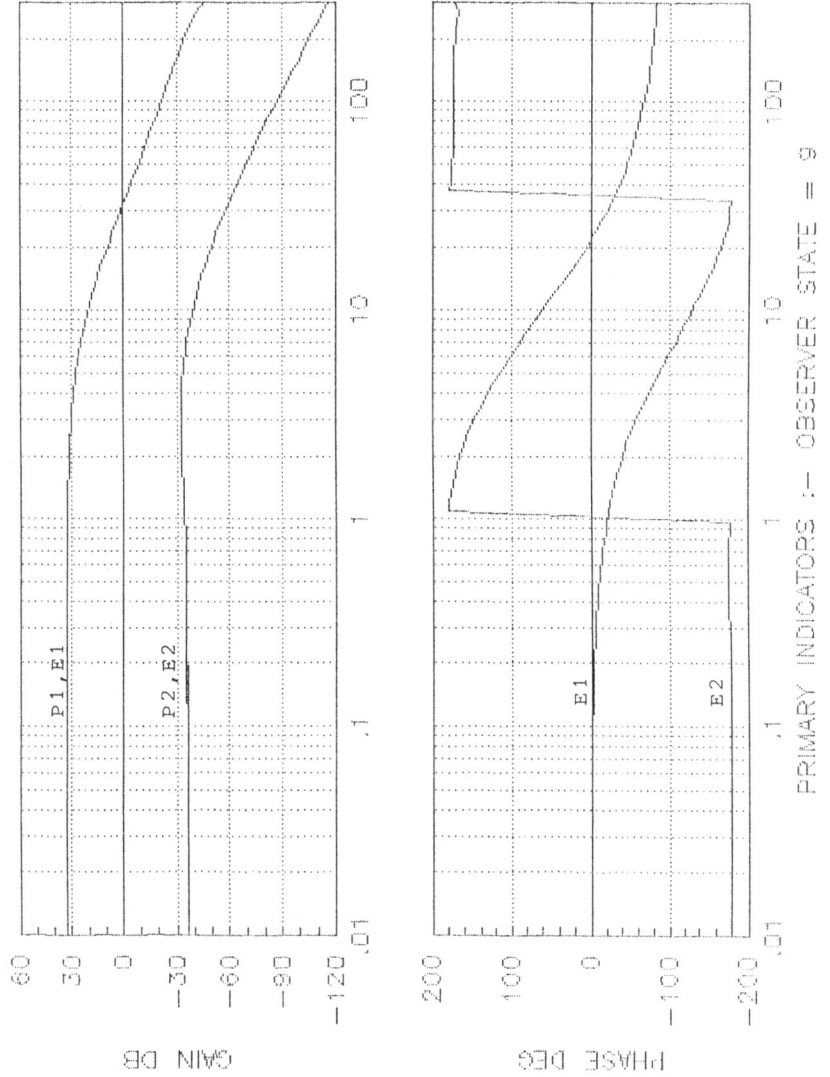

Fig. 7.11.8   The primary indicators of TGEN with
a reduced-order observer of order 9.

234

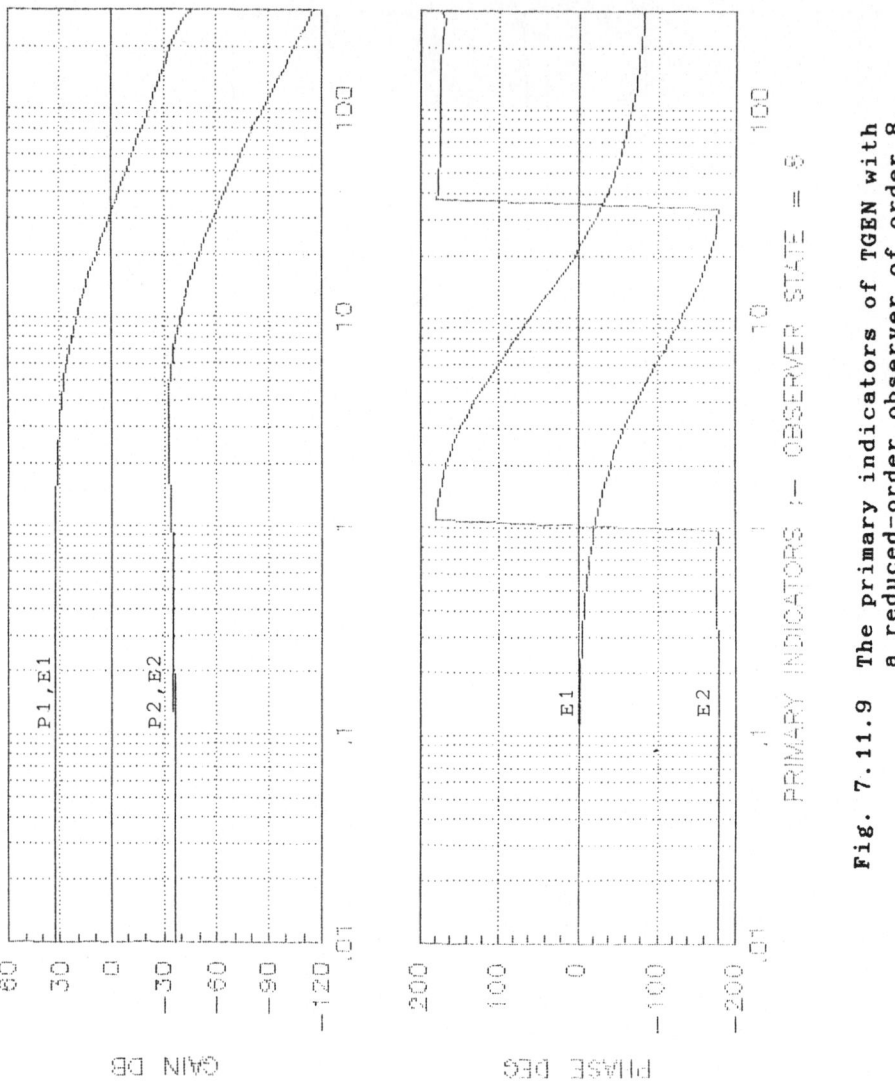

Fig. 7.11.9 The primary indicators of TGEN with a reduced-order observer of order 8.

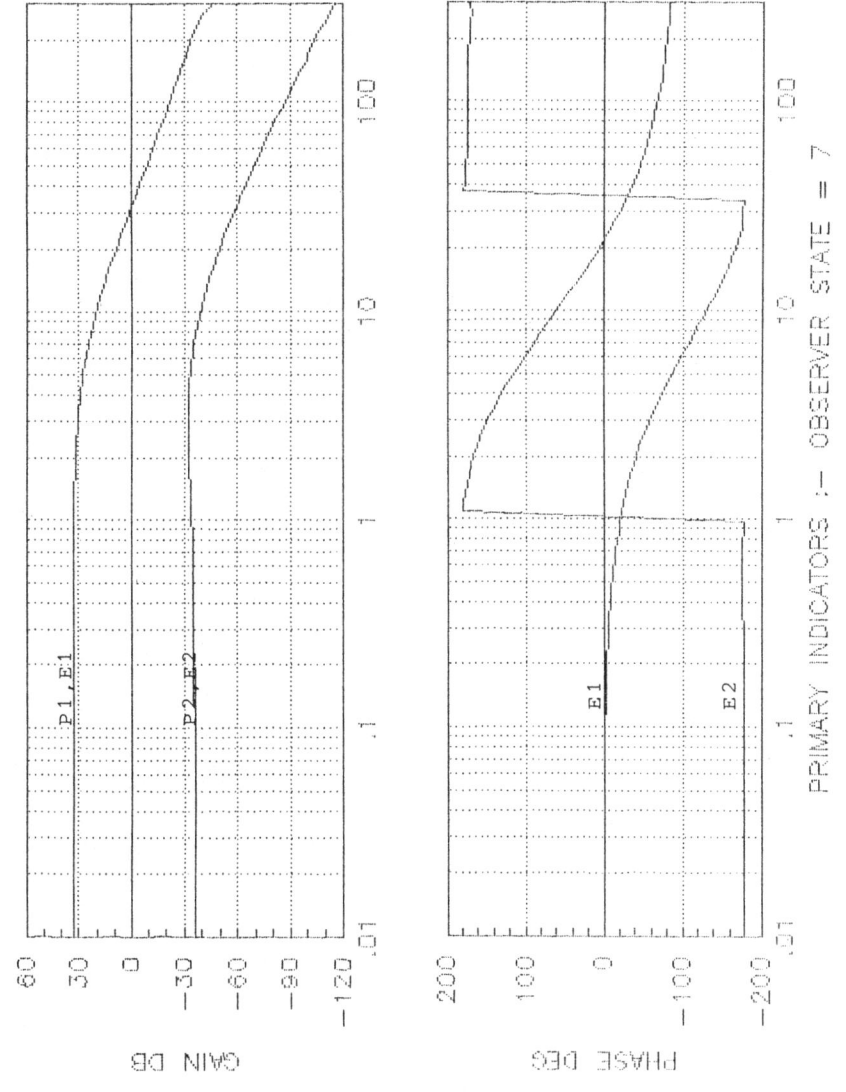

PRIMARY INDICATORS :- OBSERVER STATE = 7

Fig. 7.11.10 The primary indicators of TGEN with a reduced-order observer of order 7.

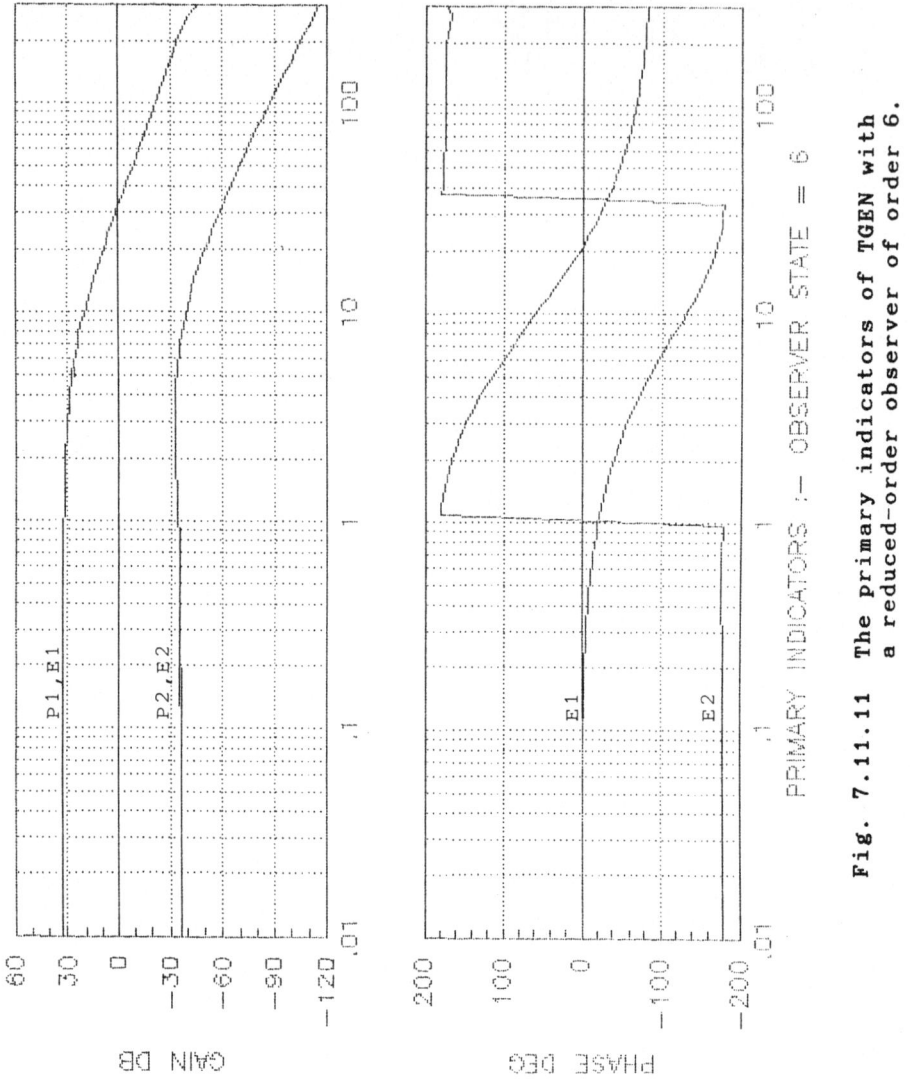

Fig. 7.11.11 The primary indicators of TGEN with a reduced-order observer of order 6.

237

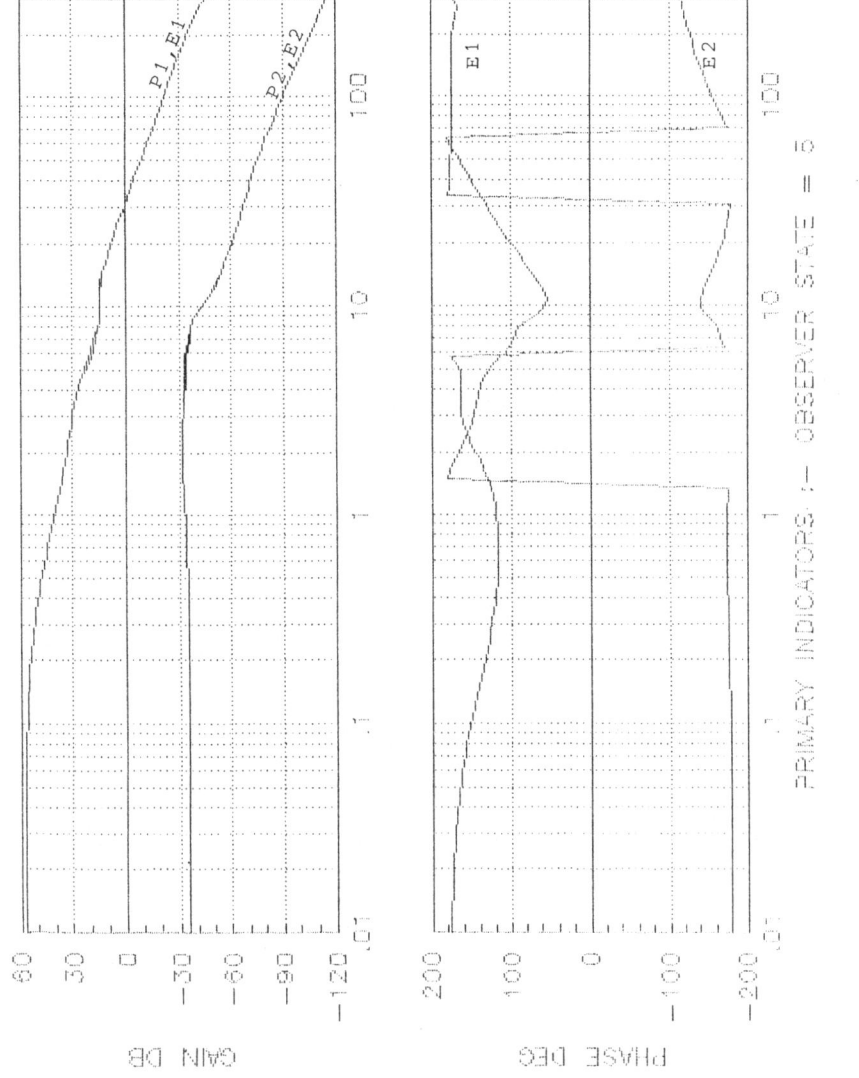

PRIMARY INDICATORS :- OBSERVER STATE = 5

Fig. 7.11.12 The primary indicators of TGEN with
a reduced-order observer of order 5.

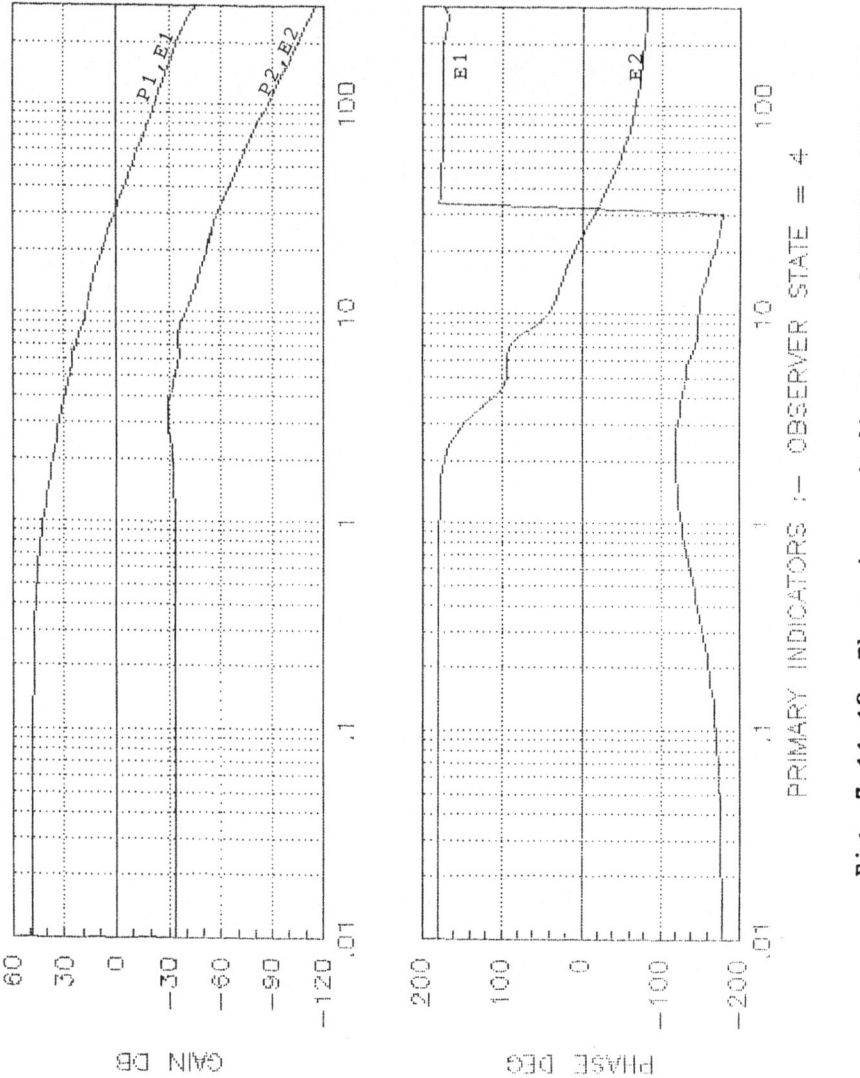

Fig. 7.11.13 The primary indicators of TGEN with a reduced-order observer of order 4.

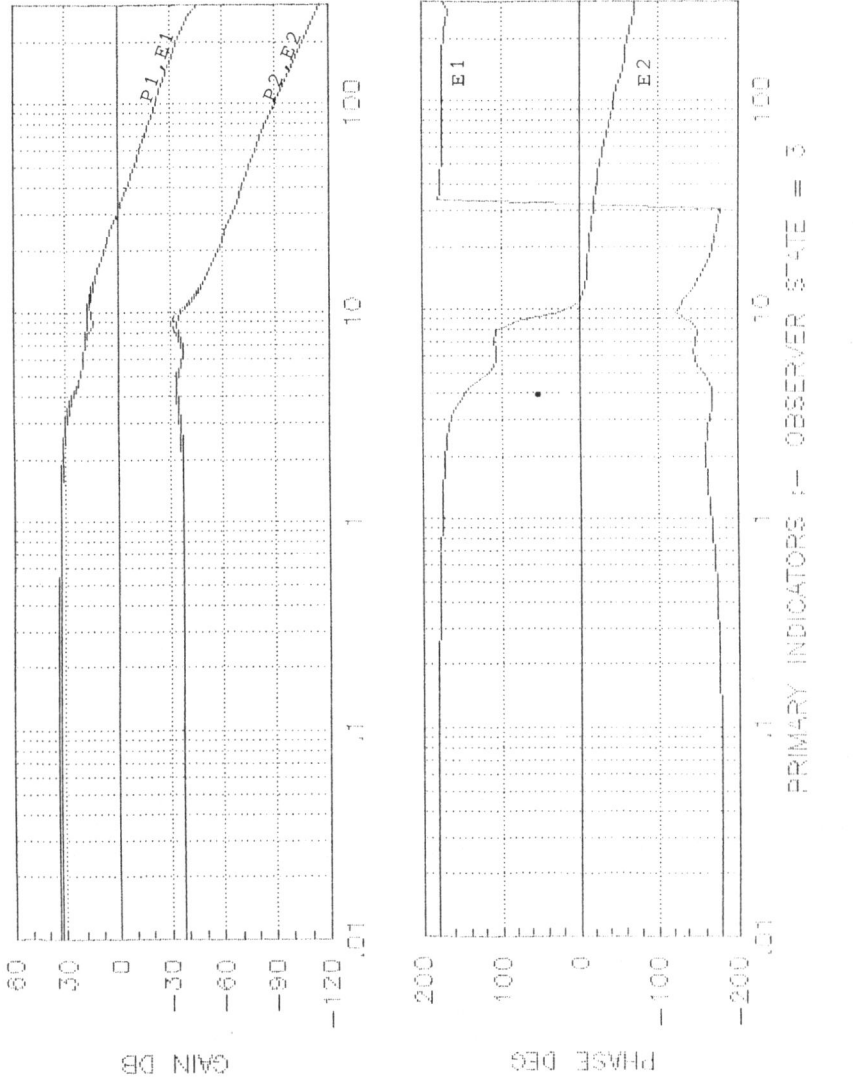

Fig. 7.11.14   The primary indicators of TGEN with
a reduced-order observer of order 3.

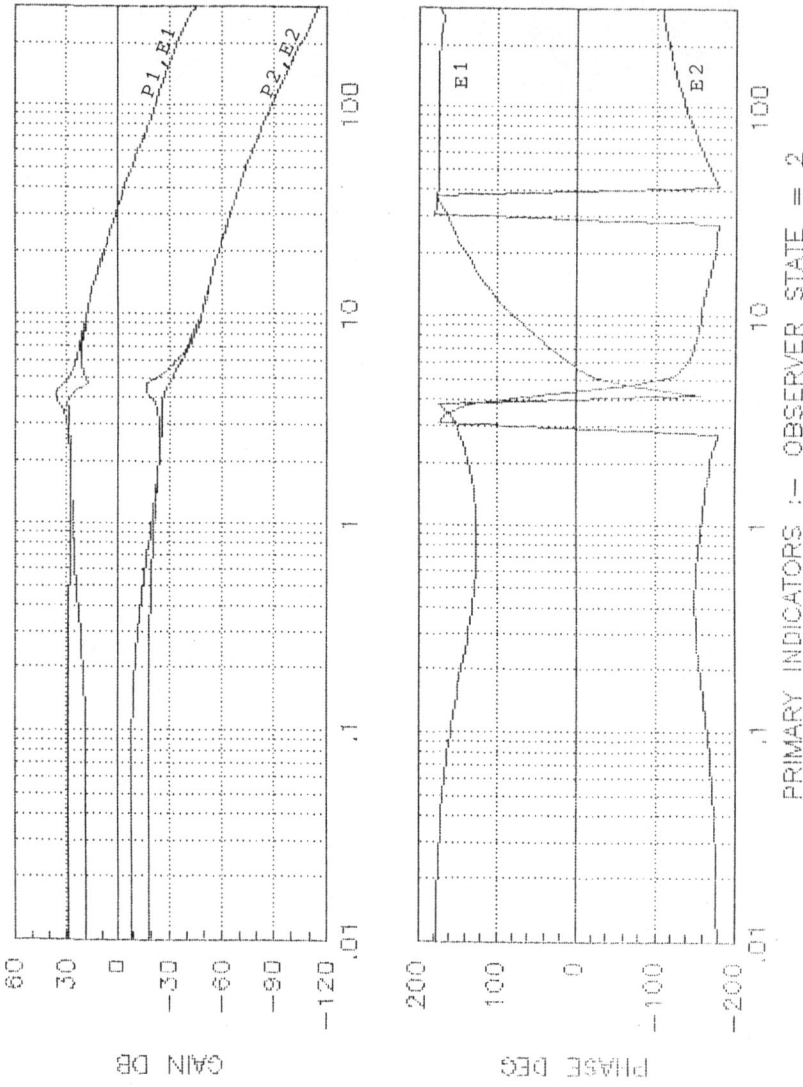

PRIMARY INDICATORS :- OBSERVER STATE = 2

Fig. 7.11.15   The primary indicators of TGEN with
a reduced-order observer of order 2.

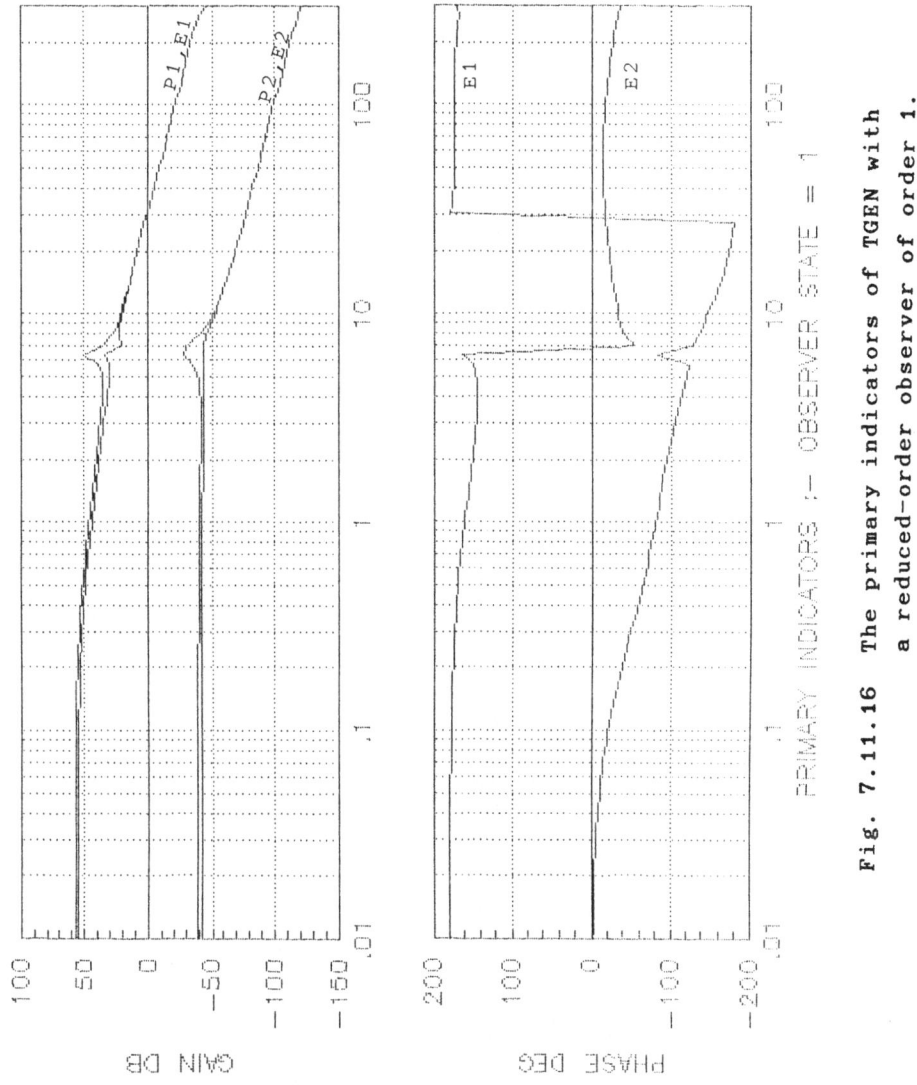

Fig. 7.11.16   The primary indicators of **TGEN** with
a reduced-order observer of order 1.

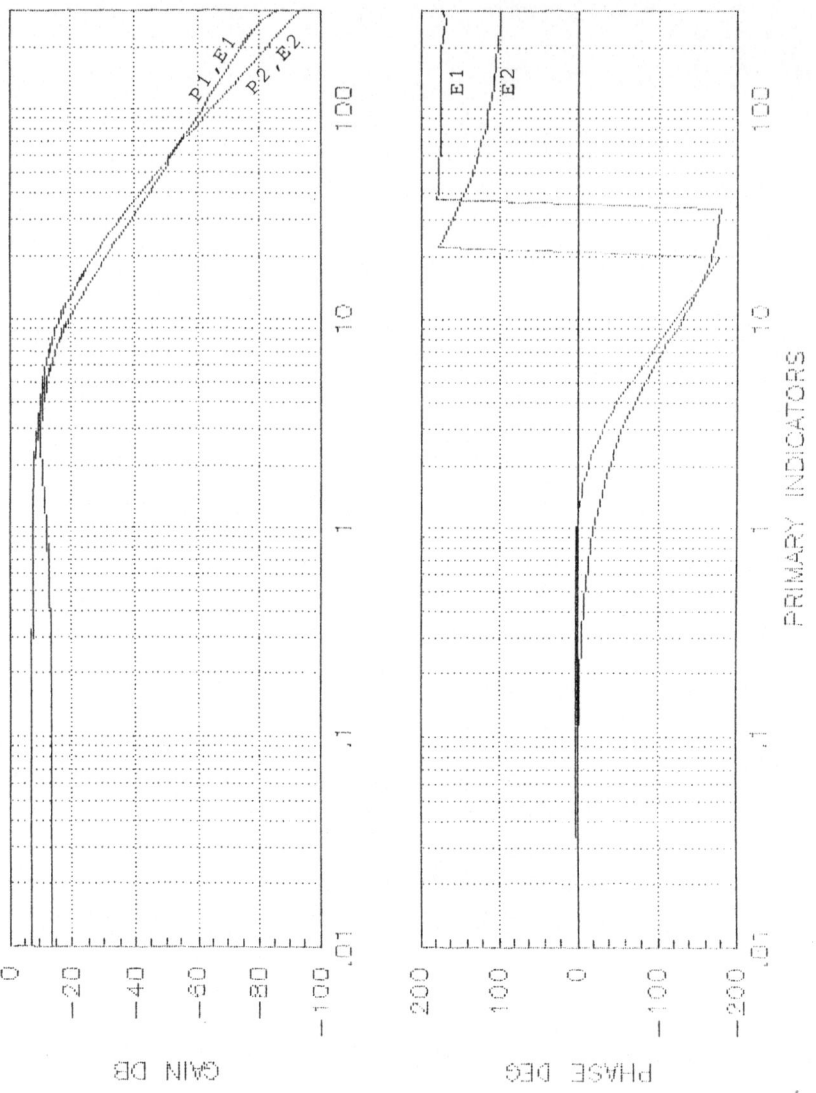

Fig. 7.11.17 The primary indicators after high frequency region design.

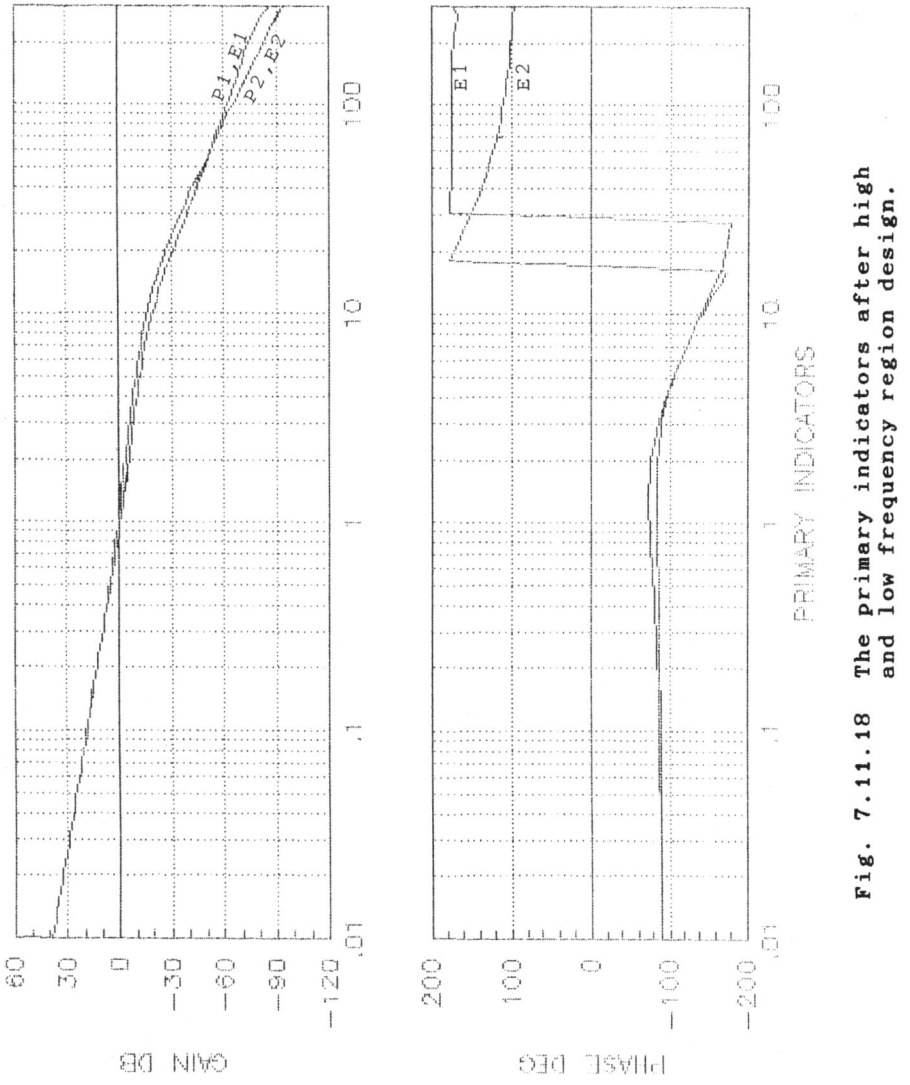

Fig. 7.11.18   The primary indicators after high and low frequency region design.

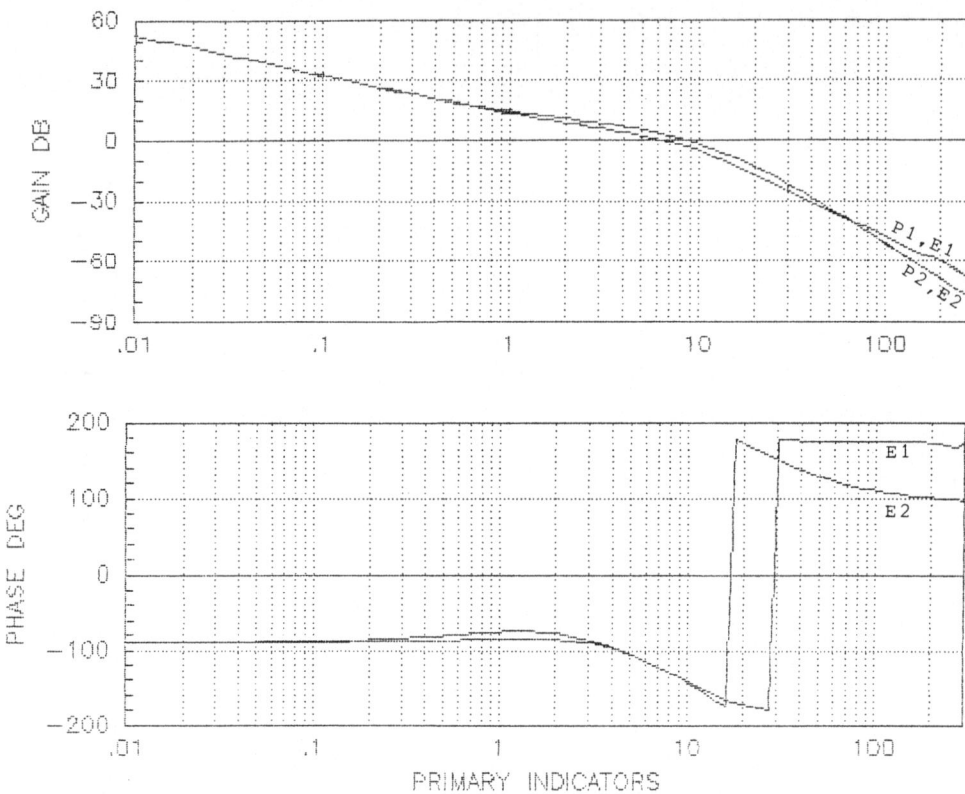

Fig. 7.11.19   The primary indicators of the
final design.

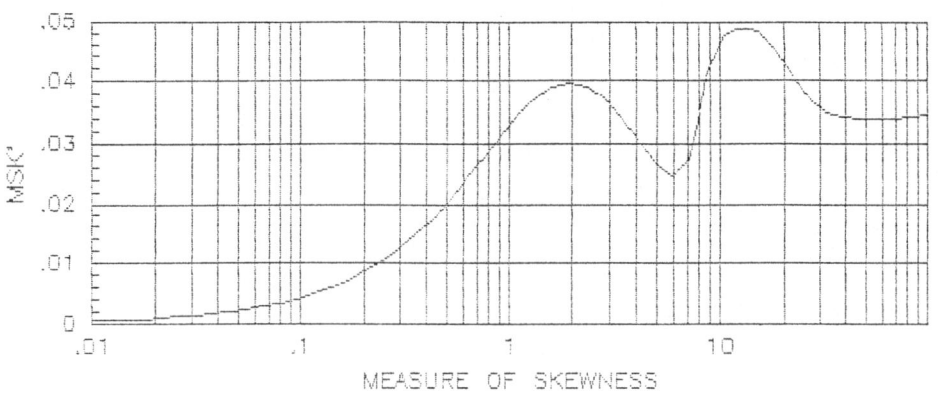

Fig. 7.11.20   The normality indicator MS of
the final design.

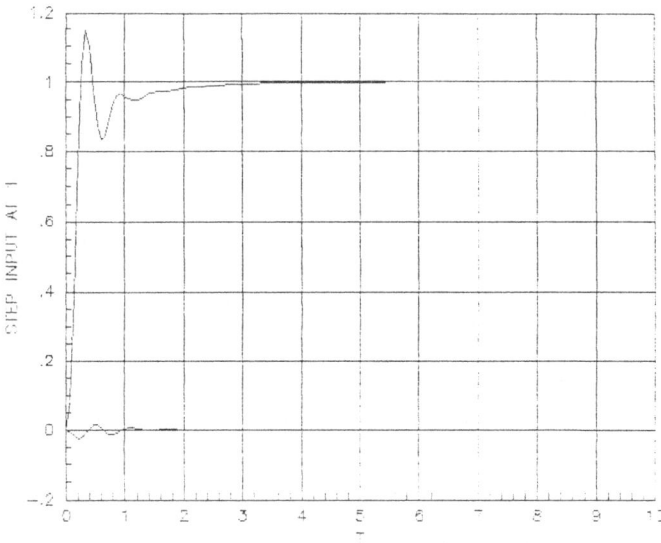

Fig. 7.11.21(a)   Closed-loop step response of the
                  final design (step at input 1).

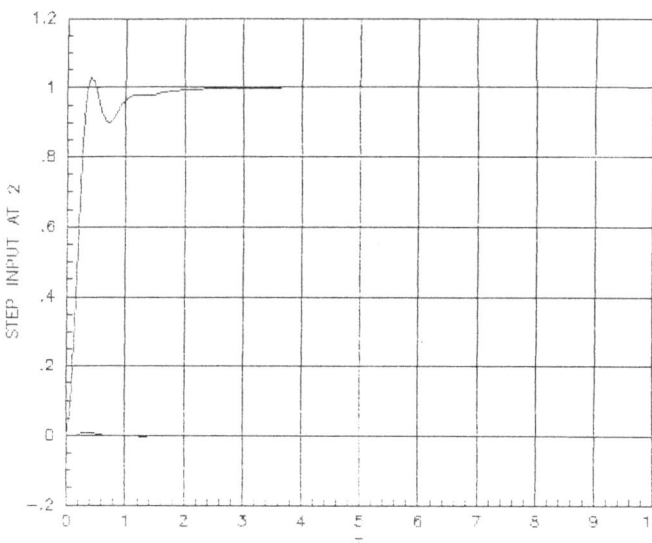

Fig. 7.11.21(b)   Closed-loop step response of the
                  final design (step at input 2).

246

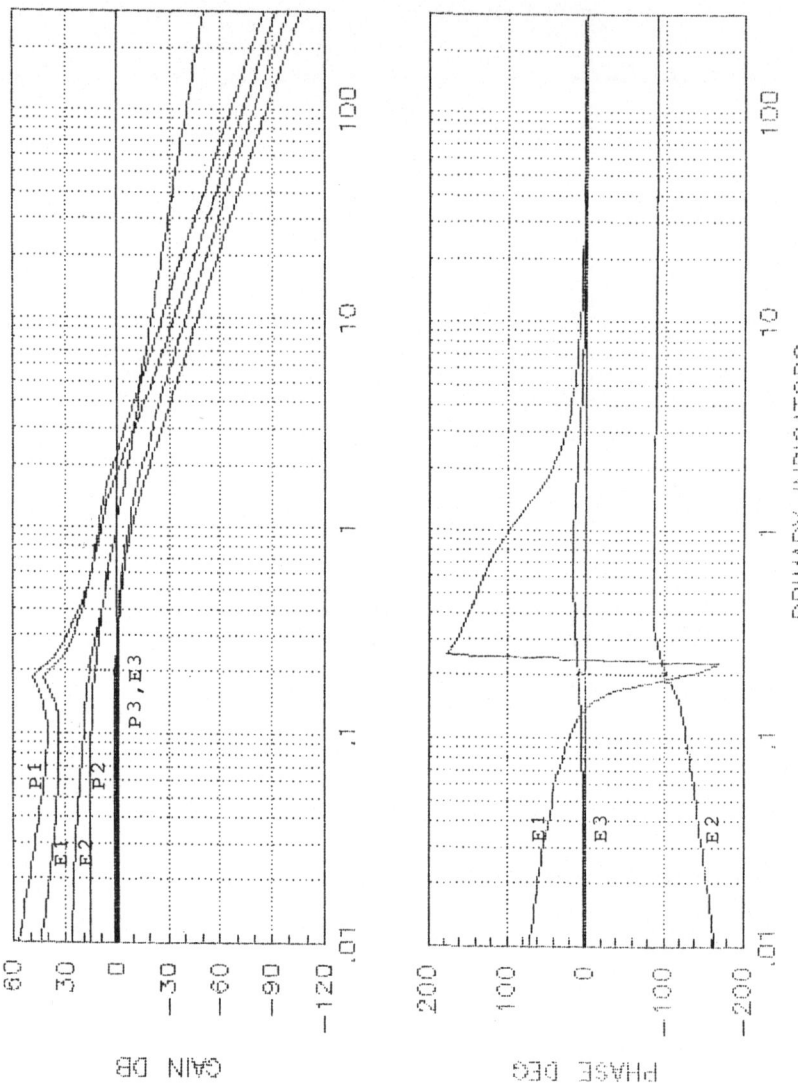

Fig. 7.12.1 The primary indicators of AIRC.

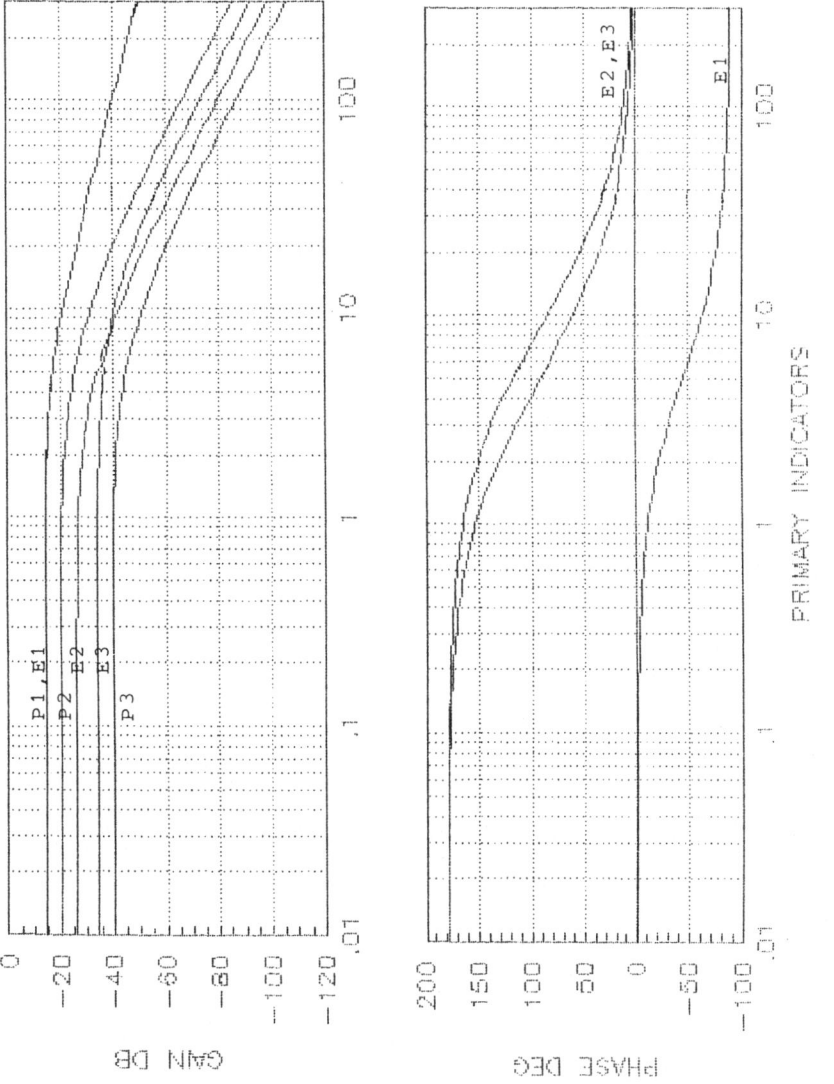

Fig. 7.12.2 The primary indicators with full
state feedback.

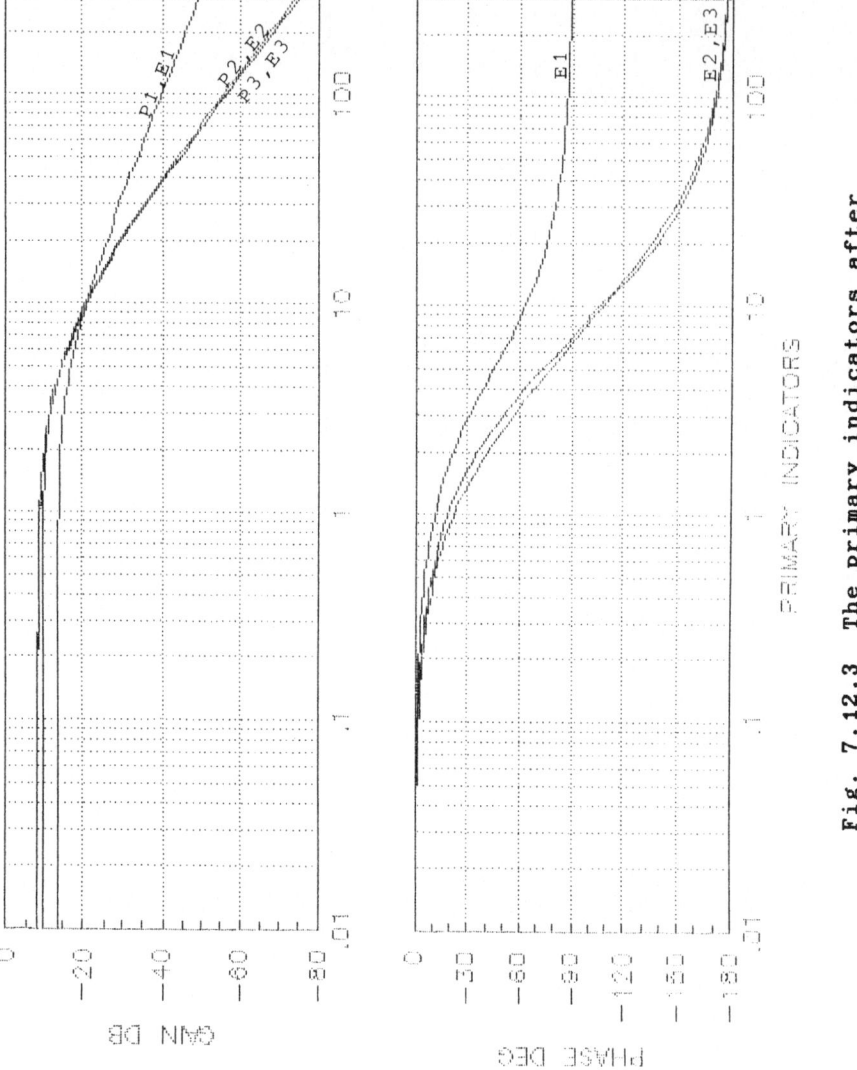

Fig. 7.12.3   The primary indicators after
high frequency region design.

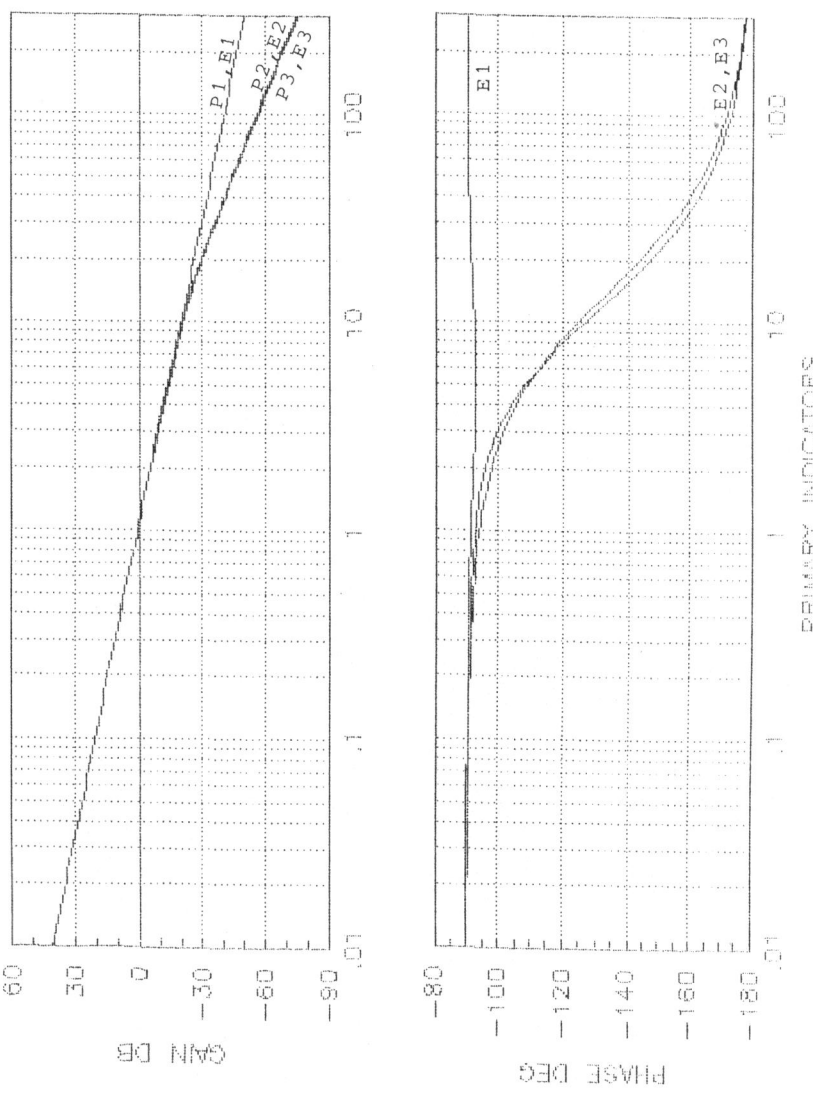

Fig. 7.12.4   The primary indicators after high
and low frequency region design.

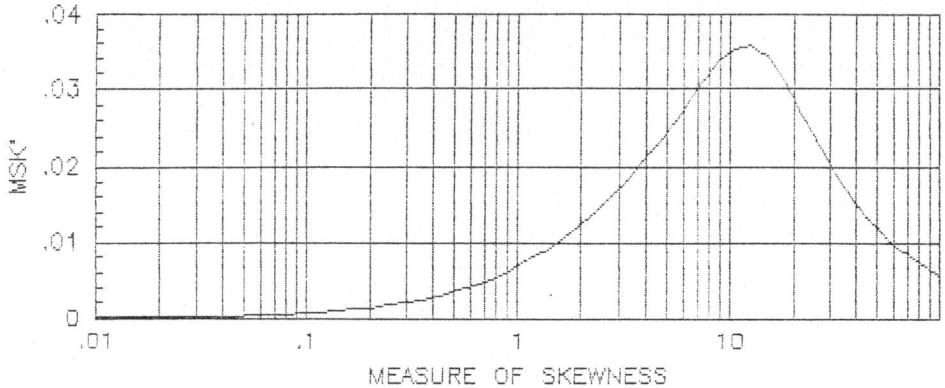

Fig. 7.12.5   The normality indicator MS of
the final design.

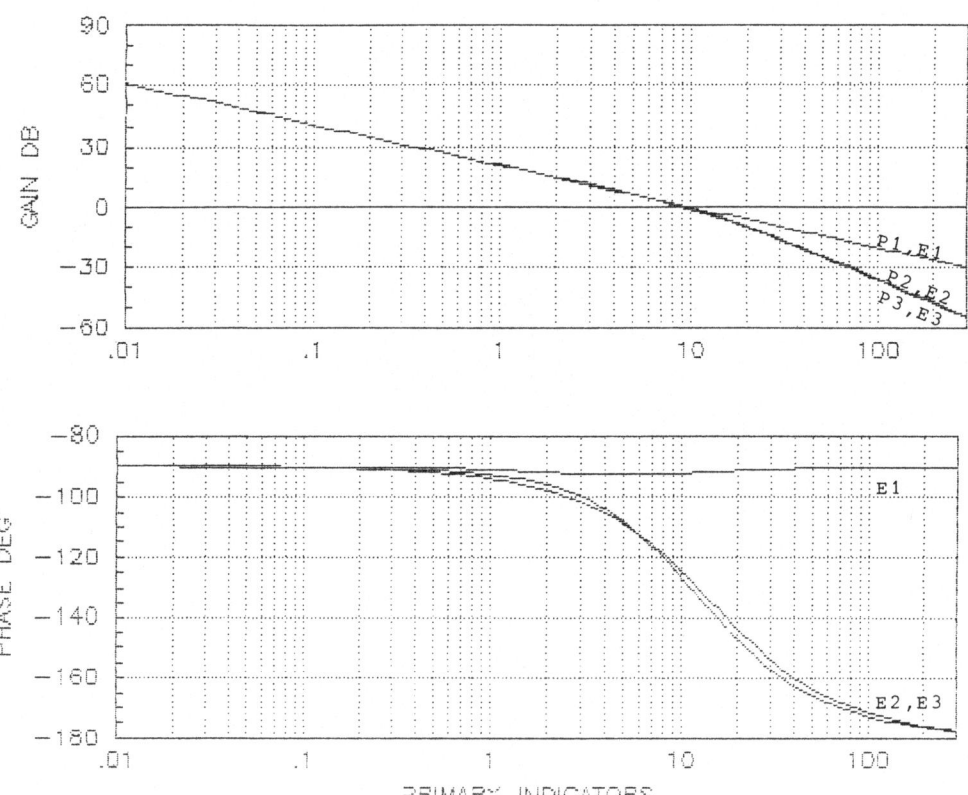

Fig. 7.12.6   The primary indicators of the
final design.

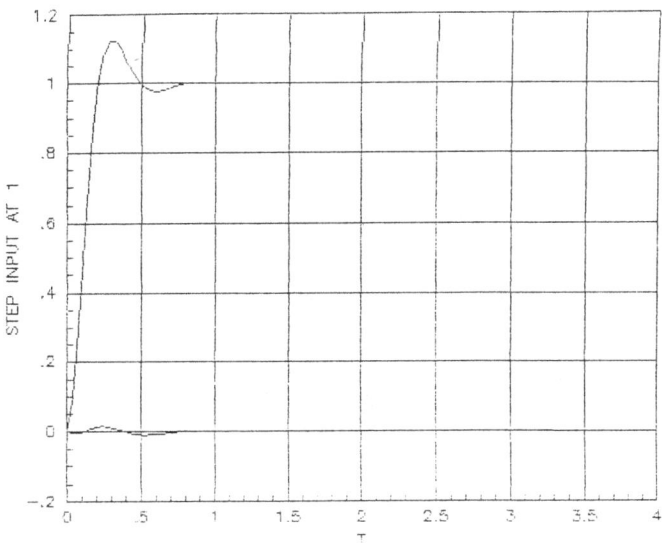

Fig. 7.12.7(a) Closed-loop step response of the
final design (step at input 1).

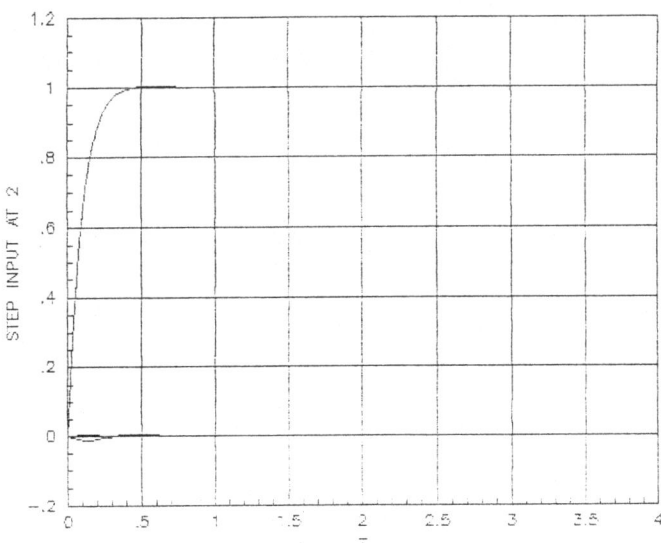

Fig. 7.12.7(b) Closed-loop step response of the
final design (step at input 2).

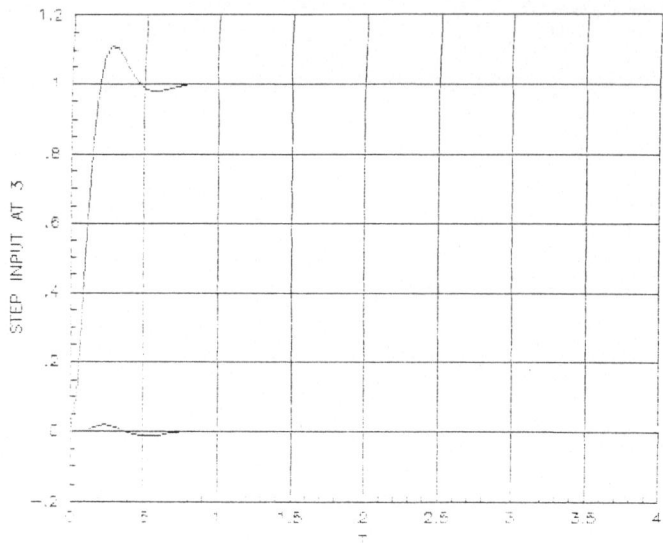

Fig. 7.12.7(c)   Closed-loop step response of the
                 final design (step at input 3).

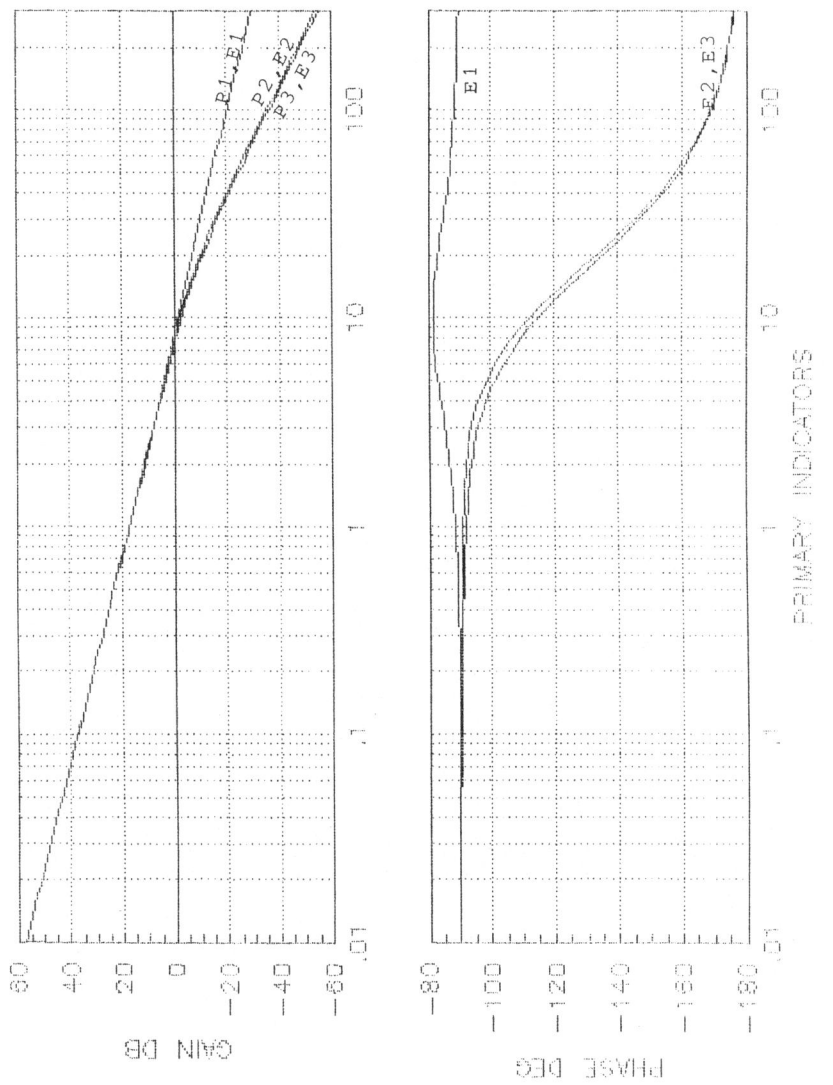

Fig. 7.12.8 The primary indicators of the
final design with a new set of poles.

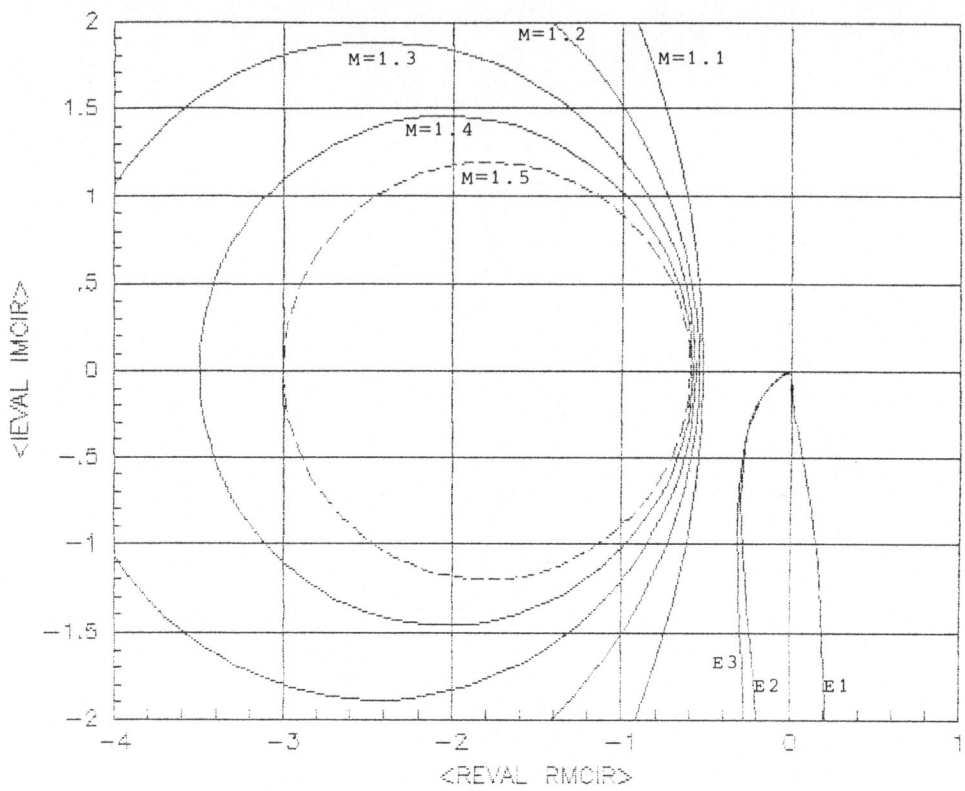

Fig. 7.12.9 Generalised Nyquist diagrams of the
final compensated system.

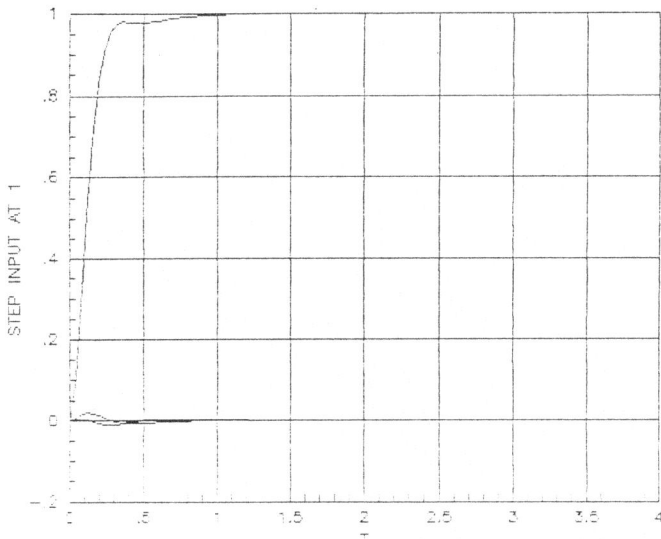

Fig. 7.12.10(a)   Closed-loop step response of the final compensated system (step at input 1).

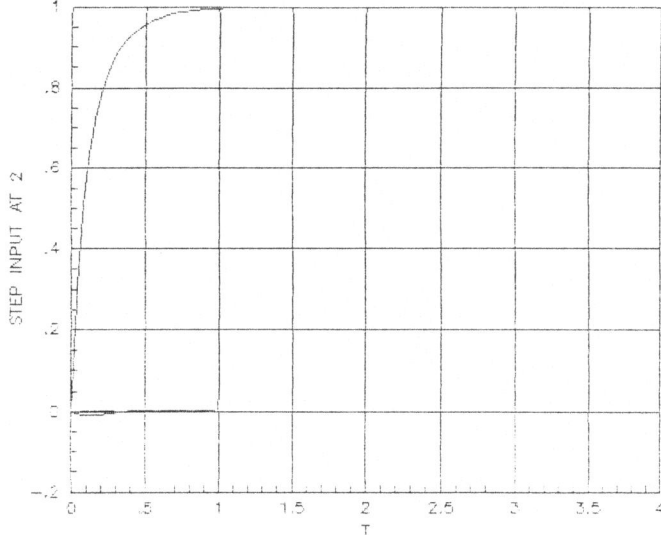

Fig. 7.12.10(b)   Closed-loop step response of the final compensated system (step at input 2).

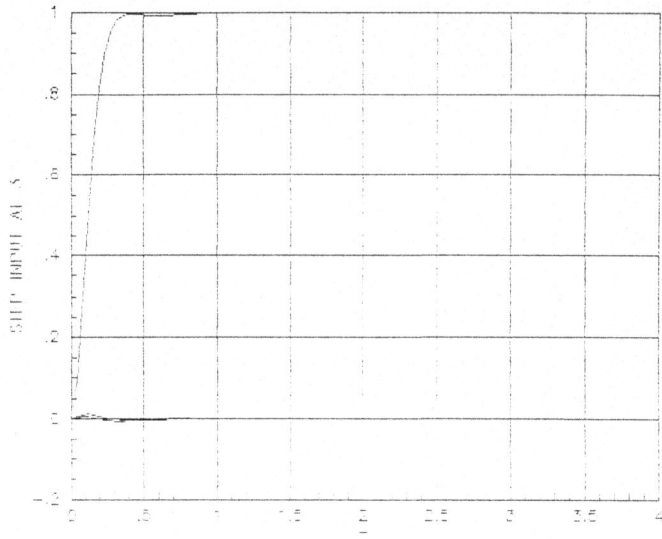

Fig. 7.12.10(c)   Closed-loop step response of the
                   final compensated system
                   (step at input 3).

CHAPTER EIGHT

DEVELOPMENT OF AN EXPERT SYSTEM FOR MULTIVARIABLE CONTROL

SYSTEM DESIGN USING A SYSTEMATIC DESIGN APPROACH

8.1 Summary of the Systematic Design Approach

A systematic approach to designing multivariable feedback control systems using the three design techniques developed in previous chapters is proposed. The design analysis is assessed by the fundamentally important primary indicators. The principle used is the well-known feedback principle. The design objective is achieved by a manipulation of the primary indicators of the open-loop transfer function into a suitable form.

The design problem is partitioned as follows. First, we break the design down into three separate frequency regions: high (HF), intermediate (IF) ,and low frequency (LF) region. Secondly, we divide the design method into three techniques which represent three levels of complexity and organize it into a hierarchy. The three levels are: Simple Design Technique (SDT), Reverse Frame Alignment Technique (RFAT), and Observer-Based Controller (OBC) technique. Each level has its own basic tools to manipulate the gains and phases of the primary indicators (see Fig. 8.1).

Figure 8.2 summarizes the systematic approach and shows how the design techniques are organised into a hierarchy. The overall approach is considered to give a powerful, systematic and intuitively appealing approach to the design of linear multivariable feedback controllers which satisfactorily handles stability, performance and robustness, and which is well-adapted for use in an expert system context. The systematic nature of the design techniques facilitates the development of an expert system. Design rules have been specifically developed in conjunction with the development of each technique. As the design has been broken down into separate frequency regions, rules for dealing with each frequency region are developed. What results is a knowledge base which is extremely modular and the rule bases can be written, debugged and updated separately. This has the significant advantages of fast development time, quick debugging and easy maintenance of the knowledge base. In the rest of this chapter, we present an approach to building an expert system which is based on this systematic design approach.

8.2 Development of the Knowledge Base

Our design problem has been divided into a number of sub-problems. This is done by breaking the design down into separate frequency regions : high, intermediate and low frequency. Also, the design knowledge has been divided into different levels of complexity and organised into a hierarchy as follows:

Level One: Simple Design Technique

Level Two: Reverse Frame Alignment Technique

Level Three: Observer-Based Controller

The advantages of organizing the design knowledge into different levels of complexity are stated below:

a. The designer can be taken through the lower levels of complexity
    systematically before he tries to tackle his problem with a more
    complex approach.

b. The manner in which the design 'fails' at each level can give important
    clues about the possible success of greater levels of design complexity.

c. The explanation of how a design solution has been reached can be
    presented in a manner reflecting the level of complexity required
    by the design.

d. The designer can rapidly determine whether his solution has reached an unacceptable level of complexity and thus decide whether he should modify his specifications rather than continuing his search with a more complex design approach.

This general technique of problem partitioning (decomposition) should embody the following two rules:

(i)   minimum coupling between components, and

(ii) maximum cohesion of functions within a component.

The partitioning enables well-defined sub-procedures to be built. This has the significant advantage that the knowledge base is extremely modular. Thus, it is one key to addressing the complexity issue.

In our development, we are not trying to automate the design procedure in the manner Taylor and James have attempted (Taylor and Frederick, 1984; James et al., 1985). Our approach emphasizes interactive design and a designer's intuition. We believe that the engineering judgement should still be left to the designer. The expert system is simply there to help the designer to make judgements, and to carry out routine tasks. Once the design techniques had been used successfully to design a number of control systems, a preliminary set of design production rules were developed. As intended, the systematic nature of the design techniques facilitated the formal development of the rules. Each rule denoted a simple design event and was of the form

IF {conditions} THEN {consequents}.

The rules were grouped into modules which were formed in accordance with the design techniques. For example, four modules were identified for the Simple Design Technique. These were : High Frequency (HF) Design, High Frequency (HF) Design Checking, Low Frequency (LF) Design and Low Frequency (LF) Design Checking. The analysis of the design, which is based on the use of the primary indicators, is also simple and systematic.

8.3 Structure of the Knowledge Base in Design Rule Modules

The structure of the knowledge base is shown in Fig. 8.3. Each module can advise on the use of the appropriate commands or functions in the control system design package MATRIX$_x$. The following is a brief description of each module.

a. Supervisor Module: This is the "brain" of the knowledge base and is goal-driven. The Supervisor can call for the action of other modules in sequence.

b. Pre-design Analysis Module: The main function of this module is to analyze the model of the plant and check the open-loop stability of the system.

c. HF Design and HF Design Checking Module: The objective of these modules is to obtain good phase properties in the high frequency region.

d. IF Design and IF Design Checking Module: The objective of these modules is to obtain good gain-phase properties in the intermediate frequency region.

e. LF Design and LF Design Checking Module: The objective of these modules is to obtain good gain properties in the low frequency region.

f. OBC and OBC Checking Module: These modules will advise the user on building an observer-based controller and will do the subsequent checking.

g. Post-design Analysis Module: This module will advise the user on the analysis of the compensated plant (e.g. closed-loop step response for the system).

8.4 Implementation of the Knowledge Base in an Expert System Shell

8.4.1 Implementation of the expert system in Expertech Xi

The arrangement of the expert system for use in design is shown in

Fig. 8.4. On the expert systems side, an expert system shell called Xi (Expertech, 1985) has been used as our implementation tool. The shell runs on an IBM PC/G terminal acting as a stand-alone personal computer. On the control engineering side, the package $MATRIX_x$ (Integrated Systems, 1984) has been used to provide the control system analysis and design facilities. Algorithms for obtaining the various sub-controllers have been programmed in the command language of $MATRIX_x$. The user has to use the package in conjunction with the expert system and act as an interface between the two. This limitation is very difficult to remove at the present stage as the two packages being used are entirely different.

Our implemented expert system is called Multivariable Analytical and Interactive Design (MAID). The development of this prototype system MAID is believed to have assisted the overall development effort, because experience was gained during its implementation. Also, the initial ideas related to the structuring of the domain knowledge were carefully examined and tested out. This has proved the case for a more comprehensive frame-based approach which has been adopted in later stages of this development.

8.4.2 Features of the expert system

MAID can aid a designer by guiding him through a design process. An appropriate module will be invoked and the designer will be asked for a piece of information or will be given advice at each stage of the design. The conceptual knowledge of design has also been incorporated into the expert system. This was performed by attaching a relevant concept to each condition and consequent of a rule. A list of facts concerning the design in the current database can be obtained at the end of each design session. The user can also query the system as to how each entry in the database was obtained. Thus, the designer remains an integral part of the design process and he can use his skill, intuition and experience to carry out the design.

8.4.3 The advantages and disadvantages of using Expertech Xi

In the initial development stage, production rules were used to represent the design knowledge. Expertech Xi can cope with this kind of representation easily. Later, it is found that a more powerful way to represent our design knowledge is in *frames*, as we shall explain later in Section 8.5.3. Expertech Xi, however, poses a restriction on the way knowledge is represented in an expert system for there is no natural means of handling frames or declarative knowledge.

The editor of Xi is very slow and hence modification of a knowledge base is very tedious. However, Xi, which is cheap to purchase and easy to use, is a useful tool for the investigation of expert system techniques and development of prototype systems.

8.4.4 Assessments of MAID for control system design

MAID has been tested on a number of design examples. The expert system advises a user in the design of a control system using a systematic procedure similar to that used in the examples presented in this dissertation. An example of a design session on the system GROC used in Section 5.5.4 is given in Appendix G. The expert system has also been used by a final year undergraduate and an M.Phil student in their projects on control system design. The following conclusions may be drawn about the expert system:

a. MAID has been found useful as a designer's assistant, which can provide guidance and explanations where necessary. It allows the user to query the system and reminds him of the facilities to use in the design package.

b. MAID is a prototype system for the investigation of expert system techniques in control system design. The design knowledge base at present is limited and incomplete. Further work is needed to extend the knowledge base and include knowledge about other design methods.

c. The production rule type of representation found in MAID is limited and inadequate to cope with the more comprehensive and powerful frame-based approach. This has led us to consider re-developing MAID and representing the design knowledge using a frame-based representation, and this will be discussed in the next section.

Concerning the suitability of the expert systems approach to control system design, we have found that the expert system is potentially very useful as an educational tool. As noted by a final year undergraduate in his project report, MAID has helped him to use MATRIX$_x$ in carrying out multivariable control system design, and has turned his (limited) knowledge of control theory directly to design work. He has also found learning more rapid and interesting, for he could ask questions such as 'how' and 'why' during the design.

The use of an expert system can provide a 'quick start' on using the package MATRIX$_x$ for multivariable feedback system design by any non-expert user, who may know about the subject but has little experience with design. The expert system can lead him and remind him of the routine tasks he has to perform.

Our approach to incorporating an expert system into control system design is different from the approach currently being used by some other researchers such as Taylor and Frederick (1984), Birdwell et al. (1985), James et al. (1985), Nolan (1986) and Trankle et al. (1986). These authors aim to use the expert system as an intelligent front-end which controls the entire environment including facilities for the graphics and numerical calculations (see Fig. 8.5). However, another approach is for the expert system to act as a design assistant or command spy (Larsson and Persson, 1986) which provides advice, comments as well as warnings, and watches over the design process (see Fig. 8.6). Our approach is the second type, except that there is no interface between the control engineering environment and the expert system. The latter approach is more flexible than the former in a significant sense, in that it suggests that the expert system may be developed more independently from the control engineering enviroment. As the languages used in developing most control software are usually very different from those used in developing an expert system, interfaces between the components should be more easy to implement with the second approach.

To conclude, we have built an appropriate rule base incorporated with design knowledge using a simple expert system shell. The power of $MATRIX_x$ has been enhanced with the guidance provided by the expert system. In addition, it has been found that such an expert system is potentially very useful for educational purposes.

8.5 A Frame-based Approach to Knowledge Representation

8.5.1 Introductory remarks

Knowledge representation is still a central topic of research in artificial intelligence. A survey of current techniques reveals that there are two main schools of thought on the representation of the expert's knowledge. The two approaches are : production rules and structured objects. Most early and conventional expert systems used the familiar production rule representation. The basic building blocks of the expert system are IF-THEN rules. Below are the advantages often claimed for the use of this production rule representation:

a. It allows for a uniform representation of knowledge.

b. It represents naturally occuring chunks of knowledge.

c. It allows for incremental growth of the knowledge base. New rules

can be added easily.

d. It represents knowledge in a declarative manner. Interaction

among the rules may lead to unplanned and interesting results.

However, the disadvantages of this kind of representation are as follows:

a. It is an inefficient way of representing knowledge and results in little

flexibility to handle the search especially if the knowledge base

becomes large.

b. The expressive power is inadequate for representing concepts and

relationships among objects.

c. It is difficult to see the consequences of adding a new rule to the

system. It may lead to an undesirable interaction and result in the

knowledge base containing contradictory and circular rules.

The other approach aims to represent knowledge by structured objects. Examples are semantic networks and frames. The knowledge base is typically a hierarchical network of objects, concepts or events. The framelike (or node-and-link) structure represents their interrelations. The use of networks avoids the exhaustive search that may arise in the use of a rule-based representation. It also allows for a very concise, neat and clear way of representing knowledge. In our latter approach to representing design knowledge, a frame-based representation is adopted which will be described later on.

8.5.2 Definition of a frame (Waterman, 1985; Harmon and King, 1985)

The idea of frames was first developed by Marvin Minsky in 1975. Since then the concept has evolved (Pang and Boyle, 1986) and frames have become a very popular way of representing knowledge.

Definition of a frame

A frame is a generic data structure containing any desired number of categories of information called *slots* (see Fig.8.7), where this information is associated with the subject of the frame. Each frame defines a semi-independent body of knowledge, which can be both procedural and declarative. Frames may be linked together to form a hierarchical classification of domain knowledge and allow for inheritance.

Each frame has a name which is similar to the concept of a header in a list.

Definition of a slot

A slot is a component of knowledge which may have a number of entries. Each entry may be either the name of another frame or a *primitive.*

If a slot is filled by the name of a frame, it indicates an inheritance relationship. Thus frames can be defined as specializations of more general frames. At the bottom of the frame hierarchy, we have primitive frames whose attribute and procedure slots contain only primitives.

Slots can be of three basic types: attribute slots, procedure slots and connection slots. The attribute slots relate to the declarative part of the frame knowledge. They describe the information represented in the frame. The type and number of frame attributes will depend on the domain whose knowledge is being represented.

The procedure slots define how the information required by the frame should be obtained and what actions should be taken if the designer selects that frame.

The connection slot defines the position of the current frame within the knowledge base.   Thus the connection slot is used to form the knowledge base taxonomy.

Definition of a primitive

A primitive serves as a basic piece of information for the current domain. Primitives are not further defined within the chosen domain.

As such, a primitive can be a rule, a conclusion, a piece of information required and so on. A primitive can have a formal component which describes the formal part of the primitive, a conceptual component which is able to express any principles behind the formal component of the primitive and a value (see Fig.8.8). One reason for breaking the domain expertise down into primitives is Lenat's experience with Eurisko (see (Schrage, 1986)). His expert system was, in his own words, disastrous until he decided to break his rules down into their fundamental attributes.

The slots required in representing design knowledge

In a frame, we have identified three basic types of slot. Figure 8.9 shows the slots we have identified as necessary for representing design knowledge. The diagram also provides a brief description of the function of each slot.

8.5.3 Advantages of the frame-based approach

Frames unify both the procedural and declarative expression of knowledge. Other advantages of the frame-based approach are as follows:

a.   The knowledge base of a frame-based system is extremely modular. This provides a natural way of representing components of expertise. Also, the knowledge base can be maintained more easily.

b.   Each frame can represent appropriate knowledge as default values.

c. Each frame is an object which represents an independent body of knowledge.

d. Once we have organised the domain expertise into frames, it is relatively simple to represent the procedural aspects as rules within the knowledge base.

e. Frames can be defined as specializations of more general frames, leading to a hierarchical classification of the domain knowledge.

f. Frames can be linked together to have inheritance relationships.

g. Uncertainty in the design process calls for flexibility and this can be provided if the frames are semi-independent.

h. The flexibility of the frames allows the experienced designer to vary the sequence in which the frames are used and therefore provides him with a more powerful structure for handling unanticipated types of design problem.

8.6 Structure of the Knowledge Base in a Frame-based System

The design knowledge is represented using a frame-based structure. Figure 8.10 shows the objects and object classes in the design knowledge base, linked together in hierarchies. The class/sub-class relationship is indicated by a solid line while the class/member relationship is indicated by a dotted line.   For example, the SIMPLE.DESIGN.TECHNIQUE is a sub-class of the DESIGN.TECHNIQUES,   and   the   HIGH.FREQ.SUB-CONTROLLER   is   a   member   of   the SIMPLE.DESIGN.TECHNIQUE.   Each object in the knowledge base is a frame, which has a structure as shown in Fig. 8.9.

Each frame has a goal and this is represented in the slot containing concept   attributes.   For   example,   the   goal   of   the   frame   HIGH.FREQ.SUB-CONTROLLER   is   in   the   slot   denoted   concept   attribute,   which   contains   the primitive " To balance the gains and align the input and output gain frames at high frequencies " (see Frame C-3 in Appendix C). At the bottom of the frame hierarchy, we have primitive frames (e.g. HFS, LFS1, RFA1 as shown in Fig.8.10). Each fundamental sub-controller used for cascading with a system is a primitive frame. Each primitive frame has a specific goal and the design process consists of a sequence of operations that aims to shape the primary indicators of the open-loop transfer function of the system.

Besides a goal, each primitive frame has a number of slots for the attribute values. The user will be asked about the relevant information when filling in the slots. For example, a HIGH.FREQ.SUB-CONTROLLER has to obtain the condition number(s) at high frequencies before the advice on gain balancing is given to the user. This attribute is represented by the evidence attribute slot in the frame. The primitive frame of a sub-controller will probably have some effects which are other than intended. For example, the HIGH.FREQ.SUB-CONTROLLER will probably make worse the gain divergences at intermediate frequencies. These side-effects have to be foreseen before the operation. Hence, a primitive frame for checking is associated with each primitive frame of a sub-controller. In the design knowledge base, the class SUB-CONTROLLER.CHECK has members which are primitive frames for checking. For example, the HFS.CHECK will be invoked after every operation of the HIGH.FREQ.SUB-CONTROLLER.

8.7 Design Knowledge Base for the Design Techniques in Frames
(IntelliCorp, 1984)

The knowledge base for the three design techniques as shown in Fig.8.10 is now described. Examples of some of the frames representing these design techniques are given in Appendix C.

8.7.1 Simple Design Technique

The SIMPLE.DESIGN.TECHNIQUE frame is linked to the HIGH.FREQ.SUB-CONTROLLER and LOW.FREQ.SUB-CONTROLLER frames (denoted by HFS and LFS respectively) plus frames of other basic types (e.g. scalar sub-controller SCALAR). The goal of the frame is to shape the primary indicators at high frequencies using the HFS and at low frequencies using the LFS.

The HFS is a primitive frame which has two specific goals. It aims to align the input and output gain frames of the system at high frequencies. It also tries to balance up the principal loci at high frequencies if the system has the same roll-off rates there. Otherwise, it will balance up the gains at bandwidth frequency. The associated checking frame, HFS.CHECK, will check whether these objectives are achieved and examine if any tradeoff is needed between the high and intermediate frequency region (see Frame C-2 to C-4 in Appendix C for more details).

The LFS has two member primitive frames, LFS1 and LFS2. The choice depends on whether the system has the same roll-off rates at low frequencies and contains any right-half plane poles. However, the main goal is to incorporate integral action so that the system has high gains at low frequencies.

8.7.2 Reverse Frame Alignment technique

The REVERSE.FRAME.ALIGNMENT frame is linked to the HFS, RFA and LFS frames together with SCALAR and OPTIMIZER. The goal of the RFA frame is to obtain phase compensation or gain adjustment at intermediate frequencies. The RFA frame has a number of member primitive frames (e.g. RFA1, RFA2 etc.). Each of them has a different operation on the primary indicators. For example, the RFA1 is a RFA sub-controller with a phase lead network on its diagonal. It is used to introduce phase lead at intermediate frequencies. In trying to fill in the evidence attribute slot, the user will be asked about the frequency at which the RFA sub-controller is formed and the amount of phase lead that is required. Since the RFA1 is a member of the sub-class RFA, it inherits the three Member attributes of the RFA as additional Own attributes. Further details can be seen from Frame C-5 to C-7 in Appendix C.

8.7.3 Observer-based Controller

The goal of the OBSERVER.BASED.CONTROLLER frame is to use a full-order observer and conclude the design with the Simple Design Technique or Reverse Frame Alignment technique. The final objective is to use a reduced-order observer-based controller instead of a full-order observer-based controller.

The FULL-ORDER.OBC and REDUCED-ORDER.OBC are sub-classes of the OBSERVER-BASED.CONTROLLER. Hence, the Member attributes of the OBSERVER-BASED.CONTROLLER are inherited as Member attributes of the FULL-ORDER.OBC and REDUCED-ORDER.OBC. These Member attributes will then become Own attributes of their members. For example, the OBSERVER-BASED.CONTROLLER class has a Member attribute " To obtain stable pole positions " and the FULL-ORDER.OBC is a sub-class of the OBSERVER-BASED.CONTROLLER. Hence, the Member attribute above will also become a Member attribute of the FULL-ORDER.OBC. Since STATE.FEEDBACK is a member of the FULL-ORDER.OBC, the Member attribute " To obtain stable pole positions " will become a Own attribute of STATE.FEEDBACK (see Frame C-8 to C-10 in Appendix C for more details).

8.8 Implementation of the Expert System

The knowledge base was developed with an aim of ultimately

implementing it in KEE (Knowledge Engineering Environment), an expert system developemnt tool marketed by Intellicorp. KEE was chosen mainly because it supports frame-based knowledge representation. Also, it provides a flexible artificial intelligence programming environment which incorporates features like object-orientated programming, rule-based programming, Lisp functional programming as well as data-driven reasoning and class property. In addition, it provides a powerful graphical interface for the user.

Figure 8.10 has already shown how the knowledge base would appear if implemented in KEE. An example of a frame associated with the object HIGH.FREQ.SUB-CONTROLLER, HFS, if implemented in KEE is given in Appendix D. However, the knowledge base can also be implemented in any expert system shell with suitable tailoring. For example, the knowledge base has been represented using production rules and implemented in an expert system shell Expertech Xi (see Section 8.4). This has resulted in a prototype system called Multivariable Analytical and Interactive Design (MAID), which can act as a designer's assistant.

8.9 Specification Considerations in Control System Design

There are a number of difficulties in dealing with the specifications in control system design. First, the specifications are very often incomplete and unclear at the beginning of a design process. Sometimes the designer may wish to re-adjust the specifications in face of the unacceptable cost he has to pay to meet the original specifications. In addition to the uncertainties involved in specifications, the very large diversity of specifications is another difficulty. The specification sheet for a multivariable control system can be very complicated with many items or it can be very simple, specifying just non-interactive control over the outputs. Because of the diversity in the nature of control systems and requirements, it is very difficult to define a "canonical" form of the specification. The consistency checking within a given specification is another difficulty when dealing with specifications. That is why a designer has to be able to handle trade-offs between competing specifications. Some ways to overcome these difficulties will be suggested below.

The impreciseness of the specifications is in the nature of the design problem. The systematic design approach already presented allows for "experimental" procedures to be performed on the system so that a designer will have a better idea of his specified requirements. The designer may wish to use his intuition to adjust the specifications during the course of design.

Furthermore, the specifications may be classified into various types for easy treatment. A way of classification is given by Nye and Tits (1986), which is based on the importance of each item in the specifications. Here, another way of classification is suggested which is directly related to the systematic design approach. Very often, the specifications are given in either time-domain (e.g. rise time, overshoot) or frequency-domain quantities (e.g. bandwidth, gain at low frequencies). If we can translate the closed-loop time-domain specifications into open-loop frequency-domain specifications, we can define all those specifications given in open-loop frequency-domain terms *primary specifications*. All the constraints are then represented as forbidden regions in the open-loop Bode diagrams.

During the design process, a designer aims to shape the primary indicators so that they will lie outside the forbidden regions as required by the primary specifications. If the objective can be achieved, he will expect to meet the specifications. Otherwise, he has to contract the forbidden regions i.e. to relax the specifications. Thus, the primary indicators are used in conjunction with the primary specifications in the systematic design approach.

However, there are two main difficulties with this method. First, some specifications are neither specified in time-domain nor frequency-domain quantities e.g. constraints on the sign and the magnitude of some controller gains. In this case, the designer has to be aware of those constraints and handle them separately. Secondly, the translation of constraints into open-loop frequency-domain specifications may not be exact and simple. Hence, the designer has to rely on some approximation methods and check his results after each design.

275

Fig. 8.1 Design tools for each design level

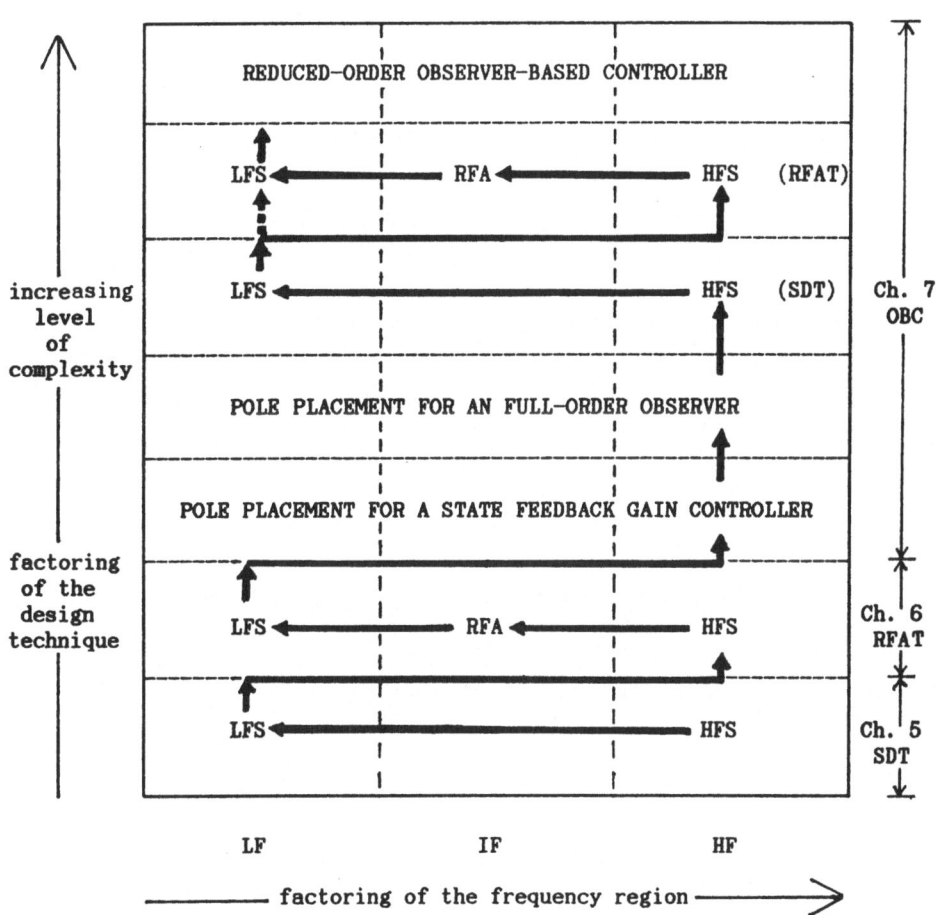

Fig. 8.2 A Systematic Approach to Multivariable Feedback
Control System Design

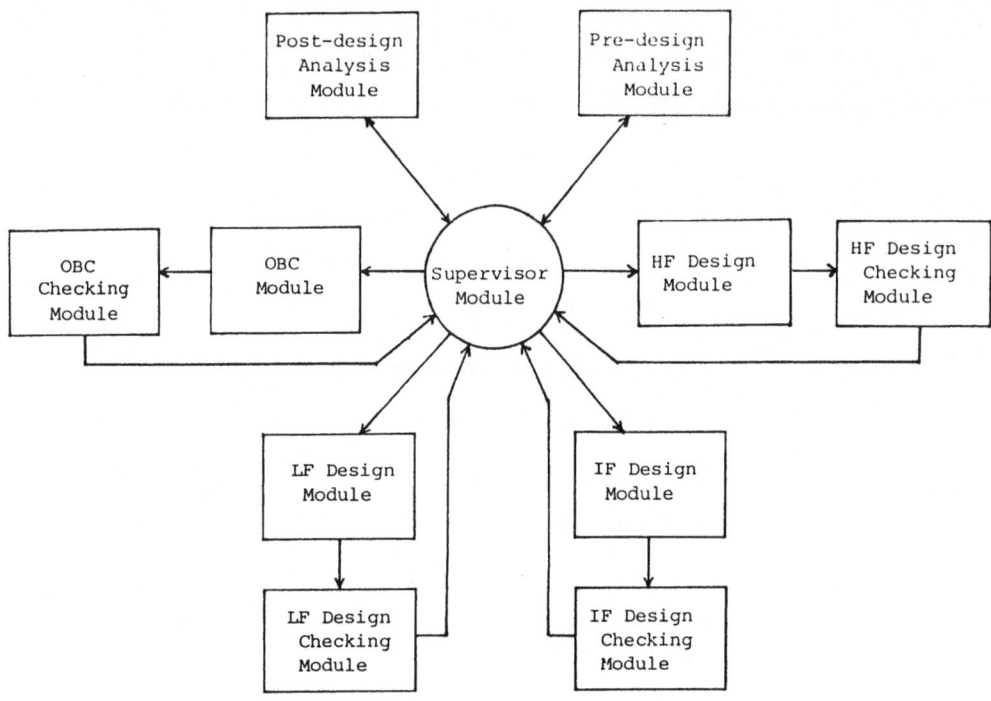

Fig. 8.3 The structure of the knowledge base

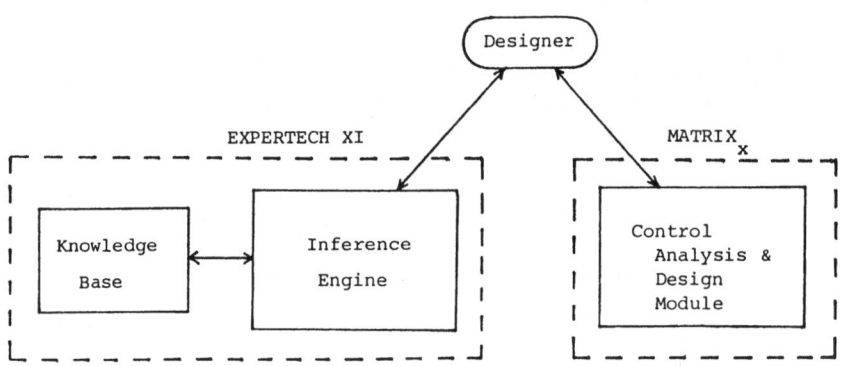

Fig. 8.4 The arrangement of the expert system for use
in control system design

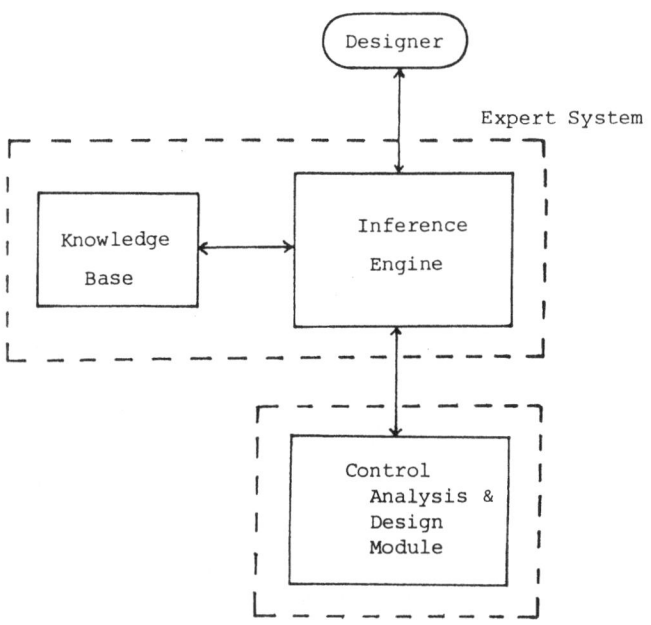

Fig. 8.5 The first approach to the use of expert systems

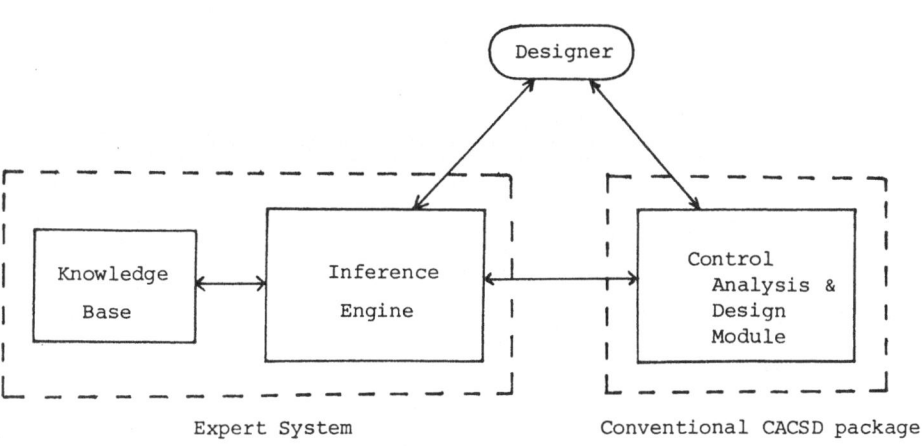

Fig. 8.6 The second approach to the use of expert systems

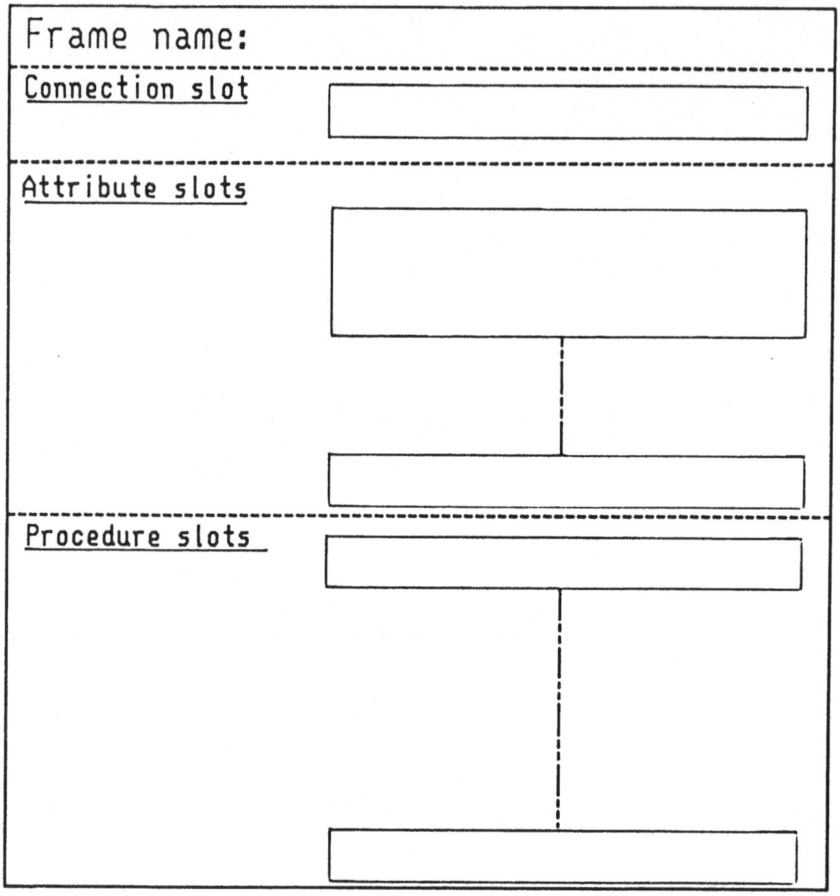

Fig. 8.7   Structure of a frame

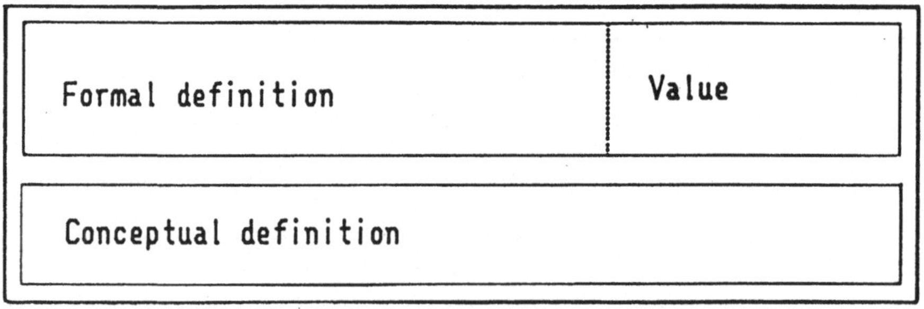

Fig. 8.8   Definition of a primitive

Frame name:	Similar to the concept of a header in a list
Connection slot	Defines the position of the frame within the knowledge base
Attribute slots	
Design attribute	Represents the frame's design actions.
Evidence attribute	Information required by the frame
Concept attribute	Conceptual information associated with the frame.
Procedure slots	
control slot	Specifies the order in which the information required by the frame should be investigated
Rule slot	Frame's production rules
Exit slot	{ compulsory subsequent frame }
Context slot	Defines the conditions under which the frame will be in context

Fig. 8.9   Slots for the frame

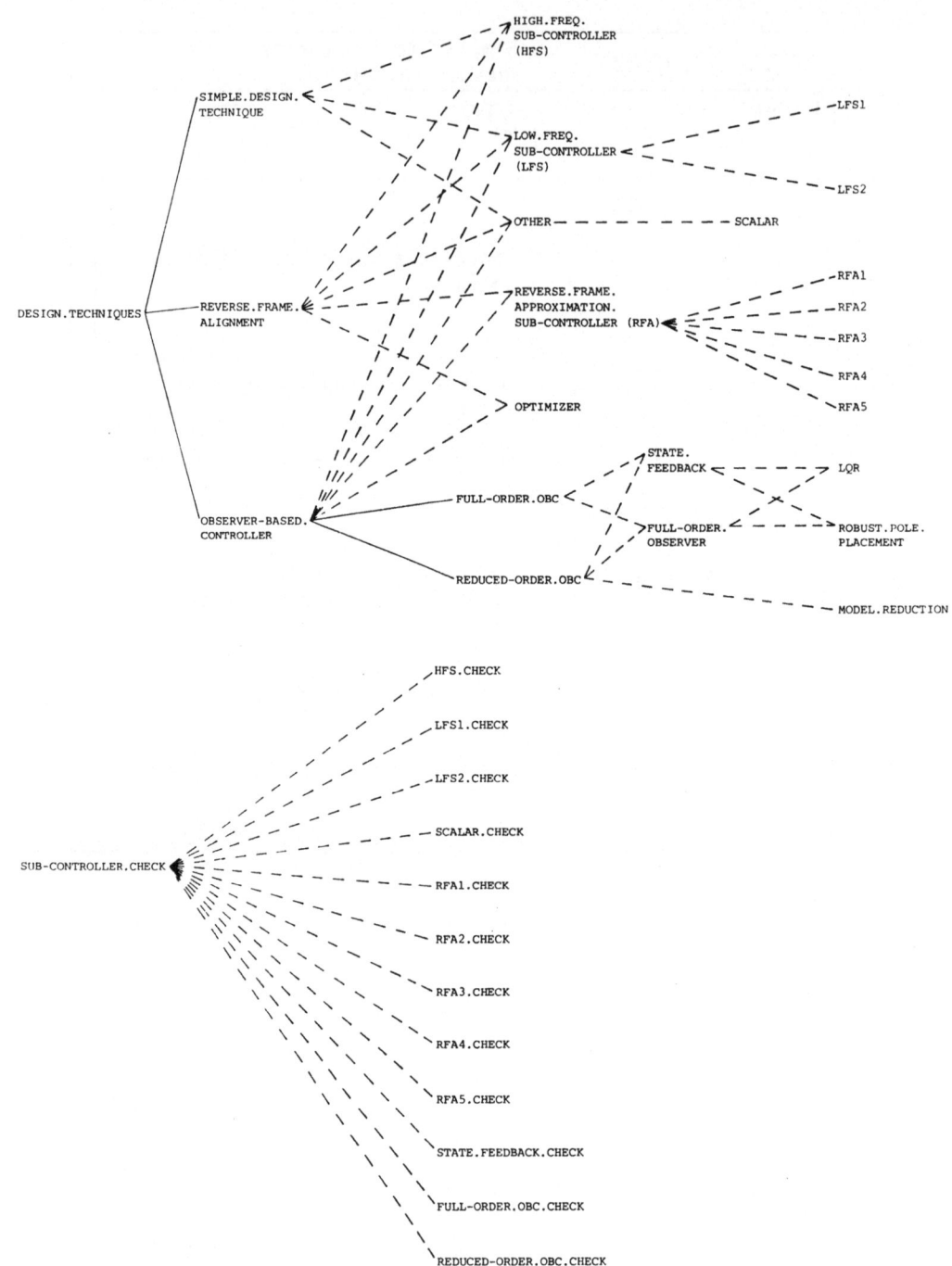

Fig. 8.10 The knowledge base for the systematic design approach

CHAPTER NINE

CONCLUSIONS

9.1 Assessment of Approach Adopted

The systematic use of primary indicators has provided a simple but fundamentally important basis for interactive design of multivariable control systems. Stability, performance and robustness can be handled in a satisfactory way. The design techniques that have been developed are based on a classical frequency response approach, which provides a key to an understanding of the system behaviour in a physical sense. Hence, practicing engineers familiar with classical control techniques should find the methodology presented here easy to follow.

The Simple Design Technique presented in Chapter 5 is a simplified version of the Reverse Frame Alignment technique. However, this technique is found to be quite effective for a number of systems and results in a simple proportional-plus-integral controller.

The Reverse Frame Alignment technique of Chapter 6 is a redevelopment of the Characteristic Locus Design Method. The new technique, which is based on an appropriate shaping of the primary indicators, is considered more systematic and powerful than the Characteristic Locus Design Method. Non-square systems can also be dealt with by this new technique. In contrast to the Characteristic Locus Design Method, the formation of the sub-controllers of the Reverse Frame Alignment technique is based on the singular value decomposition instead of the characteristic value decomposition. An optimizer for fine tuning of the parameters in an RFA sub-controller has also been developed. This is considered quite useful in practice because a designer can obtain an optimum set of parameters with respect to the setting up and weighting of the optimizer. The choice of weights can help a designer to perform trade-offs during a design.

At the top level of the design technique, the use of an observer in an observer-based controller may be considered an expensive addition to the system (Layton, 1976). However, it gives the advantage of being able to use state feedback instead of output feedback. This provides a means of accessing information from an internal model of a system. After the modification of the gain-phase characteristics of a system via an observer-based controller, the design can be concluded using either the Simple Design Technique or the Reverse Frame Alignment technique. To summarize, a systematic approach based on these three design techniques is considered to give a powerful and intuitively appealing approach to the design of linear multivariable feedback controllers.

The creation of an expert system based on this systematic design approach is a natural development of it. There are, of course, many approaches to control system design. An expert system based on these three techniques is just one approach. However, it is believed that a step has been taken towards the goal of providing an expert system design environment for multivariable feedback control system design.

9.2 Proposals for Future Work

To obtain a fully operational system, a more complete knowledge base for the systematic design approach proposed here must be developed. In addition, the expert system must be tested thoroughly to ensure that the recommendations for designs for practical systems agree with those of an expert. This is important because criticism has been made of expert systems (Bell, 1985) due to inadequate testing. Also, the design techniques need to be further developed to deal with time-varying, non-linear and discrete-time systems. The design knowledge base developed and represented using frames has not yet been implemented in any frame-based system. It is hoped that implementation in a frame-based expert system development environment such as KEE will be carried out soon. Finally, the knowledge base should be extended to include knowledge of other design methods thus drawing on a more extensive knowledge of control engineering.

APPENDICES

<u>Appendix A</u>   To prove that $\alpha = f(MS)$ is a concave, monotonic increasing

function

The matrix considered is complex and of order $m \times m$ where $m \geq 2$
(see Section 3.4.9).

<u>Proof</u>:

From [3.4.9.7], we have

$$\alpha = \delta \ (1+x+x^2+...+x^{m-1}) \tag{A.1}$$

From [A.1] & [3.4.9.6], we obtain

$$MS = \alpha.x$$
$$= \delta(x + x^2 +...x^m) \tag{A.2}$$

Differentiating α with respect to x in [A.1], we have

$$\frac{d\alpha}{dx} = \delta \ (1 + 2x + 3x^2 +... \ (m-1)x^{m-2}) \tag{A.3}$$

$\frac{d\alpha}{dx} > 0$ for positive values of x. $(m \geq 2)$

Differentiating MS with respect to x in [A.2], we have

$$\frac{dMS}{dx} = \delta \ (1 + 2x + ...mx^{m-1}) \tag{A.4}$$

$\frac{dMS}{dx} > 0$ for positive values of x. $(m \geq 2)$

Hence,

$$\frac{d\alpha}{dMS} = (\frac{d\alpha}{dx})/(\frac{dMS}{dx}) > 0 \ .$$

and f is a monotonic increasing function

To prove that it is concave, we need to prove $\dfrac{d^2\alpha}{d(MS)^2} < 0$.

But,

$$\frac{d^2\alpha}{d(MS)^2} = \frac{d}{dMS}(\frac{d\alpha}{dMS}) = \frac{d}{dx}(\frac{d\alpha}{dMS})/\frac{dMS}{dx} \quad . \tag{A.5}$$

Therefore, it remains to prove $\dfrac{d}{dx}(\dfrac{d\alpha}{dMS}) < 0$.

From [A.3] & [A.4],

$$\frac{d\alpha}{dMS} = 1 - (mx^{m-1})/(1 + 2x + 3x^2 + \ldots mx^{m-1})$$

$$\frac{d}{dx}(\frac{d\alpha}{dMS}) = - [h(x) / (1 + 2x + \ldots + mx^{m-1})^2]$$

where

$$h(x) = mx^{m-2}[(m-1) + 2(m-2)x + \ldots + k(m-k)x^{k-1} + \ldots + (m-1)x^{m-2}]$$

for $m \geq k \geq 1$.

$h(x) > 0$ for positive values of x. $(m \geq 2)$

Hence,

$$\frac{d}{dx}(\frac{d\alpha}{dMS}) < 0$$

and

$$\frac{d^2\alpha}{d(MS)^2} < 0 \quad . \quad (m \geq 2)$$

Therefore, $\alpha = f(MS)$ is a concave, monotonic increasing function. $\qquad \square$

Appendix B   System Models

B-1   : GFLO
Description   : Pressurised flow-box in a paper-making machine
Reference   : MacFarlane and Belletrutti, 1973
No. of inputs   : 2
No. of outputs   : 2
No. of states   : 2
State space model :

$$
D = \begin{bmatrix} 0. & 0. \\ 0. & 0. \end{bmatrix}
\qquad
B = \begin{bmatrix} 0.0336 & 1.0380 \\ 0.0010 & 0.0000 \end{bmatrix}
$$

$$
C = \begin{bmatrix} 1. & 0. \\ 0. & 1. \end{bmatrix}
\qquad
A = \begin{bmatrix} -0.3950 & 0.0115 \\ -0.0110 & 0.0000 \end{bmatrix}
$$

B-2   : GROC
Description   : Nuclear rocket engine
Reference   : Davison and Chow, 1977
No. of inputs   : 2
No. of outputs   : 2
No. of states   : 4
State space model :

$$
D = \begin{bmatrix} 0. & 0. \\ 0. & 0. \end{bmatrix}
\qquad
B = \begin{bmatrix} 65.0000 & 0.0000 \\ 0.0000 & 0.0000 \\ 0.0000 & 0.0000 \\ 0.0000 & 0.4000 \end{bmatrix}
$$

$$
C = \begin{bmatrix} 0. & 0. & 1. & 0. \\ 0. & 0. & 0. & 1. \end{bmatrix}
$$

$$
A = \begin{bmatrix}
-65.0000 & 65.0000 & -19.5000 & 19.5000 \\
0.1000 & -0.1000 & 0.0000 & 0.0000 \\
1.0000 & 0.0000 & -0.5000 & -1.0000 \\
0.0000 & 0.0000 & 0.4000 & -0.4000
\end{bmatrix}
$$

B-3   : REAC
Description   : Chemical reactor
Reference   : Munro, 1972
No. of inputs : 2
No. of outputs : 2
No. of states : 4
State space model :

$$
D =
\begin{bmatrix}
0. & 0. \\
0. & 0.
\end{bmatrix}
$$

$$
B =
\begin{bmatrix}
0.0000 & 0.0000 \\
5.6790 & 0.0000 \\
1.1360 & -3.1460 \\
1.1360 & 0.0000
\end{bmatrix}
$$

$$
C =
\begin{bmatrix}
1. & 0. & 1. & -1. \\
0. & 1. & 0. & 0.
\end{bmatrix}
$$

$$
A =
\begin{bmatrix}
1.4000 & -0.2080 & 6.7150 & -5.6760 \\
-0.5810 & -4.2900 & 0.0000 & 0.6750 \\
1.0670 & 4.2730 & -6.6540 & 5.8930 \\
0.0480 & 4.2730 & 1.3430 & -2.1040
\end{bmatrix}
$$

B-4   : AUTO
Description   : Automobile gas turbine
Reference   : Edmunds et al., 1983
No. of inputs : 2
No. of outputs : 2
No. of states : 12
State space model :

$$
D =
\begin{bmatrix}
0. & 0. \\
0. & 0.
\end{bmatrix}
$$

$$
B =
\begin{bmatrix}
0. & 0. \\
1. & 0. \\
0. & 0. \\
0. & 0. \\
0. & 1. \\
0. & 0. \\
0. & 0. \\
1. & 0. \\
0. & 0. \\
0. & 0. \\
0. & 0. \\
0. & 1.
\end{bmatrix}
$$

$$
C =
\begin{bmatrix}
0.2640 & 0.8060 & -1.4200 & -15.0000 & 0.0000 & 0.0000 & 0.0000 & 0.0000 & 0.0000 & 0.0000 & 0.0000 & 0.0000 \\
0.0000 & 0.0000 & 0.0000 & 0.0000 & 0.0000 & 4.9000 & 2.1200 & 1.9500 & 9.3500 & 25.8000 & 7.1400 & 0.0000
\end{bmatrix}
$$

$$
A =
\begin{bmatrix}
0.0000 & 1.0000 & 0.0000 & 0.0000 & 0.0000 & 0.0000 & 0.0000 & 0.0000 & 0.0000 & 0.0000 & 0.0000 & 0.0000 \\
-0.2020 & -1.1500 & 0.0000 & 0.0000 & 0.0000 & 0.0000 & 0.0000 & 0.0000 & 0.0000 & 0.0000 & 0.0000 & 0.0000 \\
0.0000 & 0.0000 & 0.0000 & 1.0000 & 0.0000 & 0.0000 & 0.0000 & 0.0000 & 0.0000 & 0.0000 & 0.0000 & 0.0000 \\
0.0000 & 0.0000 & 0.0000 & 0.0000 & 1.0000 & 0.0000 & 0.0000 & 0.0000 & 0.0000 & 0.0000 & 0.0000 & 0.0000 \\
0.0000 & 0.0000 & -2.3600 & -13.6000 & -12.8000 & 0.0000 & 0.0000 & 0.0000 & 0.0000 & 0.0000 & 0.0000 & 0.0000 \\
0.0000 & 0.0000 & 0.0000 & 0.0000 & 0.0000 & 0.0000 & 1.0000 & 0.0000 & 0.0000 & 0.0000 & 0.0000 & 0.0000 \\
0.0000 & 0.0000 & 0.0000 & 0.0000 & 0.0000 & 0.0000 & 0.0000 & 1.0000 & 0.0000 & 0.0000 & 0.0000 & 0.0000 \\
0.0000 & 0.0000 & 0.0000 & 0.0000 & 0.0000 & -1.6200 & -9.4000 & -9.1500 & 0.0000 & 0.0000 & 0.0000 & 0.0000 \\
0.0000 & 0.0000 & 0.0000 & 0.0000 & 0.0000 & 0.0000 & 0.0000 & 0.0000 & 0.0000 & 1.0000 & 0.0000 & 0.0000 \\
0.0000 & 0.0000 & 0.0000 & 0.0000 & 0.0000 & 0.0000 & 0.0000 & 0.0000 & 0.0000 & 0.0000 & 1.0000 & 0.0000 \\
0.0000 & 0.0000 & 0.0000 & 0.0000 & 0.0000 & 0.0000 & 0.0000 & 0.0000 & 0.0000 & 0.0000 & 0.0000 & 1.0000 \\
0.0000 & 0.0000 & 0.0000 & 0.0000 & 0.0000 & 0.0000 & 0.0000 & 0.0000 & -188.0000 & -111.6000 & -116.4000 & -20.8000
\end{bmatrix}
$$

B-5   : GHEL
Description   : CH-47 tandem rotor helicopter
Reference   : Bloch and Postlethwaite, 1981; Doyle and Stein, 1981
No. of inputs   : 2
No. of outputs : 2
No. of states   : 4
State space model :

$$
B = \begin{bmatrix} 0.1400 & -0.1200 \\ 0.3600 & -8.6000 \\ 0.3500 & 0.0090 \\ 0.0000 & 0.0000 \end{bmatrix}.
$$

$$
D = \begin{bmatrix} 0. & 0. \\ 0. & 0. \end{bmatrix}
$$

$$
C = \begin{bmatrix} 0.0000 & 1.0000 & 0.0000 & 0.0000 \\ 0.0000 & 0.0000 & 0.0000 & 57.3000 \end{bmatrix}
$$

$$
A = \begin{bmatrix} -0.0200 & 0.0050 & 2.4000 & -32.0000 \\ -0.1400 & -0.4400 & -1.3000 & -30.0000 \\ 0.0000 & 0.0180 & -1.6000 & 1.2000 \\ 0.0000 & 0.0000 & 1.0000 & 0.0000 \end{bmatrix}
$$

B-6   : TGEN
Description   : Nuclear-powered turbo-generator
Reference   : Limebeer and Maciejowski, 1982; Hung and MacFarlane, 1982
No. of inputs   : 2
No. of outputs : 2
No. of states   : 10
State space model :

$$
D = \begin{bmatrix} 0. & 0. \\ 0. & 0. \end{bmatrix}
$$

$$
B = \begin{bmatrix} 0.0000 & 0.0000 \\ 0.0000 & 0.0000 \\ 0.0000 & 0.0000 \\ 0.0000 & 0.0000 \\ 0.0000 & 0.0000 \\ 0.0000 & 0.0000 \\ 0.0000 & 0.0000 \\ 0.0000 & 0.0000 \\ 1.6660 & 0.0000 \\ 0.0000 & 10.0000 \end{bmatrix}
$$

$$
C = \begin{bmatrix} 1.0000 & 0.0000 & 0.0000 & 0.0000 & 0.0000 & 0.0000 & 0.0000 & 0.0000 & 0.0000 & 0.0000 \\ -0.4910 & 0.0000 & -0.6320 & 0.0000 & 0.0000 & -0.2074 & 0.0000 & 0.0000 & 0.0000 & 0.0000 \end{bmatrix}
$$

$$
A = 1.0D+03 *
$$

$$
\begin{bmatrix}
0.0000 & 0.0010 & 0.0000 & 0.0000 & 0.0000 & 0.0000 & 0.0000 & 0.0000 & 0.0000 & 0.0000 \\
0.0000 & -0.0001 & -0.0010 & -0.0118 & -0.0118 & -0.0631 & -0.0343 & -0.0343 & -0.0276 & 0.0000 \\
0.3241 & -0.0012 & -0.0291 & 0.0001 & 0.0028 & -0.9677 & -0.6781 & -0.6781 & 0.0000 & -0.1293 \\
-0.1273 & 0.0005 & 0.0114 & -0.0010 & 0.0131 & 0.3801 & 0.2663 & 0.2663 & 0.0000 & 1.0548 \\
-0.1860 & 0.0007 & 0.0167 & 0.0009. & -0.0171 & 0.5555 & 0.3893 & 0.3893 & 0.0000 & -0.8749 \\
0.3419 & 0.0011 & 1.0527 & 0.7565 & 0.7565 & -0.0298 & 0.0002 & 0.0033 & 0.0000 & 0.0000 \\
-0.0307 & -0.0001 & -0.0947 & -0.0680 & -0.0680 & 0.0027 & -0.0027 & 0.0049 & 0.0000 & 0.0000 \\
-0.3023 & -0.0010 & -0.9309 & -0.6689 & -0.6689 & 0.0263 & 0.0024 & -0.0096 & 0.0000 & 0.0000 \\
0.0000 & 0.0000 & 0.0000 & 0.0000 & 0.0000 & 0.0000 & 0.0000 & 0.0000 & -0.0017 & 0.0000 \\
0.0000 & 0.0000 & 0.0000 & 0.0000 & 0.0000 & 0.0000 & 0.0000 & 0.0000 & 0.0000 & -0.0100
\end{bmatrix}
$$

B-7   : NSRE
Description   : Non-square model of a chemical reactor
Reference   : Munro, 1972
No. of inputs   : 2
No. of outputs   : 3
No. of states   : 4
State space model :

D   =

$$\begin{bmatrix} 0. & 0. \\ 0. & 0. \\ 0. & 0. \end{bmatrix}$$

C   =

$$\begin{bmatrix} 1. & 0. & 0. & 0. \\ 0. & 1. & 0. & 0. \\ 0. & 0. & 1. & -1. \end{bmatrix}$$

B   =

$$\begin{bmatrix} 0.0000 & 0.0000 \\ 5.6790 & 0.0000 \\ 1.1360 & -3.1460 \\ 1.1360 & 0.0000 \end{bmatrix}$$

A   =

$$\begin{bmatrix} 1.4000 & -0.2080 & 6.7150 & -5.6760 \\ -0.5810 & -4.2900 & 0.0000 & 0.6750 \\ 1.0670 & 4.2730 & -6.6540 & 5.8930 \\ 0.0480 & 4.2730 & 1.3430 & -2.1040 \end{bmatrix}$$

B-8   : AIRC
Description   : Aircraft
Reference   : Kouvaritakis et al., 1979
No. of inputs   : 3
No. of outputs   : 3
No. of states   : 5
State space model :

D   =

$$\begin{bmatrix} 0. & 0. & 0. \\ 0. & 0. & 0. \\ 0. & 0. & 0. \end{bmatrix}$$

C   =

$$\begin{bmatrix} 1. & 0. & 0. & 0. & 0. \\ 0. & 1. & 0. & 0. & 0. \\ 0. & 0. & 1. & 0. & 0. \end{bmatrix}$$

B   =

$$\begin{bmatrix} 0.0000 & 0.0000 & 0.0000 \\ -0.1200 & 1.0000 & 0.0000 \\ 0.0000 & 0.0000 & 0.0000 \\ 4.4190 & 0.0000 & -1.6650 \\ 1.5750 & 0.0000 & -0.0732 \end{bmatrix}$$

A   =

$$\begin{bmatrix} 0.0000 & 0.0000 & 1.1320 & 0.0000 & -1.0000 \\ 0.0000 & -0.0538 & -0.1712 & 0.0000 & 0.0705 \\ 0.0000 & 0.0000 & 0.0000 & 1.0000 & 0.0000 \\ 0.0000 & 0.0485 & 0.0000 & -0.8556 & -1.0130 \\ 0.0000 & -0.2909 & 0.0000 & 1.0532 & -0.6859 \end{bmatrix}$$

Appendix C   Examples of the Design Knowledge Base represented using Frames

FRAME NAME   : DESIGN.TECHNIQUES

CONNECTION SLOT   : Nil

ATTRIBUTE SLOTS

    Design attribute   : D1 - SIMPLE.DESIGN.TECHNIQUE
    (Own)   D2 - REVERSE.FRAME.ALIGNMENT
                         D3 - OBSERVER-BASED.CONTROLLER

    Evidence attribute :
    (Own)   E1 - SIMPLE.DESIGN.TECHNIQUE has not been attempted.
                         E2 - SIMPLE.DESIGN.TECHNIQUE has been attempted but
                              the design is not satisfactory.
                         E3 - SIMPLE.DESIGN.TECHNIQUE & REVERSE.FRAME.ALIGNMENT
                              has been attempted but the design is not
                              satisfactory.

    Concept attribute : To manipulate the primary indicators of the open-loop
    (Own)   transfer function matrix into a suitable form.

PROCEDURE SLOTS

    Control slot   : Nil

    Rule slot   : If (E1) is (Yes), then (D1) is (Yes).
                         If (E2) is (Yes), then (D2) is (Yes).
                         If (E3) is (Yes), then (D3) is (Yes).

    Exit slot   : SIMPLE.DESIGN.TECHNIQUE /
                         REVERSE.FRAME.ALIGNMENT /
                         OBSERVER-BASED.CONTROLLER

    Context slot   : Nil

Frame C-1

FRAME NAME   : SIMPLE.DESIGN.TECHNIQUE

CONNECTION SLOT   : DESIGN.TECHNIQUES

ATTRIBUTE SLOTS

  Design attribute   : D1 - HIGH.FREQ.SUB-CONTROLLER
  (Own)   D2 - LOW.FREQ.SUB-CONTROLLER
                             D3 - OTHER

  Evidence attribute : Nil

  Concept attribute : To design for good phase properties at high frequencies
  (Own)   and good gain properties at low frequencies.

PROCEDURE SLOTS

  Control slot   : D1 $\longrightarrow$ D2 $\longrightarrow$ D3

  Rule slot   : Nil

  Exit slot   : DESIGN.TECHNIQUES

  Context slot   : Nil

Frame C-2

```
FRAME NAME              : HIGH.FREQ.SUB-CONTROLLER (HFS)

CONNECTION SLOT         : SIMPLE.DESIGN.TECHNIQUE
                          REVERSE.FRAME.ALIGNMENT
                          OBSERVER-BASED.CONTROLLER

ATTRIBUTE SLOTS

  Design attribute    : D1 - Gain balancing at high frequencies.
  (Own)                 D2 - Gain balancing at bandwidth frequency.

  Evidence attribute  : E1 - High frequency roll-off rates.
  (Own)                 E2 - Condition number at high frequencies.
                        E3 - Condition number at bandwidth frequency.

  Concept attribute   : To balance the gains and align the input and output
  (Own)                      gain frames at high frequencies.

PROCEDURE SLOTS

  Control slot        : E1

  Rule slot           : If (E1) is (the same), then (D1) is (Yes),
                             else (D2) is (Yes).
                         If (D1) is (Yes), then evaluate (E2),
                             else evaluate (E3).

  Exit slot           : HFS.CHECK

  Context slot        : Nil
```

Frame C-3

FRAME NAME   : HIGH.FREQ.SUB-CONTROLLER.CHECK (HFS.CHECK)

CONNECTION SLOT   : HFS

ATTRIBUTE SLOTS

   Design attribute   : D1 - To obtain other condition number(s)
   (Own)   (usually smaller).
                    : D2 - To change the signs of the gain parameters.
                      D3 - To drop using the chosen type of controlller.

   Evidence attribute : E1 - Gain divergences worsen at intermediate
   (Own)   frequencies.
                      E2 - Wrong gain-phase characteristics at high
                            frequencies.
                      E3 - To keep trying the chosen type of sub-controller.

   Concept attribute : To check the operation of the HFS.
   (Own)

PROCEDURE SLOTS

   Control slot   : E1 $\longrightarrow$ E2

   Rule slot   : If (E1) is (Yes) and (E3) is (Yes), then (D1) is (Yes).
                      If (E2) is (Yes) and (E3) is (Yes), then (D2) is (Yes).
                      If (D1) is (No) and (D2) is (No), then exit slot is
                            connection slot of HFS, else exit slot is HFS.

   Exit slot   : Connection slot of HFS / HFS

   Context slot   : Nil

Frame C-4

<u>FRAME NAME</u>   : REVERSE.FRAME.ALIGNMENT

<u>CONNECTION SLOT</u>   : DESIGN.TECHNIQUES

<u>ATTRIBUTE SLOTS</u>

    Design attribute   : D1 - HIGH.FREQ.SUB-CONTROLLER
    (Own)   D2 - REVERSE.FRAME.APPROXIMATION.SUB-CONTROLLER
                         D3 - LOW.FREQ.SUB-CONTROLLER
                         D4 - OTHER
                         D5 - OPTIMIZER

    Evidence attribute : Nil

    Concept attribute : C1 - To use HFS at high frequencies, RFA at
    (Own)   intermediate frequencies and LFS at
                              low frequencies.
                    C2 - To use the optimizer when fine tuning of
                              the parameters at intermediate frequencies
                              is needed.

<u>PROCEDURE SLOTS</u>

    Control slot   : D1 $\longrightarrow$ D2 $\longrightarrow$ D3 $\longrightarrow$ D4 or
                    : D1 $\longrightarrow$ D2 $\longrightarrow$ D5 $\longrightarrow$ D3 $\longrightarrow$ D4

    Rule slot   : Nil

    Exit slot   : DESIGN.TECHNIQUES

    Context slot   : Nil

<u>Frame C-5</u>

FRAME NAME : REVERSE.FRAME.APPROXIMATION.SUB-CONTROLLER (RFA)

CONNECTION SLOT : REVERSE.FRAME.ALIGNMENT
OBSERVER-BASED.CONTROLLER

ATTRIBUTE SLOTS

Design attribute : D1 - RFA1
(Own) D2 - RFA2
D3 - RFA3
D4 - RFA4
D5 - RFA5

Evidence attribute :
(Own) E1 - Phase lead.
E2 - Phase lag.
E3 - Lead-lag.
E4 - Lag-lead.
E5 - Conjugate-pole cancellation.
(Member) E6 - Frequency at which the RFA is formed.
E7 - To decide whether the frequency at which the RFA
is formed needs to be adjusted automatically.
E8 - To choose a locus.

PROCEDURE SLOTS

Control slot : Nil

Rule slot : If (E1) is (Yes), then D1.
If (E2) is (Yes), then D2.
If (E3) is (Yes), then D3.
If (E4) is (Yes), then D4.
If (E5) is (Yes), then D5.

Exit slot : RFA1 / RFA2 / RFA3 / RFA4 / RFA5

Context slot : Nil

Frame C-6

FRAME NAME   : RFA1

CONNECTION SLOT   : RFA

ATTRIBUTE SLOTS

   Design attribute   : D1 - To form a RFA sub-controller with phase advance.
   (Own)

   Evidence attribute : E1 - Amount of phase.
   (Own)   E2 - Frequency at which the phase is added.
                    E3 - Frequency at which the RFA is formed.
                    E4 - To decide whether the frequency at which the RFA
                            is formed needs to be adjusted automatically.
                    E5 - To choose a locus.

   Concept attribute : Phase compensation at intermediate frequencies.
   (Own)

(*Note that E3, E4 and E5 are inherited from the RFA frame*)

PROCEDURE SLOTS

   Control slot   : Nil

   Rule slot   : Nil

   Exit slot   : RFA1.CHECK

   Context slot   : Nil

Frame C-7

FRAME NAME   : OBSERVER-BASED.CONTROLLER

CONNECTION SLOT   : DESIGN.TECHNIQUES

ATTRIBUTE SLOTS

  Design attribute   : D1 - FULL-ORDER.OBC
  (Own)   D2 - REDUCED-ORDER.OBC
                    D3 - SIMPLE.DESIGN.TECHNIQUE
                    D4 - REVERSE.FRAME.ALIGNMENT

  Evidence attribute : E1 - Full-order observer gives a satisfactory frequency
  (Own)   response.
                  E2 - Full-order observer with SDT gives a satisfactory
                   result.
                  E3 - Full-order observer with Reverse Frame Alignment
                   gives a satisfactory result.
                  E4 - Reduced-order observer gives a satisfactory
                   frequency response.

  Concept attribute   :
  (Own)   C1 - To use a full-order observer and conclude the
                   design with the SDT or the Reverse Frame
                   Alignment technique.
                  C2 - The final aim is to use a reduced-order observer-
                   based controller.
  (Member)   C3 - To obtain stable pole positions.
                  C4 - The gains of the controller should not be
                   unacceptably high.

PROCEDURE SLOTS

  (*Here we give the flowchart instead of the control and rule slots*)

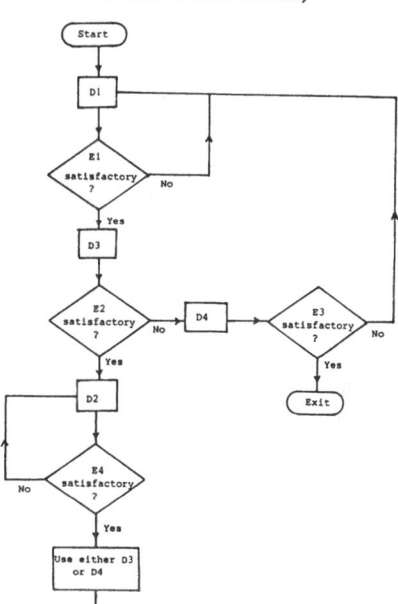

Exit slot   : DESIGN.TECHNIQUES

Context slot   : Nil

Frame C-8

FRAME NAME   : FULL-ORDER.OBC

CONNECTION SLOT   : OBSERVER-BASED.CONTROLLER

ATTRIBUTE SLOTS

 Design attribute : D1 - STATE.FEEDBACK
 (Own)   D2 - FULL-ORDER.OBSERVER

 Evidence attribute : Nil

 Concept attribute :
 (Own)   C1 - To obtain a satisfactory frequency response.
 (Member)   C2 - To obtain stable pole positions.
         C3 - The gains of the controller should not be
            unacceptably high.

(Note that C2 and C3 are inherited from the frame OBSERVER-BASED.CONTROLLER)

PROCEDURE SLOTS

 Control slot   : D1 $\longrightarrow$ D2

 Rule slot   : Nil

 Exit slot   : OBSERVER-BASED.CONTROLLER

 Context slot   : Nil

Frame C-9

FRAME NAME   : STATE.FEEDBACK

CONNECTION SLOT   : FULL-ORDER.OBC

ATTRIBUTE SLOTS

   Design attribute   : D1 - ROBUST.POLE.PLACEMENT
   (Own)   D2 - LQR

   Evidence attribute : E1 - Robust pole placement has not been used.
   (Own)   E2 - State feedback gain controller has unacceptable
                              high gains.

   Concept attribute : C1 - To obtain stable pole positions.
   (Own)   C2 - The gains of the controller should not be
                              unacceptably high.

(Note that C1 and C2 are inherited from FULL-ORDER.OBC)

PROCEDURE SLOTS

   Control slot   : E1 ⟶ E2

   Rule slot   : If (E1) is (Yes), then (D1) is (Yes).
                     If (E2) is (Yes), then (D2) is (Yes).

   Exit slot   : FULL-ORDER.OBC

   Context slot   : Nil

Frame C-10

Appendix D   Example of a Frame in KEE

Unit: HIGH.FREQ.SUB-CONTROLLER (HFS) in knowledge base MAID
Created by Grantham Pang on 1-July-86 00:00:00
Sub-class: SIMPLE.DESIGN.TECHNIQUE
          REVERSE.FRAME.ALIGNMENT
          OBSERVER-BASED.CONTROLLER
Member: Nil

OwnSlot: GAIN.BALANCING.AT.HIGH.FREQ from HFS
Inheritance: OVERRIDE
ValueClass: (WORD)
Value: Unknown
COMMENT: Value assigned by rules

OwnSlot: GAIN.BALANCING.AT.BAND.FREQ from HFS
Inheritance: OVERRIDE
ValueClass: (WORD)
Value: Unknown
COMMENT: Value assigned by rules

OwnSlot: HIGH.FREQ.ROLL.OFF.RATES from HFS
Inheritance: OVERRIDE
ValueClass: (WORD)
Value: Unknown
COMMENT: Value assigned by user

OwnSlot: COND.NUMBER.AT.HIGH.FREQ from HFS
Inheritance: OVERRIDE
ValueClass: (NUMBER)
Value: Unknown
COMMENT: Value assigned by user
RANGE: (1 1000)

OwnSlot: COND.NUMBER.AT.BAND.FREQ from HFS
Inheritance: OVERRIDE
ValueClass: (NUMBER)
Value: Unknown
COMMENT: Value assigned by user
RANGE: (1 1000)

OwnSlot: CONDITION.NUMBER from HFS
Inheritance: OVERRIDE
ValueClass: ((COND.NUMBER.AT.HIGH.FREQ)(COND.NUMBER.AT.BAND.FREQ))
Value: Unknown
COMMENT: Value assigned by rules

OwnSlot: CHECKING.TYPE from HFS
Inheritance: OVERRIDE
ValueClass: (SUB-CONTROLLER.CHECK)
Value: Unknown
COMMENT: Value assigned by rules

OwnSlot: RULES from HFS
Inheritance: OVERRIDE
Value: [(RULE 1 (IF (HIGH.FREQ.ROLL.OFF.RATES IS SAME))
                (THEN(GAIN.BALANCING.AT.HIGH.FREQ IS YES)
                    (CONDITION.NUMBER IS COND.NUMBER.AT.HIGH.FREQ)
                    (CHECKING.TYPE IS HFS.CHECK)))

        (RULE 2 (IF (HIGH.FREQ.ROLL.OFF.RATES IS NOT.SAME))
                (THEN(GAIN.BALANCING.AT.BAND.FREQ IS YES)
                    (CONDITION.NUMBER IS COND.NUMBER.AT.BAND.FREQ)
                    (CHECKING.TYPE IS HFS.CHECK)))]

Appendix E   Proof of Theorem 3.4.2

The conditions for partial gain balancing and partial alignment will first be stated.

Let $G \in C^{m \times m}$ have a singular value decomposition (SVD)

$$G = Y \cdot \Sigma \cdot U^*$$

$$= [\, y_1 \ \ y_2 \ \cdots \ y_m \,]
\begin{bmatrix} \sigma_1 & . & 0 \\ & . & \\ & . & \\ 0 & & \sigma_m \end{bmatrix}
\begin{bmatrix} u_1^* \\ \vdots \\ u_m^* \end{bmatrix}
\qquad [E.1]$$

where the singular values $(\sigma_1, \ldots, \sigma_m)$ are arranged in descending order of magnitude. Next, the singular values and the corresponding singular vectors are arranged in the following manner. Let the first $i_1 (\geq 1)$ singular values be equal and let

$$\tilde{Y}_1 = [\, y_1 \ \cdots \ y_{i_1} \,] , \qquad \tilde{\Sigma}_1 = \sigma_1 I_{i_1} , \qquad \tilde{U}_1^* = \begin{bmatrix} u_1^* \\ \vdots \\ u_{i_1}^* \end{bmatrix} .$$

Similarly, if the next i_2 singular values are equal (but distinct from σ_1), then we put

$$\tilde{Y}_2 = [\, y_{i_1+1} \ \cdots \ y_{i_1+i_2} \,] , \qquad \tilde{\Sigma}_2 = \sigma_{i+1} I_{i_2} , \qquad \tilde{U}_2^* = \begin{bmatrix} u_{i_1+1}^* \\ \vdots \\ u_{i_1+i_2}^* \end{bmatrix} \qquad \text{etc.}$$

In this way, the SVD [E.1] can be written in a block form

$$G = \tilde{Y} \cdot \tilde{\Sigma} \cdot \tilde{U}^*$$

$$= [\, \tilde{Y}_1 \ \ \tilde{Y}_2 \ \cdots \ \tilde{Y}_r \,]
\begin{bmatrix} \tilde{\Sigma}_1 & . & 0 \\ & . & \\ & . & \\ 0 & & \tilde{\Sigma}_r \end{bmatrix}
\begin{bmatrix} \tilde{U}_1^* \\ \vdots \\ \tilde{U}_r^* \end{bmatrix}
\qquad [E.2]$$

with $i_1 + i_2 + \cdots + i_r = m$.

Hence, we have a " block SVD " in which each Σ_j is a scalar matrix containing i_j equal singular values and $\tilde{U}_j$, $\tilde{Y}_j$ define the input and output singular subspaces corresponding to the i_j equal singular values in Σ_j. We say that G is partially gain balanced and partially aligned (PGB & PA) if for any k and ℓ ($k \leq r$, $\ell \leq r$),

$$\tilde{U}_k^* \cdot \tilde{Y}_\ell = 0 \qquad [E.3]$$

for all $\ell \neq k$.

That is, the input subspace spanned by $\tilde{U}_k$ is orthogonal to all the other output subspaces except that spanned by $\tilde{Y}_k$. Alternatively, the condition [E.3] can be stated as

$$\tilde{U}^* \cdot \tilde{Y} = \begin{bmatrix} Z_1 & & 0 \\ & \cdot & \\ & & \cdot \\ 0 & & \cdot \ Z_r \end{bmatrix} \qquad [E.4]$$

That is, the product of the singular vector frames is block diagonally structured, where the dimensions of the diagonal blocks Z_j are equal to the dimensions of the Σ_j's, respectively.

Lemma 1

If G is PGB & PA, then in [E.4], each of the diagonal sub-blocks Z_j is unitary.

Proof of Lemma 1:

From [E.4],

$$\tilde{Y} = \tilde{U} \cdot \begin{bmatrix} Z_1 & & 0 \\ & \cdot & \\ & & \cdot \\ 0 & & \cdot \ Z_r \end{bmatrix} \qquad [E.5]$$

so that

$$\tilde{Y}_j = \tilde{U}_j \cdot Z_j \qquad [E.6]$$

($j = 1, \ldots, r$) .

Multiplying both sides of [E.6] by the respective conjugate transpose, we have

$$\tilde{Y}_j^* \cdot \tilde{Y}_j = Z_j^* \cdot \tilde{U}_j^* \cdot \tilde{U}_j \cdot Z_j$$

Hence,

$$I = Z_j^* \cdot Z_j$$

and Z_j is unitary.   □

Proof of Theorem 3.4.2

(<==)

      Suppose G is PGB & PA, then G can be written in the form [E.2] where the singular vector subspaces satisfy condition [E.4]. Substituting [E.5] into [E.2], we have

$$G = [\ \tilde{U}_1 \ \ \tilde{U}_2 \ \cdots \ \tilde{U}_r \ ] \begin{bmatrix} Z_1 & \cdot & & 0 \\ & \cdot & \cdot & \\ 0 & & \cdot & Z_r \end{bmatrix} \begin{bmatrix} \Sigma_1 & \cdot & & 0 \\ & \cdot & \cdot & \\ 0 & & \cdot & \Sigma_r \end{bmatrix} \begin{bmatrix} \tilde{U}_1^* \\ \vdots \\ \tilde{U}_r^* \end{bmatrix} \qquad [E.7]$$

By Lemma 1, each Z_j is unitary and hence has a CVD given by

$$Z_j = W_j \cdot e^{j\theta_j} \cdot W_j^* \qquad [E.8]$$

where

W_j is unitary , and $e^{j\theta_j}$ is a diagonal phase matrix.

Substituting [E.8] into [E.7] and noting that W_j^* commutes with Σ_j (since Σ_j is scalar), we have that

$$G = [(\tilde{U}_1 W_1) \ \cdots \ (\tilde{U}_r W_r)] \begin{bmatrix} \Sigma_1 e^{j\theta_1} & \cdot & & 0 \\ & \cdot & \cdot & \\ 0 & & \cdot & \Sigma_r e^{j\theta_r} \end{bmatrix} \begin{bmatrix} (\tilde{U}_1 W_1)^* \\ \vdots \\ (\tilde{U}_r W_r)^* \end{bmatrix} \qquad [E.9]$$

The above is a CVD of G with an eigenvector frame which is unitary. Hence, G is normal.

(==>)   (Hung and MacFarlane, 1982)

      If G is normal, then it is unitarily similar to its diagonal matrix of eigenvalues. That is,

$$G = W \cdot \Lambda \cdot W^*$$

$$= W \cdot | \ \Lambda \ | \cdot \arg(\Lambda) \cdot W^*$$

$$= W \cdot | \; \Lambda \; | \cdot U^* \qquad\qquad [E.10]$$

where

W is unitary, Λ is a diagonal matrix with eigenvalues on its diagonal, and

U is defined to be $W \cdot \arg(\Lambda^*)$.

Note that [E.10] may be taken as an SVD of G.

If G is normal and has distinct singular values, then G is aligned. However,

if G has some equal singular values, say the first i ($1 \leq i \leq m$) singular

values are equal, then

$$G = \begin{bmatrix} & \vdots & \\ W_1 & \vdots & W_2 \\ & \vdots & \end{bmatrix} \begin{bmatrix} \sigma & & & & \\ & \ddots & & & \\ & & \sigma & & \\ & & & \sigma_{i+1} & \\ & & & & \ddots & \\ & & & & & \sigma_m \end{bmatrix} \begin{bmatrix} U_1^* \\ \text{-----} \\ U_2^* \end{bmatrix} \qquad [E.11]$$

The singular vectors of output subspace W_1 need not be aligned with singular

vectors of input subspace U_1^* because we can have

$$G = \begin{bmatrix} & \vdots & \\ W_1 X & \vdots & W_2 \\ & \vdots & \end{bmatrix} \begin{bmatrix} \sigma & & & & \\ & \ddots & & & \\ & & \sigma & & \\ & & & \sigma_{i+1} & \\ & & & & \ddots & \\ & & & & & \sigma_m \end{bmatrix} \begin{bmatrix} X^* U_1^* \\ \text{-----} \\ U_2^* \end{bmatrix} \qquad [E.12]$$

where X is unitary.

However, W_2 and U_2 are aligned. Hence, G is partially gain-balanced and

partially aligned. $\qquad\qquad\qquad\qquad\qquad\qquad\qquad\qquad\qquad\qquad\quad$ □

Appendix F   Effect of Scaling the Units on the Sensitivity of Eigenvalues

During a design process, it is always assumed that appropriate units for the inputs and outputs have been determined. However, it is not obvious how this should be performed and the procedure adopted usually depends on the experience and intuition of the designer. Munro (1985) has discussed the scaling of the system input and output variables with an aim of achieving a modified system which is diagonal dominant. Here, a discussion on the scaling of units and its effect on the eigenvalues and singular values is given.

Consider the 2-input, 2-output ship propulsion control system below:

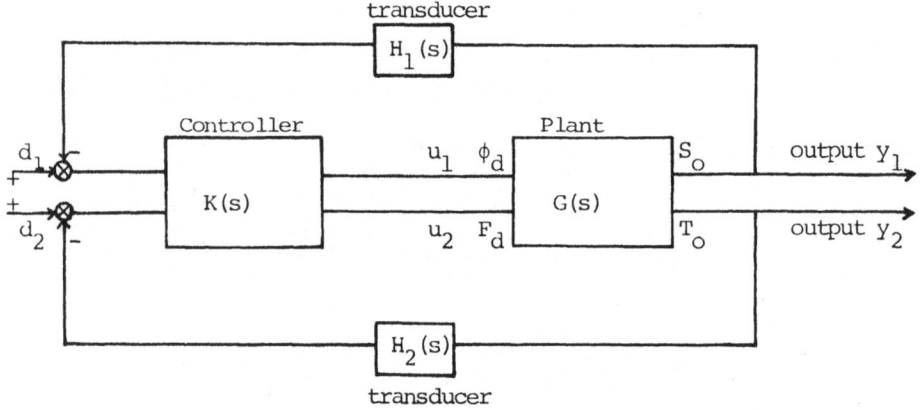

Fig. F.1   Feedback control of a ship propulsion system

The two inputs (u_1, u_2) are

     ϕ_d : propeller pitch angle (degrees)

     F_d : gas turbine fuel rate (per cent)

The two outputs (y_1, y_2) are

     S_o : propeller speed (rev/min)

     T_o : propeller shaft torque (Nm)

d_1 and d_2 are the desired propeller speed and shaft torque. Note that the units of the outputs are rev/min and Nm. If the units of the outputs S_o and T_o are changed to rev/sec and kNm respectively, this would be equivalent to cascading a pre- and post-modifier to the plant as in Fig. F.2.

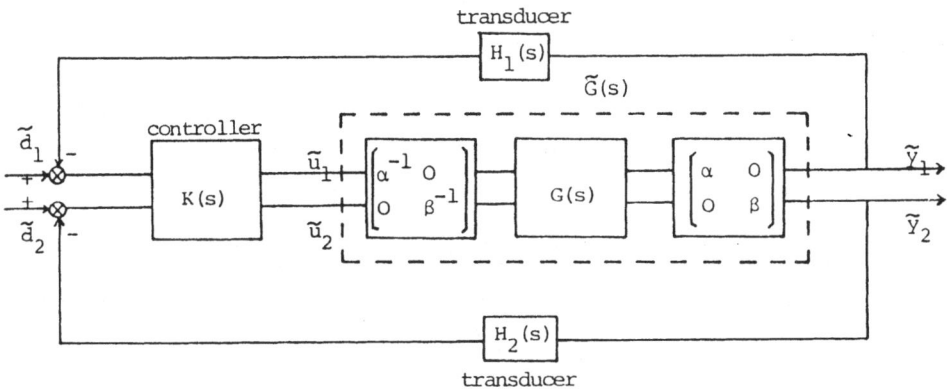

Fig. F.2 Feedback control of a ship propulsion system with scaling of units

where $\alpha = 60,$ $\beta = 0.001,$ $\tilde{y}_1 = \alpha \cdot y_1,$ $\tilde{y}_2 = \beta \cdot y_2,$

$\tilde{u}_1 = \alpha \cdot u_1,$ $\tilde{u}_2 = \beta \cdot u_2,$ $\tilde{d}_1 = \alpha \cdot d_1,$ $\tilde{d}_2 = \beta \cdot d_2$

and

$$\tilde{G}(s) = diag\{\alpha, \beta\} \cdot G(s) \cdot diag\{\alpha^{-1}, \beta^{-1}\} \ . \qquad [F.1]$$

Let

$$G(s) = \begin{bmatrix} g_{11}(s) & g_{12}(s) \\ g_{21}(s) & g_{22}(s) \end{bmatrix}$$

then

$$\tilde{G}(s) = \begin{bmatrix} g_{11}(s) & \alpha\beta^{-1}g_{12}(s) \\ \alpha^{-1}\beta \cdot g_{21}(s) & g_{22}(s) \end{bmatrix}$$

It is interesting to note that the scaling factors α and β appear only in the off-diagonal elements of $\tilde{G}(s)$. Since $G(j\omega)$ and $\tilde{G}(j\omega)$ are similar matrices at all frequencies ω, they have the same eigenvalues and their generalised Nyquist diagrams are unchanged. However, the singular values of $G(j\omega)$ and $\tilde{G}(j\omega)$ would be different. In fact, by varying α and β, a considerable change on the singular values may be obtained. It was shown that (Theorem 3.4.13) the divergences between the characteristic and principal gains give an indication of the robustness of a system. Hence, there must be a choice of α and β which will give minimal divergence of gain and hence minimize the sensitivity of the generalised Nyquist diagrams.

Another view of input and output scalings is the following. It is known that the use of input and output scalings can change the Gershgorin bands on the closed-loop Nyquist array of a system (Edmunds et al., 1983). As Gershgorin bands on this array indicate how robust the control is for changes in the characteristics of the sensors, therefore input and output scalings can affect inferences about the robustness of a system.

Hence, a designer should check the units being used, and examine whether the design problem can be better-posed by scaling the units. There are two ways which this scaling may be carried out:

a. Before carrying out a design, a scaling of units may be performed so that the eigenloci are as robust as possible.

b. Before completing a design, a pair of diagonal scaling matrices may be included as pre- and post-compensators to the plant. The objective is to minimize the gain divergences with respect to the scaling parameters.

Example F.1

This example will illustrate the effect of scaling the units before carrying out a design. The system REAC in Section 5.5.5 is considered; let its open-loop transfer function matrix be G(s). After a few trials on the values of α and β, a ratio of $\alpha/\beta = 1.4$ is found to be suitable and the gain divergences have been reduced as shown by the primary indicators in Fig. F.3. The indicator MS is given in Fig. F.4 and this skewness measure is found to be below 0.06 at high frequencies whereas it is about 0.39 for the originial system of REAC (Fig. F.5). Note that the actual values of α and β have not yet been fixed and only their ratio is determined. For simplicity, $\alpha = 1.4$ and $\beta = 1$ are chosen. The design on $\tilde{G}(s)$ is then carried out where $\tilde{G}(s)$ is expressed as in [F.1]. A high frequency sub-controller is obtained as follows:

$$ K_\infty = \begin{bmatrix} 0 & 1.0854 \\ -1 & 0.0000 \end{bmatrix} $$

A low frequency sub-controller with a scalar gain of 10 is obtained as follows:

$$ K_L(s) = 10 \cdot (2 \cdot I_2/s + I_2) $$

The primary indicators and indicator MS are given in Figs. F.7 & F.8 and these are found to be very similar to the design using the SDT (see Figs. 5.5.3 & 5.5.6). However, the inputs and outputs of the plant are now different from the previous design.

Example F.2

This example is used to illustrate the effect of scaling the units before completing a design. The system in Section 5.5.2 is considered. The primary indicators of the compensated system are shown in Fig. 5.2.3. It can be seen from the indicator MS (Fig. 5.2.5) that the skewness of the system has a peak of 0.43 at about 1 rad/s. However, a scaling of the units with $\alpha/\beta = 0.4$ is chosen and the indicator MS after scaling (Fig. F.9) shows that the peak value is now around 0.11. The primary indicators are given in Fig. F.10 and it can be seen that a more robust design has been obtained by scaling the units.

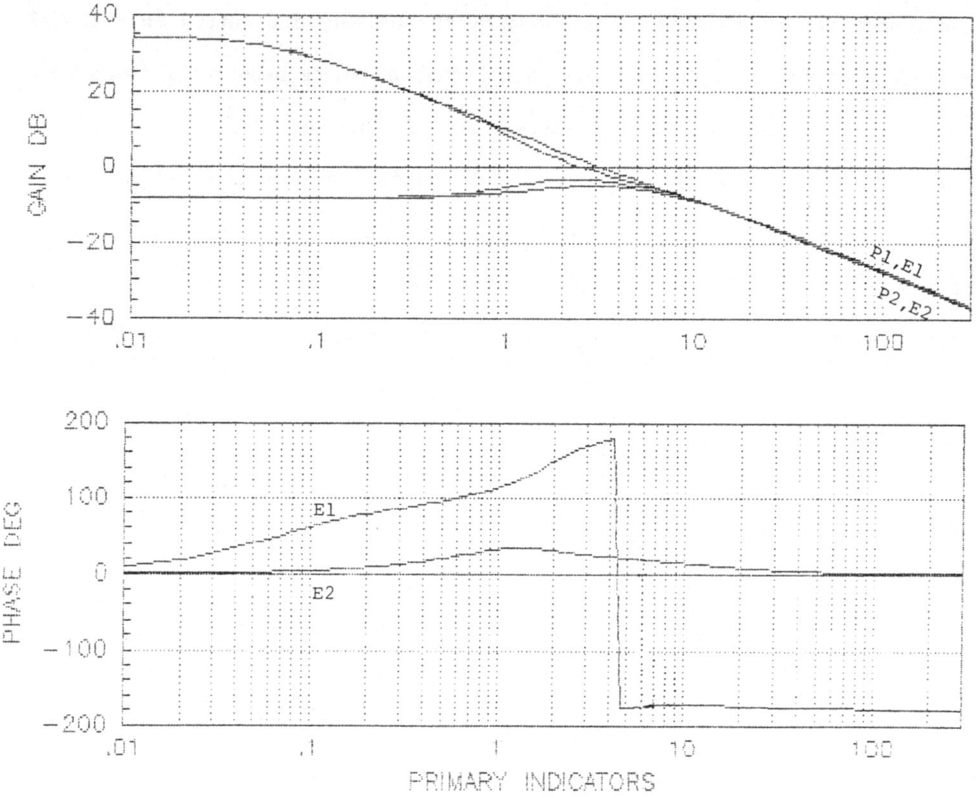

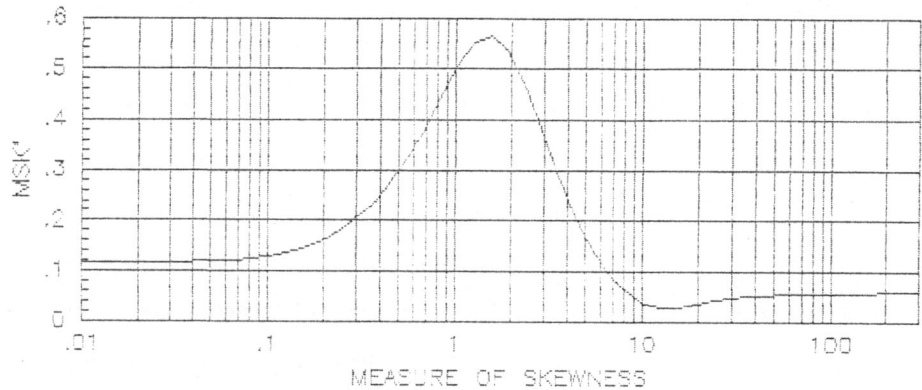

Fig. F.3   The primary indicators of REAC with unit scalings (α/β = 1.4)

Fig. F.4   Measure of skewness of REAC with unit scalings (α/β = 1.4)

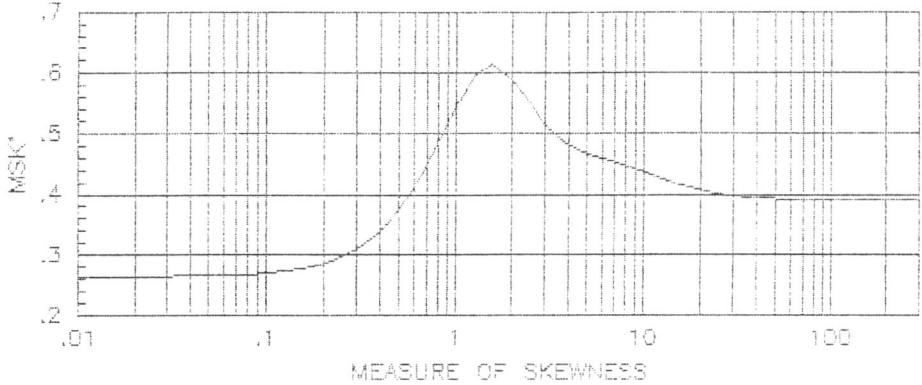

Fig. F.5   Measure of skewness of the original system REAC

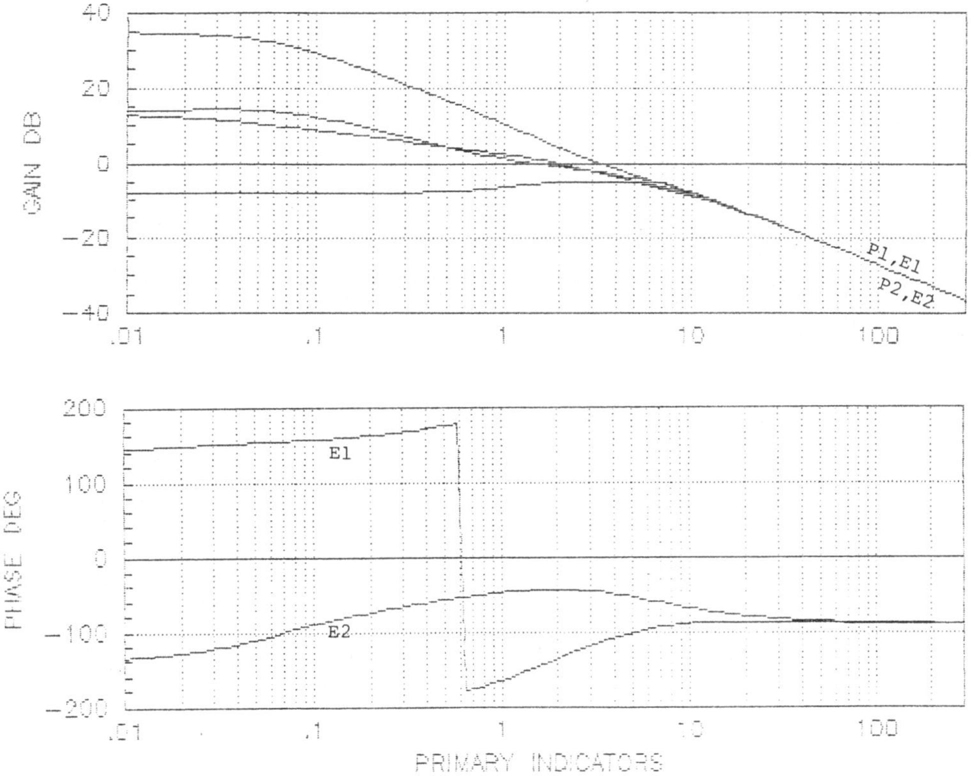

Fig. F.6   The primary indicators of G(s) after high frequency
           region design

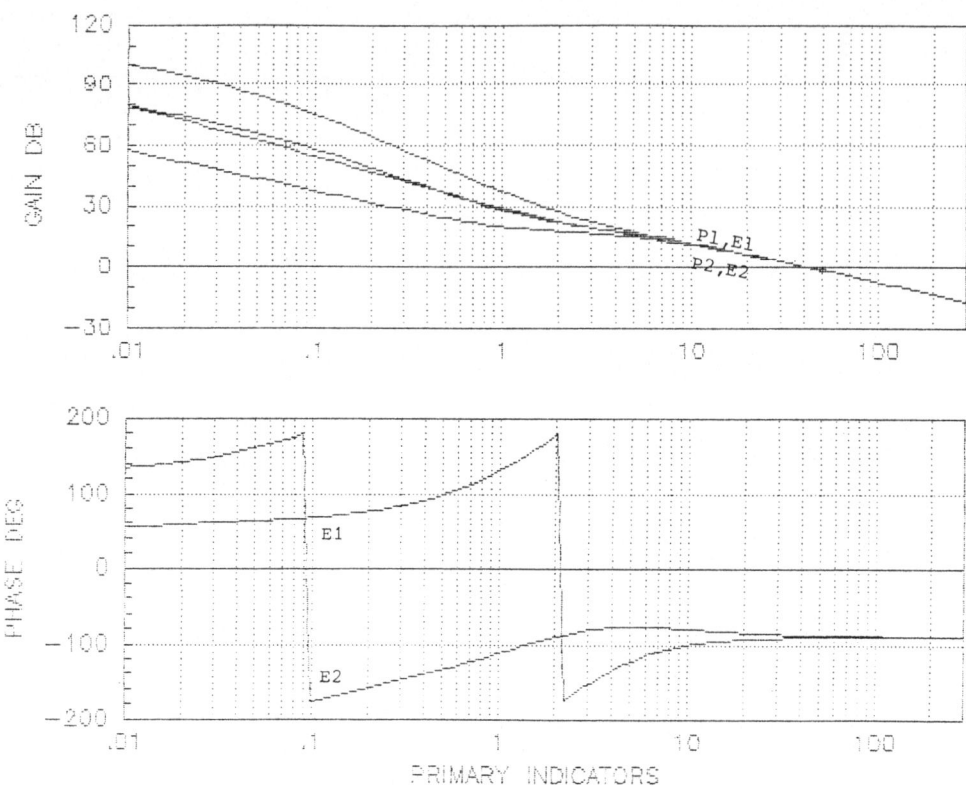

Fig. F.7   The primary indicators of the final design of $\tilde{G}(s)$

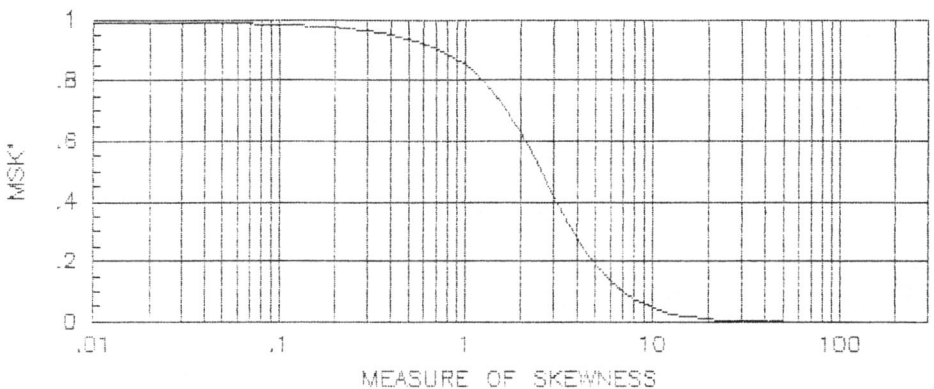

Fig. F.8   The indicator MS of the final design of $\tilde{G}(s)$

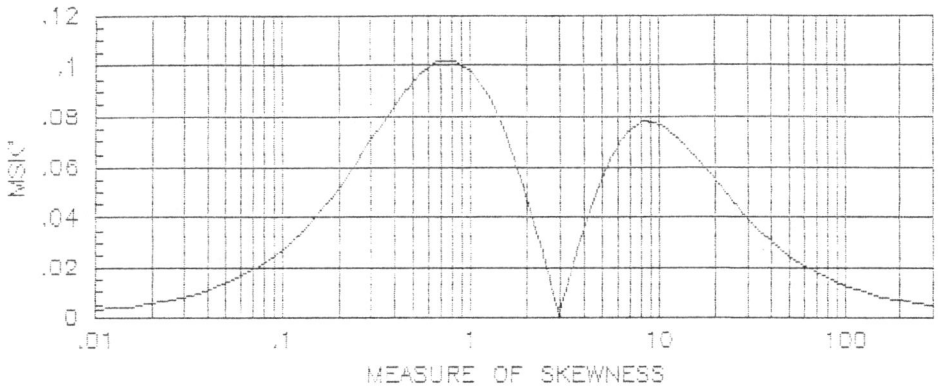

Fig. F.9   The indicator MS of the final design with
unit scalings (α/β = 0.4)

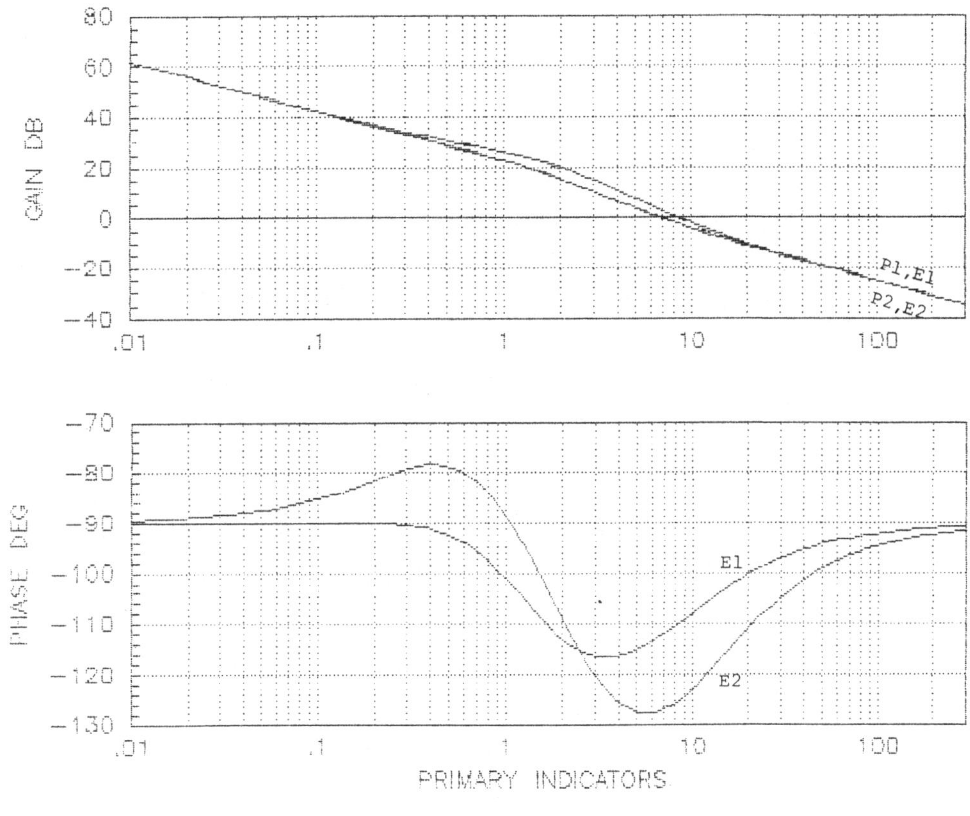

Fig. F.10   The primary indicators of the final design with
unit scalings (α/β = 0.4)

Appendix G   Example of a design session on GROC using MAID

       The design of the system GROC using the Simple Design Technique is
given in Section 5.5.4. The following is an example of a design session on
GROC using the prototype expert system MAID. The responses from the user are
shown in bold. In the example, the user would respond WHY to a question where
he want to query MAID why that question is asked. The subsequent explanations
from MAID are shown in *italics*. Conceptual information requested by the user
is shown enclosed by {{ brackets }} and the names of the function that the
user should use in MATRIX$_x$ are enclosed by ((brackets)). Any comments which
are not part of the dialogue are enclosed by [*   *]

    [* Start of the design session *]

> **Simple Design Technique ?**

Report
Investigating the high frequency region

Are the high frequency roll-off rates the same ? ((PRIMARY))
    [* The user should look at the primary indicators of the
        system (Fig. 5.4.1) *]

> **help high frequency design**
    [* Instead of answering the question, the user queries the
        system for concepts relating to the high frequency design *]

  {{ 1/ THE CHARACTERISTIC LOCI SHOULD EXHIBIT CORRECT GAIN-PHASE
     CHARACTERISTICS AT HIGH FREQUENCIES.

     2/ THE PRINCIPAL GAINS ARE MANIPULATED TO ACHIEVE CLOSED-LOOP
     PERFORMANCE.

     3/ THE CHARACTERISTIC AND PRINCIPAL GAINS ARE MANIPULATED
     FOR GOOD ROBUSTNESS PROPERTIES.

     The aim of the High Frequency Design Module is to balance up the
     principal gains as well as to align the input and output gain frames
     of the system so that the system is normal, and thus robust at high
     frequencies. The characteristic loci are manipulated so that they
     satisfy the Nyquist Stability Criterion.

     The gains may be balanced up over the high frequency region if the
     orders of infinite zeros of the loci are the same (i.e. the high
     frequency roll-off rates are the same). }}
>
Are the high frequency roll-off rates the same ? ((PRIMARY))
    [* The previous question is asked again *]
> **no**

What is the condition number at the bandwidth frequency ? ((HFCOND))
    [* A bandwidth of 10 rad/s is assumed *]
> **2.49**

Report

Balance the gains at the bandwidth frequency . ((HFCOMPX))
The condition number at the bandwidth frequency is = 2.49.
The input and output gain frames will be aligned at high frequencies.

Report
Checking the system after the high frequency region design.

Does the system exhibit the correct gain-phase characteristics ?
((PRIMARY))
     [* The user should look at Fig. 5.4.2 *]
> yes

Has the high frequency sub-controller caused unacceptable robustness at
intermediate frequencies ? ((MS))

> why

JUSTIFICATIONS
I am trying to determine intermediate freq after high freq design
using rule: 10

rule 10
if unaceptable deterioration at intermediate frequencies is yes
then intermediate freq after high freq design is interm freq unsatisfactory
     and report The high frequency sub-controller has caused an unacceptable
     and report   deterioration in robustness at intermediate frequencies.
     and report The designer should reduce the condition number used in the
     and report   high frequency design stage in order to reduce the gain
     and report   divergences. ((HFCOMPX)).
     and command reset unaceptable deterioration at intermediate frequencies
     and command reset intermediate freq after high freq design
     and check intermediate freq after high freq design

>
Has the high frequency sub-controller caused unacceptable robustness at
intermediate frequencies ? ((MS))

> no

Report
Investigating the low frequency region .

Does the system have any right half plane poles ? CLPOLE

> no

Is there integral action in all the loci ? ((PRIMARY))

> why

JUSTIFICATIONS
I am trying to determine presence of integral action
using rule: 14

rule 14
if right half plane poles observed is no
     and integral action in all loci is yes

then presence of integral action is low frequency frame inappropriate
and report The program will not check the Low Freq Design Module any
and report   further since there is already integral action in
and report   all the loci

>
Is there integral action in all loci ?   ((PRIMARY))

> **no**

Is there integral action in any one of the loci ? ((PRIMARY))

> **yes**

Report
Low frequency sub-controller should be used ((LFINTE)). The effect of
integral action can be controlled by the parameter in the low frequency
sub-controller.
     [* The primary indicators after the high and low frequency region design
        are given in Fig. 5.4.3 *]

Has the low frequency sub-controller caused unacceptable robustness at
intermediate frequencies ? ((MS))
     [* The user should look at Fig. 5.4.5 for the normality indicator MS*]
> **no**

Is the compensated system closed-loop stable ? ((GNYQZ)) ((CLPOLE))

> **why**

JUSTIFICATIONS
I am trying to determine post design modification for zero rhp poles
using rule: 25

rule 25
if the compensated system is closed-loop stable is no
then post design modification for zero rhp poles is reduce gain for stability
and report Reduce the gain of the system by introducing a scalar sub-
and report   controller until the system is stable with an acceptable
and report   phase margin ((GSGAIN)).
>
Is the compensated system closed-loop stable ? ((GNYQZ)) ((CLPOLE))

> **yes**

Report
Increase the gain of the system by introducing a scalar sub-controller
which gives improvement in performance with an acceptable phase margin.
[GSGAIN]

     [* The primary indicators after a scalar gain of 10
        are given in Fig. 6.4.4 *]

Is there excessive gain divergence at intermediate frequencies ? ((PRIMARY))

> **no**

Does system require further characteristic phase compensation ? ((PRIMARY))

> no

Report
The design has been completed using the
Simple Design Technique .

Closed-loop step responses should be examined ((CLSTEP)).

DESIGN COMPLETE.

   [* A list of the current results can be obtained *]

> data

CURRENT DATABASE ITEMS
characteristic phase compensation is no
excessive gain divergence_at intermediate frequencies is no
post design modification for zero rhp poles is increase gain for perform
the compensated system_is closed loop stable is yes
trade-off caused by low freq controller is no
condition number at low freq = 1
low freq design is balance gains at low frequency
integral action in all loci is no
right-half plane poles observed is no
unaceptable deterioration at intermediate frequencies is no
system closed-loop stable after high freq design is yes
the system has the correct phase characteristics is yes
high frequency design interm is balance gains at the bandwidth frequency
condition number at bandwidth frequency = 2.49
high frequency roll-off rates are the same is no

    [* End of the design session *]

REFERENCES

ÅSTRÖM K.J., J.J.ANTON & K.-E. ÅRZÉN (1986). Expert control. Automatica, 22, No. 3, 277-286.

BELL M.Z. (1985). Why expert systemsfail. J. Opl. Res. Soc., 36 , No.7 , 613-619.

BIRDWELL J.D. et al. (1985). Expert systems techniques in a computer-aided control system analysis and design environment. 3rd IFAC/IFIP International Symposium, CADCE'85, Denmark.

BLOCH A.M. & I. POSTLETHWAITE (1981). Robustness in multivariable control system design. Report CUED/F-CAMS/TR-215, Engineering Dept., Cambridge University.

BODE H.W. (1940). Relations between attenuation and phase in feedback amplifier design. Bell System Technical Journal, 19, 421-454.

BODE H.W. (1945). Network Analysis and Feedback Amplifier Design. Van Nostrand, Princeton.

BÜNZ D. & K. GÜTSCHOW (1985). CATPAC-An interactive software package for control system design. Automatica, 21, No.2, 209-213.

CHU E.K-W. & N.K. NICHOLS (1984). Robust pole assignment by output feedback. Proc. 4th IMA Conference on Control Theory, Cambridge, U.K..

DAVISON E.J. & S.H. WANG (1975). On pole assignment in linear multivariable systems using output feedback. IEEE Trans. Aut. Control, AC-20, 516-518.

DAVISON E.J. & S.G. CHOW (1977). Perfect control in linear time-invariant multivariable systems: The control inequality principle. In Fallside, F., (Ed.) Control System Design by Pole-Zero Assignment. Academic Press.

DESOER C.A. & Y.T. WANG (1980). On the generalized Nyquist stability criterion. IEEE Trans. Aut. Control, AC-25, 187-196.

DOYLE J.C. & G. STEIN (1981). Multivariable feedback design: concepts for a classical/modern synthesis. IEEE Trans. Aut. Control, AC-26, 4-16.

EDMUNDS J.M. (1978). Cambridge linear analysis and design programs. Report CUED/F-CAMS/TR-170, Engineering Dept., Cambridge University.

EDMUNDS J. M., S.E. JEANES, & J.M. MACIEJOWSKI (1983). CLADP: The Cambridge linear analysis and design programs - user reference manual. Engineering Dept., University of Cambridge.

EXPERTECH (1985). Expertech Xi - User Manual, Expertech Ltd.

FAHMY M.M. & J. O'REILLY (1982). IEEE Trans. Aut. Control, AC-27, 690-693.

FLETCHER L.R., J.KAUTSKY, G.K.G. KOLKA & N.K. NICHOLS (1985). Some necessary and sufficient conditions for eigenstructure assignment. Int. J. Control, 42, 1457-1468.

GLOVER K. (1984). All optimal Hankel-norn approximations of linear multivariable systems and their L∞ -error bounds. Int. J. Control, 39, 1115-1193.

GOURISHANKAR V. & K. RAMAR (1976). Int. J. Control, 23, 493-504.

HABER R.N. & L. WILKINSON (1982). Perceptual components of computer displays. IEEE Computer Graphics and Applications. Vol. 2, No. 3, 23-25.

HARMON P. & D. KING (1985). Expert Systems. Wiley Press, New York.

HASSAN M.M. & M.H. AMIN (1985). Recursive eigenstructure assignment in linear systems. Submitted to Int. J. Control.

HENRICI P. (1962). Bounds for iterates, inverses, spectral variation and field of values of non-normal matrices. Numerische Math., Vol.4, 24-40.

HIGHAM N.J. (1984). Computing the polar decomposition - with applications. Numerical Analysis Report No.94, Dept. of Mathematics, University of Manchester

HUNG Y.S. & A.G.J. MACFARLANE (1981). On the relationships between the unbounded asymptote behaviour of multivariable root loci, impulse response and infinite zeros. Int. J. Control, 34, 31-69.

HUNG Y.S. & A.G.J. MACFARLANE (1982). Multivariable Feedback: A Quasi-Classical Approach. Lecture Notes in Control and Information Sciences 40, Springer-Verlag, Berlin.

INTEGRATED SYSTEMS (1984). MATRIX$_x$: The User's Guide. Integrated Systems Inc..

INTELLICORP (1984). The knowledge engineering environment. IntelliCorp GmbH Knowledge Systems Division.

JAMES J.R., D.K. FREDERIK & J.H. TAYLOR (1985). The use of expert-systems programming techniques for the design of lead-lag compensators. International Conference, Control 85, Cambridge U.K..

JUANG J.C. & T.T. LEE (1984). On optimal pole assignment in a specified region. Int. J. Control, 40, 65-79,

KAUTSKY J., N.K. NICHOLS & P. VAN DOOREN (1985). Robust pole assignment in liar state feedback. Int. J. Control, 41, 1129-1155.

KLEIN G & B.C. MOORE (1977). Eigenvalue-generalized eigenvector assignment with state feedback. IEEE Trans. Aut. Control, AC-22, 140-141.

KOUVARITAKIS B. (1974). Characteristic locus methods for multivariable feedback systems design. Ph.D. thesis, The University of Manchester, Institute of Science and Technology.

KOUVARITAKIS B., W. MURRAY & A.G.J. MACFARLANE (1979). Characteristic frequency-gain design study of an automatic flight control system. Int. J. Control, 29, 787-796.

LARSSON J.E. & P. PERSSON (1986). Knowledge representation by scripts in an expert interface. Proc. of the American Control Conference. Seattle, U.S.A..

LAYTON J.M. (1976). Multivariable control theory. Peter Peregrinus Ltd., U.K..

LIMEBEER D.J.N. & J.M. MACIEJOWSKI (1982). Two tutorial examples of multivariable control system design. Report CUED/F-CAMS/TR-229, Engineering Dept., Cambridge University.

MACFARLANE A.G.J. & J.J. BELLETRUTTI (1973). The characteristic locus design method. Automatica, 9, 575-588.

MACFARLANE A.G.J. & KOUVARITAKIS B. (1977). A design technique for linear multivariable feedback systems. Int. J. Control, 25, 837-874.

MACFARLANE A.G.J. & D.J.N. LIMEBEER (1981). Notes on the Analysis and Design of Multivariable Feedback Systems. Internal Notes, Engineering Dept., University of Cambridge.

MACFARLANE A.G.J. & I. POSTLETHWAITE (1977). The generalized Nyquist stability criterion and multivariable root loci. Int. J. Control, 25, 81-127.

MACFARLANE A.G.J. & D.F.A. SCOTT-JONES (1979). Vector gain. Int. J. Control, 29, 65-91

MACFARLANE A.G.J. & Y.S. HUNG (1984). Gains, phases and angles. Proc. IFAC World Congress, Budapest, Hungary.

MACFARLANE A.G.J. & Y.S. HUNG (1985). Indicators for multivariable feedback systems. International Conference Control 85, Cambridge, U.K., 9-11 July, 1985.

MOORE B.C. (1976). On the flexibility offered by state feedback in multivariable systems beyond closed loop eigenvalue assignment. IEEE Trans. Aut. Control, October, 689-692.

MUNRO N. (1972). Design of controllers for open-loop unstable multivariable system using inverse Nyquist array. Proc. IEE, 119, No. 9, 1377-1382.

MUNRO N. (1985). Recent extensions to the inverse Nyquist array design method. IEEE Conference on Decision and Control, Iowa, U.S.A..

MUNRO N. & S. NOVIN-HIRBOD (1979). Pole assignment using full-rank output-feedback compensators. Int. J. Systems Sci., 10, 285-306.

MURRAY W. (1972). Numerical Methods for Unconstrained Optimization, Academic Press, London.

NELDER J.A. & R. MEAD (1965). A simplex method for function minimization. Computer Journal, 7, 308-313.

NOLAN P.J. (1986). An intelligent assistant for control system design. Proc. of the 1st International Conference on the Application of Artificial Intelligence in Engineering Problems, University of Southampton, U.K..

NYE W.T. & A.L. TITS (1986). An application-orientated, optimization-based methodology for interactive design of engineering systems. Int. J. Control, 43, 1693-1721.

NYQUIST H. (1932). Regeneration theory. Bell Syst. Tech. Journ., 11, 126-147.

O'REILLY J. (1983). Observer for Linear Systems. Academic Press.

OWENS D.H. (1975). Dyadic expansion, characteristic loci and multivariable-control-systems design. Proc. IEE, Vol.122, March, pp.315-320.

PANG G.K.H. & J-M BOYLE (1986). An Expert System for Analytical and Interactive Design of Control Systems. To appear in The Second International Expert Systems Conference, London.

PATEL R.V. & N. MURNO (1982). Multivariable System Theory And Design. Pergamon Press, Oxford.

PORTER B. & J.J. D'AZZO (1978). Int. J. Control, 27,487.

POSTLETHWAITE I. & A.G.J. MACFARLANE (1979). A Complex Variable Approach to the Analysis of Linear Multivariable Feedback Systems. Lecture Notes in Control and Information Sciences 12. Springer-Verlag, New York.

SAGE A.P. (1981). Linear Systems Control. Pitman, London.

SCHRAGE M. (1986). Artificial intelligence: needing common sense. IEEE Expert, Spring 1986.

SHAH S.L., D.G.FISHER & D.E. SEBORG (1975). Eigenvalue/eigenvector assignment for multivariable systems and further results for output feedback control. Electronics Letters, 11, No.6, 388-389.

SHIEH L.S., H.M. DIB & B.C. MCINNIS (1986). Linear quadratic regulators with eigenvalue placement in a vertical strip. IEEE Trans Aut. Control, AC-31, 241-243.

SILVERMAN L.M. (1980). Approximation of linear systems. Department of Electrical Engineering, University of Southern California.

SMITH M.C. (1981). On the generalized Nyquist stability criterion. Int. J. Control, 34, 885-920.

SMITH M.C. (1982). A generalized Nyquist/root-locus theory for multi-loop feedback systems. Ph.D. thesis, Engineering Dept., Cambridge University.

SOLHEIM, O.A. (1972). Design of optimal control systems with prescribed eigenvalues. Int. J. Control, 15,143-160.

SPRINATHKUMAR S. & R.P. RHOTEN (1975). Eigenvalue/eigenvector assignment for multivariable systems. Electronics Letters, 11, No.16, 124-125.

STEWART G.W. (1973). Introduction to Matrix Computations. Academic Press, New York.

TAYLOR J.H. & D.K. FREDERICK (1984). An expert system architecture for computer aided control engineering. Proc. IEEE, December issue.

TRANKLE T.L. & L.Z. MARKOSIAN (1985). An expert system for control system design. International Conference Control 85, Cambridge, U.K..

TRANKLE T.L., P. SHEU & U.H. RABIN (1986). Expert system architecture for control system design. Proc. of the American Control Conference. Seattle, U.S.A.

WATERMAN D.A. (1985). A Guide to Expert System. Addison-Wesley.

WILKINSON J.H. (1965). The Algebraic Eigenvalue Problem. Clarendon Press, Oxford.

INDEX

Lecture Notes in Control and Information Sciences

Edited by M. Thoma

Lecture Notes in Control and Information Sciences

Edited by M. Thoma and A. Wyner

Lecture Notes in Control and Information Sciences

Edited by M. Thoma and A. Wyner